S0-BXE-548

Normal and Pathological Development of Energy Metabolism

Normal and Pathological Development of Energy Metabolism

Edited by

F. A. Hommes

Department of Pediatrics
University of Groningen
The Netherlands

C. J. Van den Berg

Study Group Inborn Errors
University of Groningen
The Netherlands

RC620.5
A1
N6
1975

1975

ACADEMIC PRESS
London New York San Francisco
A Subsidiary of Harcourt Brace Jovanovich, Publishers

325398

ACADEMIC PRESS INC. (LONDON) LTD.
24/28 Oval Road
London NW1

United States Edition published by
ACADEMIC PRESS INC.
111 Fifth Avenue
New York, New York 10003

Copyright © 1975 by
ACADEMIC PRESS INC. (LONDON) LTD.

All Rights Reserved

No part of this book may be reproduced in any form by photostat, microfilm or any other means, without written permission from the publishers

Library of Congress Catalog Card Number: 75-29715
ISBN: 0–12–354560–9

PRINTED IN GREAT BRITAIN BY
T. AND A. CONSTABLE LTD., EDINBURGH

PARTICIPANTS

Th. Akerbaan, Laboratory of Biochemistry, B.C.P. Jansen Institute, Plantage Muidergracht 12, Amsterdam, The Netherlands.
P.G. Barth, Department of Pediatrics, Free University, De Boelelaan 1117, Amsterdam, The Netherlands.
N.Z. Baquer, Courtauld Institute of Biochemistry, The Middlesex Hospital, Medical School, London W1R 5PR, England.
R. Berger, Laboratory of Developmental Biochemistry, Department of Pediatrics, University of Groningen, Bloemsingel 10, Groningen, The Netherlands.
J.P. Blass, Mental Retardation Center, University of California, Medical School, Los Angeles, California 90024, U.S.A.
W. Blom, Sophia Children's Hospital, Erasmus University, Gordelweg 160, Rotterdam, The Netherlands.
L. Cathelineau, Hôpital Trousseau, 24 Avenue de Dr. Arnold Metter, 75 Paris XII, France.
R. Charles, Laboratory for Anatomy and Embryology, University of Amsterdam, Mauritskade 61, Amsterdam, The Netherlands.
J.B. Clark, Department of Biochemistry, St. Bartholomew's Hospital, University of London, Charterhouse Square, London EC1 M6BQ, England.
J.E. Cremer, Toxicology Research Unit, Medical Research Council Laboratories, Woodmansterne Road, Carshalton, Surrey, England.
C.J. de Groot, Department of Pediatrics, University of Groningen, Groningen, The Netherlands.
J. Fernandes, Sophia Children's Hospital, Erasmus University, Gordelweg 160, Rotterdam, The Netherlands.
G.E. Gaull, New York State Institute for Basic Research in Mental Retardation, 1050 Forrest Hill Road, Staten Island, New York 10314, U.S.A.
R. Gitzelmann, Kinderspital, Steinwiesstrasse 75, 8032 Zürich, Switzerland.
O. Greengard, Cancer Research Institute, New England Deaconess Hospital, 194 Pilgrim Road, Boston, Massachusetts 02215, U.S.A.
B. Hansen, Kinderspital, Steinwiesstrasse 75, 8032 Zurich, Switzerland.
A. Hatzfeld, Institut de Pathologie Moléculaire, Université de Paris, 24 Rue de Faubourg Saint Jacques, Paris, France.
F.A. Hommes, Laboratory of Developmental Biochemistry, Department of Pediatrics, Bloemsingel 10, Groningen, The Netherlands.
G.J.M. Hoogwinkel, Laboratory for Medical Chemistry, University of Leiden, Wassenaarseweg 72, Leiden, The Netherlands.
J.H.P. Jonxis, Department of Pediatrics, University of Groningen, Oostersingel 59, Groningen, The Netherlands.
P. Koepp, Universitäts-Kinderklinik, Martinistrasse 52, 2 Hamburg 20, Germany.
J.F. Koster, Laboratory of Biochemistry I, Medical Faculty, Erasmus University, Postbox 1738, Rotterdam, The Netherlands.
A.M. Kroon, Laboratory of Physiological Chemistry, University of Groningen, Bloemsingel 10, Groningen, The Netherlands.

J.M. Land, Department of Biochemistry, St. Bartholomew's Hospital, Medical Center, University of London, Charterhouse Square, London EC1 M6BQ, England.

J.P. Leroux, Clinique Medical Infantile, Hôpital des Enfants-Malades, 149 Rue de Sèvres, 75730 Paris Cedex 15, France.

I. Lombeck, Kinderklinik der Stadtischen Krankenanstalten, Moorenstrasse 5, 4 Dusseldorf 1, Germany.

D. Matheson, Department of Biological Psychiatry, University of Groningen, Oostersingel 59, Groningen, The Netherlands.

L.A.H. Monnens, Department of Pediatrics, St. Radboud Hospital, Geert Grooteplein 20, Nijmegen, The Netherlands.

W.J. Rutter, Department of Biophysics and Biochemistry, University of California, San Francisco, California 94122, U.S.A.

J.M. Saudubray, Clinique Medicale Infantile, Hôpital des Enfants-Malades, 149 Rue de Sèvres, 75730 Paris Cedex 15, France.

E.D.A.M. Schretlen, Department of Pediatrics, St. Radboud Hospital, Geert Grooteplein 20, Nijmegen, The Netherlands.

R. Sengers, Department of Pediatrics, St. Radboud Hospital, Geert Grooteplein 20, Nijmegen, The Netherlands.

K. Snell, Department of Biochemistry, University of Surrey, Guildford, Surrey GU2 5XH, England.

J. Stern, Department of Pathology, Queen Mary's Hospital for Children, Carshalton, Surrey, England.

H.J. Sternowksy, Universitäts-Kinderklinik, Martinistrasse 52, 2 Hamburg 50, Germany.

J.M. Tager, Laboratory of Biochemistry, B.C.P. Jansen Institute, Plantage Muidergracht 12, Amsterdam, The Netherlands.

W.H.H. Tegelaars, University Children's Hospital, Binnengasthuis, Grimburgwal 10, Amsterdam, The Netherlands.

J.T. Tildon, Department of Pediatrics, University of Maryland, School of Medicine, Baltimore, Md 21201, U.S.A.

C.J. Van den Berg, Department of Biological Psychiatry, University of Groningen, Oostersingel 59, Groningen, The Netherlands.

G. Van den Berghe, Laboratoire de Chimie Physiologie, Université de Louvain, Dekenstraat 5, 3000 Louvain, Belgium.

H.H. Van Gelderen, Department of Pediatrics, University Hospital, Leiden, The Netherlands.

F.J. Van Sprang, Wilhelmina Children's Hospital, Nieuwe Gracht 137, Utrecht, The Netherlands.

G. Verellen, Clinique Universitaire St. Raphael, Medicine des Enfants, Université Catholique de Louvain, Kapucijnenvoer 35, 3000 Louvain, Belgium.

H.K.A. Visser, Sophia Children's Hospital, Erasmus University, Gordelweg 160, Rotterdam, The Netherlands.

S.K. Wadman, Wilhelmina Children's Hospital, Nieuwe Gracht 137, Utrecht, The Netherlands.

J. White, Courtauld Institute of Biochemistry, Middlesex Hospital Medical School, London W1P 5PR, England.

H. Wick, Basler Kinderspital, Römergasse 8, 400 Basel 5, Switzerland.

J.L. Willems, Department of Pediatrics, St. Radboud Hospital, Geert Grooteplein 20, Nijmegen, The Netherlands.

PREFACE

This is the second time that a meeting is held on the relation between developmental biochemistry and inborn errors of metabolism associated with brain damage.

Impressive progress has been made in the biochemistry of inborn errors of metabolism during the last few years. This progress is mainly due to the application of biochemical techniques and approaches to the study of human metabolism in patients, cultured fibroblasts or animal model systems. It can be expected that this progress will continue for a while because there is so much more known in biochemistry than is applied to or practised in medicine. However, with the advance and penetration of these biochemical methods into this field a new problem arises, namely that of communication between the student of chemical phenomena who uses a rather sophisticated and specific language and the practitioner who deals with patients, parents and social backgrounds. This problem should not be treated as of little importance. It may be expected that with the advancement and introduction of more refined biochemical approaches, requiring a more specific language to communicate with specialist colleagues, the problem will only be aggravated.

It is for this reason that we have invited a rather mixed company to participate in this meeting, consisting of developmental biochemists, biochemists, clinicians and geneticists, to expose the purists to practical clinical problems and the clinicians to the possibilities, but also the limitations, of biochemical approaches. I sincerely hope that our efforts will be fruitful. We have set the stage; it is now up to you by your contribution, formal or in the discussions, to determine whether or not a meaningful dialogue between different disciplines is possible for the benefit of some children whose fate is determined by an odd genetic coincidence.

F.A. HOMMES

CONTENTS

ENERGETIC ASPECTS OF LATE FETAL AND NEONATAL METABOLISM

F.A. Hommes

Laboratory of Developmental Biochemistry, Department of Pediatrics University of Groningen, Groningen, The Netherlands

The late fetal and neonatal period is characterized by dramatic changes in energy metabolism. During the last stages of gestation glycogen is accumulating, to be used during delivery and the very early postnatal period and to bridge the time until the onset of gluconeogenesis. Associated with these changes in the supply of substrate for energy generation, changes are observed in the machinery for energy production, that is oxidative phosphorylation and glycolysis. Energy supply and energy generation are both of paramount importance for maintaining and maturation of a normal brain function.

It is obvious that gross interference with energy supply is incompatible with life. Disturbed function which could be ascribed to these effects, probably takes place at a much more subtle or sophisticated level of organization, which requires a more detailed knowledge of the structural organization of energy generating processes. This is a difficult area of research since the basic mechanism of ATP generation in mitochondria is unknown. There are however some important differences between the fetal and adult energy generating systems which may give some clues to a better insight into a disturbed function due to an inborn error of metabolism. In our studies we have used predominantly fetal rat liver because this tissue permits the preparation of isolated hepatocytes, only to a limited extent contaminated with other cells (Hommes *et al.*, 1971). A rather homogeneous preparation which is easy to handle can therefore be studied.

MITOCHONDRIA

A critical review on the development of components of the respiratory chain in liver mitochondria has recently appeared (Pollak and Duck-Chang, 1973). It is generally agreed that the total mitochondrial oxidative capacity per cell increases considerably (Hommes and Richters, 1969; Gregson and Williams, 1969; Jakovčic *et al.*, 1971) and that this increase is predominantly due to an increase in the number of mitochondria per cell, as illustrated in Fig.1. Similar results have been reported for the brain (Gregson and Williams, 1969; Pysh, 1970). Not only does the number of mitochondria per cell increase, but also there is an increase in the activity of a number of enzymes of the tricarboxylic acid cycle (Hommes *et al.*

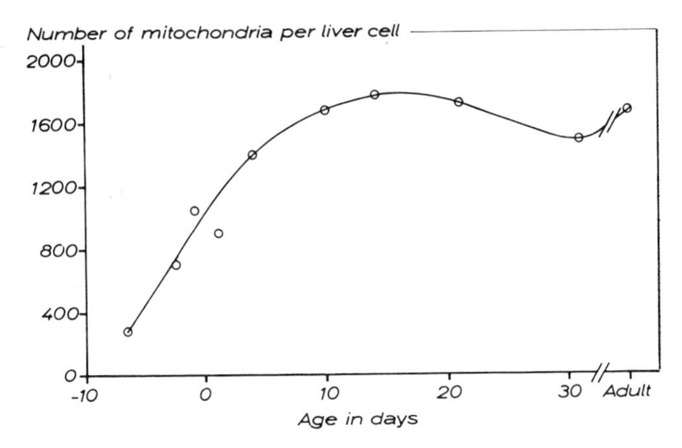

Fig.1 Number of mitochondria per liver cell as a function of age.

1969, 1971) and some enzymes closely associated with that cycle (Hemon, 1967; Kofer and Pollak, 1961; Snell and Walker, 1973).

Of particular importance in this respect is the pyruvate dehydrogenase complex because this enzyme system is an important donor for acetyl-CoA involved in the function of the tricarboxylic acid cycle. The development of this enzyme has been studied by Knowles and Ballard (1974). Its activity increases about 3 times during the period from 3 days before birth to adult, again partly due to an increase in the number of mitochondria and partly due to an increase in the activity of this enzyme per mg mitochondrial protein (Hommes *et al.*, 1974). Since others will contribute to this subject, I shall leave the implications for their comments.

Considerable attention has been paid in the past to the development of mitochondrial enzyme systems associated with energy metabolism. As long as the activity of an enzyme is low in development it may be expected that this enzyme contributes to the control of metabolism, by virtue of this low activity. Many mitochondrial reactions in the adult tissue are however controlled not only by the activity of the mitochondrial enzyme systems but also by the rate of uptake of release of metabolites by the mitochondria. Very little is known about this aspect of late fetal and neonatal energy metabolism. Williams (1961, 1966) observed a resistance to swelling of mitochondria from newborn rat liver in the presence of Ca^{++}, potassium fumarate and potassium phosphate. It was suggested that this could be due to a difference in membrane structure. Further evidence of such differences came from studies by Pollak and Munn (1970) by isopycnic density gradient centrifugation of mitochondrial preparations. They demonstrated that the mitochondria of fetal rat liver consists of two populations differing in permeability to sucrose of the inner mitochondrial membrane. The proportion of mitochondria with a matrix space readily accessible to sucrose decreases with increasing age. A higher permeability of the inner membrane of fetal rat liver mitochondria to sucrose has recently been confirmed by De Schrijver *et al.* (1974).

These results seem to be at variance with some data recently obtained in our laboratory. The fatty acid composition of the inner mitochondrial membrane of fetal and adult rat liver mitochondria has been compared (Luit *et al.*, 1974) and

Table I. Fatty acid composition of lipids of inner mitochondrial membranes of adult and fetal rat liver mitochondria. Inner mitochondrial membranes were isolated according to Parsons *et al.* (1967). The determination of fatty acid composition was carried out on lipid extracts prepared by the method of Dawson *et al.* (1960) by gas chromatography according to Lipsky *et al.* (1959) after conversion to the methylesters (Metcalfe *et al.*, 1966). The values are given as percentage of total fatty acids.

Fatty acid	Inner mitochondrial membrane of adult rat liver	Inner mitochondrial membrane of fetal rat liver
Myristinic acid (14:0)	1.0	4.4
Palmitic acid (16:0)	20.9	28.6
Palmitoleic acid (16:1)	2.8	9.2
Stearic acid (18:0)	22.4	15.1
Oleic acid (18:1)	9.1	26.2
Linoleic acid (18:2)	25.1	11.5
Arachidonic acid (20:4)	18.7	5.0

it was found that the inner membrane of fetal liver mitochondria contained less unsaturated moieties than the inner membrane of adult liver mitochondria, as illustrated in Table I. The difference is about 28%. Similar differences have been found in the membranes of rough and smooth endoplasmic reticulum (Luit *et al.*, 1974).

Current concepts of the structure of cellular membranes, i.e. the fluid mosaic model as developed by Singer and Nicolson (1972), would then suggest that a higher degree of saturation of the fatty acid moieties of the phospholipids of the membrane results in a more crystalline structure of the phospholipid part of the membrane and consequently a lower rate of transfer of metabolites across that membrane. The rate of iso-osmotic swelling - which can be taken as a measure for the uptake of metabolites - with the ammonium salts of pyruvate and malate is indeed lower in fetal rat liver mitochondria than in adult rat liver mitochondria (Luit *et al.*, 1974). This difference could also be ascribed to a difference in content of translocators involved in these transport processes. Titrations with mersalyl (Meyer and Tager, 1969) did not however show any difference in the amount of mersalyl per mg mitochondrial protein required for 50% inhibition of the rate of swelling. The amount of dicarboxylate translocator per mg mitochondrial protein seems therefore to be the same for fetal and adult rat liver mitochondria. It remains to be established whether the same holds true for transport of pyruvate by titration with α-hydroxy-4-cyano-cinnamate (Halestrap and Denton, 1974).

It is however evident that differences in fatty acid composition of the membrane can result in differences in the rate of transport of metabolites across the inner mitochondrial membrane. The implications for the control of metabolism need further investigations.

Even far less is known about possible defects in these translocators, which are proteins and therefore genetically determined. It may furnish a basis for the many poorly-understood lactic acidoses.

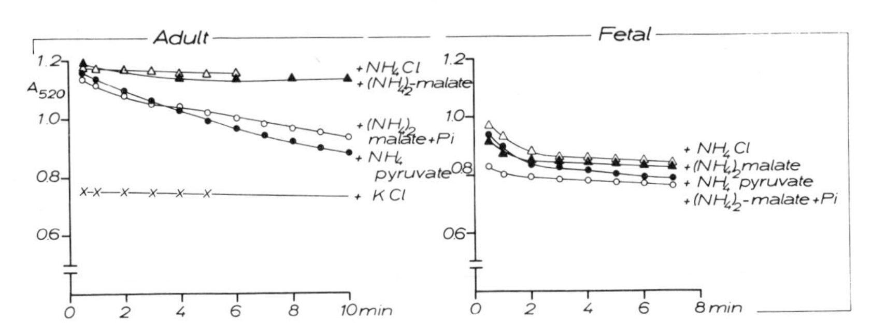

Fig. 2 Iso-osmotic swelling of adult and fetal rat liver mitochondria. Rat liver mitochondria (0.33 mg protein for adult and 0.73 mg protein for fetal) were suspended in tris-HCl buffer, 20 mM, 10 μg, EDTA, 1 mM and NH_4 Cl, 125 mM, ammonium pyruvate 125 mM or ammonium malate 100 mM. Volume 3.0 ml, pH 7.0. At time zero the mitochondria were added and the change in absorption was measured.

GLYCOLYSIS

Since the number of mitochondria per cell and the activity of a number of enzymes per mg mitochondrial protein are lower in the fetal tissue as compared to the adult tissue, a substantial contribution to total energy production is supplied by glycolysis.

Previous studies with isolated fetal hepatocytes demonstrated that under the conditions used less than 10% of the glycolytic flux was derived from glucose via the hexokinase reaction (Berger and Hommes, 1974). About 90% was contributed by the breakdown of glycogen. The reason is the rather high concentration of glucose-6-phosphate in these cells, which is a competitive inhibitor of hexokinase and thus the glucose phosphorylating system is shut down almost completely. A striking feature of the intact fetal liver during the last days of pregnancy is however the accumulation of glycogen (Shelky, 1961). The concentration of glucose-6-phosphate must therefore be lower in the intact liver to allow rapid phosphorylation of glucose. Furthermore phosphorylase must be mostly in the inactive form (Devos and Hers, 1974), the more so as glycogen synthetase is only partially active (Schaub *et al.*, 1972). The question arises therefore which factors control the rate of glycogen breakdown versus glucose phosphorylation in the fetal hepatocytes.

The contribution by the glucose phosphorylating system to the total glycolytic flux can be measured by incubating the cells in the presence of [3,4-^{14}C]-D-glucose and measuring lactate accumulation, $^{14}CO_2$ production and the specific activities of lactate and glucose.

Besides the interconversion of phosphorylase a and phosphorylase b, the rate of glycogenolysis is determined by the inorganic phosphate concentration - being a substrate for phosphorylase with an apparent K_m of 2 mM (Maddaiah and Madsen, 1966) and the intracellular glucose concentration. The latter effect is caused by binding of glucose to phosphorylase a. This complex is less active than phosphorylase a without bound glucose and furthermore more susceptible to the action of phosphorylase a phosphatase (Stalman *et al.*, 1972).

Fetal hepatocytes were therefore isolated (Hommes *et al.*, 1971) and subsequently incubated in media with different glucose and inorganic phosphate con-

Table II. Effect of glucose and inorganic phosphate concentration of the isolation medium for the preparation of isolated fetal hepatocytes on the contribution to the glycolytic flux by the glucose phosphorylating system and glycogenolysis. Isolated cells were incubated in Krebs-Ringer-phosphate medium, supplemented with glucose as indicated, in the presence of [3,4-^{14}C]-D-glucose for 20 min at 37°C. Lactate accumulation, $^{14}CO_2$ production and the specific activities of glucose and lactate were determined. Cells were isolated in medium I (Hommes *et al.*, 1970) containing 0.7 mM Pi or in medium II (Berger and Hommes, 1974) containing 5.0 mM Pi.

Exp.	Medium for isolation of cells	Additions to isolation medium	Subsequently incubated with glucose (mM)	Glycolytic flux derived from glycogenolysis as % of glycolytic flux
1	II	2 mM glucose	2	87.6
2	II	2 mM glucose	2	82.4
3	II	2 mM glucose	10	14.3
4	I	2 mM glucose	2	54.0
5	I	2 mM glucose	2	52.9
6	I	10 mM glucose	2	54.0
7	I	10 mM glucose	10	45.2
8	I	10 mM glucose + insulin, 200 μU/ml	10	49.4
9	I	10 mM glucose + insulin, 200 μU/ml	10 + insulin 200 μU/ml	45.2

centrations to determine the contributions to the total glycolytic flux by the glucose phosphorylating system and glycogenolysis. Table II shows the results. When cells are isolated in a medium containing 5 mM inorganic phosphate and subsequently incubated with 2 mM glucose, the lactate produced is predominantly derived from glycogen. Increasing the glucose concentration in the incubation medium to 10 mM resulted in a considerable decrease in the contribution to the glycolytic flux by glycogenolysis. This demonstrates that the glucose concentration does have an effect on glycogen breakdown in isolated fetal hepatocytes. Decreasing the phosphate concentration in the isolation medium resulted likewise in a decrease of the contribution by glycogenolysis to the glycolytic flux although to a lesser extent. In this case, however, the glucose concentration of the incubation medium has only a marginal effect.

The effect of insulin on glycogen synthesis is controversial (Hers *et al.*, 1970). Whatever the insulin effect *in vivo* may be, it is clear that under the present conditions insulin has only a very small effect, if any at all. A high concentration of insulin was chosen in this case because fetal plasma insulin levels are high in the rat fetus near the end of gestation (Cohen and Turner, 1972).

The plasma glucose concentration in the fetal rat near the end of gestation has been determined at 9 mM (Walker and Snell, 1973) and the inorganic phosphate concentration at 1 mM. Indeed conditions which favour limited activation of the phosphorylase system. The question arises whether these effects can be correlated with the intracellular concentrations of glucose and inorganic phosphate. These were therefore determined under conditions similar to those described in Table II. The results are shown in Table III. A high phosphate concentration of the incubation medium results in a high intracellular phosphate concentration. The intracellular glucose concentration similarly parallels the glucose

Table III. Intracellular concentration of UDPG, UTP, G-6-P, glucose and inorganic phosphate (Pi) of isolated fetal rat hepatocytes. Experimental conditions were as described in the capture of Table II. Cells were separated from the incubation medium by centrifugation over a silicon layer (Meyer, 1971). The amount of incubation medium adhering to the cells was estimated by adding [U-^{14}C]-sucrose to the incubation medium. Metabolites were estimated by standard enzymic methods (Bergmeyer, 1970). Phosphate was determined by a modification of the Fiske and Subbarow method (1925). Values are given in μM, taking into account the morphometric constants of the fetal rat liver cell (Vergonet *et al.*, 1970). n.d.: not determined.

Exp.	Medium for isolation of cells	Additions to isolation medium	Subsequently incubated with glucose (mM)	UDPG	UTP	G-6-P μM	Gluc.	Pi
1	I	2 mM glucose	2	200	291	385	2300	n.d.
2	I	2 mM glucose	10	291	506	426	2020	500
3	II	2 mM glucose	2	109	265	156	1200	1400
4	II	2 mM glucose	10	179	347	312	2600	6500
5	II	10 mM glucose	2	610	815	350	2025	8400
6	II	10 mM glucose	10	356	560	505	1960	6400

concentration of the isolation medium and the incubation medium. A direct effect of the glucose and inorganic phosphate concentrations on glycogen breakdown is thus likely. It is also evident why in the isolation medium with a low inorganic phosphate concentration the glucose concentration has little effect on the contribution by glycogenolysis to the glycolytic flux. The intracellular glucose concentration is already high. The reason for this phenomenon is unclear.

Both the glucose and inorganic phosphate concentrations are therefore of importance for the regulation of glycogen breakdown in the fetal liver. To what extent both factors contribute to the quantitative aspects needs further investigation. As hepatic glycogen mobilization seems to begin between 1 and 2 h after delivery, the hypoglycemia observed after birth therefore not only serves a purpose by inducing hormonal changes necessary for the mobilization of glycogen, but also by the low plasma glucose concentration itself.

ACKNOWLEDGEMENTS

The expert technical assistance of Mrs. G. Luit-De Haan is gratefully acknowledged. These investigations were supported in part by the Netherlands Foundation for Chemical Research (SON) with financial aid from the Netherlands Foundation for the Advancement of Pure Research (ZWO).

REFERENCES

Berger, R. and Hommes, F.A. (1974). *Biochim. biophys. Acta* **333**, 535.
Bergmeyer, H.U. (1970). "Methoden der enzymatischen Analyse". Verlag Chemie, Weinheim.
Cohen, N.M. and Turner, R.C. (1972). *Biol. Neonate* **21**, 107.
Dawson, R.M.C., Hemington, N. and Lindrag, D.B. (1960). *Biochem. J.* **77**, 226.
De Schrijver, D., Mertens-Strijthagen, J., Wattiaux-De Coninck, S. and Wattiaux, R. (1974). Abstr. Comm. 9th FEBS Meeting, Budapest, p.281.
Devos, P. and Hers, H.G. (1974). *Biochem. J.* **140**, 331.
Fiske, C.H. and Subbarow, Y. (1925). *J. biol. Chem.* **66**, 375.
Gregson, N.A. and Williams, P.L. (1969). *J. Neurochem.* **16**, 617.
Halestrap, A.P. and Denton, R.M. (1974). *Biochem. J.* **138**, 720.
Hemon, P. (1967). *Biochim. biophys. Acta* **132**, 175.
Hers, H.G., De Wulf, H., Stalmans, W. and Van den Berghe, G. (1970). *Adv. Enzyme Reg.* **8**, 171.

Hommes. F. A. and Richters, A.R. (1969). **14**, 359.
Hommes. F.A., Wilmink, C.W. and Richters, A.R. (1969). *Biol. Neonate* **14**, 69.
Hommes, F.A., Luit-De Haan, G. and Richters, A.R. (1971). *Biol. Neonate* **17**, 15.
Hommes, F.A., Oudman-Richters, A.R. and Molenaar, I. (1971). *Biochim. biophys. Acta* **244**, 191.
Hommes, F.A., Kraan, G.P.B. and Berger, R. (1973). *Enzyme* **15**, 351.
Jakovčic, S., Haddock, J., Getz, G.D., Rabinowitz, M. and Swift, H. (1971). *Biochem. J.* **121**, 341.
Knowles, S.E. and Ballard, F.J. (1974). *Biol. Neonate* **24**, 41.
Kofer, E. and Pollak, J.K. (1961). *Exp. cell. Res.* **22**, 120.
Lipsky, S.R., Landown, R.A. and Godet, M.R. (1959). *Biochim. biophys. Acta* **31**, 336.
Luit, G., Berger, R. and Hommes, F.A. (1974). *Biol. Neonate (in press).*
Maddaiah, V.T. and Madsen, N.B. (1966). *J. biol. Chem.* **241**, 3873.
Metcalfe, L.D., Smitz, A.A. and Pelko, J.R.P. (1966). *Anal. Chem.* **38**, 514.
Meijer, A.J. and Tager, J.M. (1969). *Biochim. biophys. Acta* **189**, 136.
Meijfer, A.J. (1971). "Anion translocation in mitochondria". Thesis, University of Amsterdam.
Parsons, D.F., Williams, G.R., Thompson, W., Wilson, D. and Chance, B. (1967). *In* "Round Table Discussion on Mitochondrial Structure and Compartmentation" (E. Quagliariello, S. Papa, E.C. Slater and J.M. Tager, eds), p.29. Adriatica Editrice Bari.
Pollak, J.K. and Munn, E.A. (1970). *Biochem. J.* **117**, 913.
Pollak, J.K. and Duck-Chong, C.G. (1973). *Enzyme* **15**, 139.
Pysh, J.J. (1970). *Brain Research* **18**, 325.
Schaub, J., Gutmann, I. and Lippert, H. (1972). *Horm. Metab. Res.* **4**, 110.
Shelley, H.J. (1961). *Brit. Med. Bull.* **17**, 137.
Singer, S.J. and Nicolson, G.L. (1972). *Science* **175**, 720.
Snell, K. and Walker, D.G. (1973). *Enzyme* **15**, 40.
Stlamans, W., De Barry, Th., Laloux, M., De Wulf, H. and Hers, H.G. (1972). *In* "Metabolic Interconversion of Enzymes" (O. Wieland, E. Helmreich and H. Holzer, eds), p.121. Springer Verlag, Berlin.
Vergonet, G., Hommes, F.A. and Molenaar, I. (1970). *Biol. Neonate* **16**, 297.
Walker, D.G. and Snell, K. (1973). *In* "Inborn Errors of Metabolism" (F.A. Hommes and C. Van den Berg, eds), p.97. Academic Press, London.
Williams, M.L. (1961). *Biochim. biophys. Acta* **47**, 411.
Williams, M.L. (1966). *Biochim. biophys. Acta* **118**, 221.

DISCUSSION

Tager: I am convinced that the transport of metabolites across the mitochondrial membrane in the cell is an important factor in controlling metabolism. The data you showed on the differences between fetal and adult rat liver is very suggestive. The translocators in the mitochondrial membrane catalyse an exchange diffusion process in which one ion is exchanged stoichiometrically for another ion. For instance, malate exchanges for phosphate. The rate of transport of a metabolite across the mitochondrial membrane and the direction of flux are determined, not only by the amount of translocator present in the membrane but also by the concentration of metabolites on either site of the membrane. Have you measured the endogeneous pool of metabolites in the fetal liver mitochondria? An additional comment is the following: When you use the swelling technique in ammonium salts to detect the presence of a translocator, you are actually (if you use ammonium malate for instance) measuring the rate of the reaction catalysed by two translocators, namely the dicarboxylate translocator and the phosphate translocator. Have you measured the activity of the phosphate translocator in the fetal rat liver mitochondria? Because this might be a rate limiting factor.

Hommes: We have not specifically measured the phosphate translocator. Neither did we measure the intramitochondrial content of metabolites. The mitochondria were used as isolated like the adult mitochondria and not specifically loaded with any of the substrates.

Gaull: One of the things that always puzzled me is the fact that the concentration of glucose in the blood of the human fetus is lower than that in the mother. I wonder whether this would give you worry in this regard.

Hommes: I was not aware of the fact that the blood glucose concentration in the human fetus is lower than that of the mother. At what time has this been measured? After delivery?

Gaull: No, during gestation. The amino acids and the water soluble vitamins are usually maintained at a higher concentration in the blood of the fetus by the placenta. The concentration of glucose in the blood of the human fetus is believed to be maintained by a facilitated transfer but the fetal concentration is actually lower in the fetus than in the mother and after birth it increases.

Hommes: This would suggest that the situation I just described does not apply to the human. There is one other thing I should mention: in the rat, and I do not know of any studies on the human in this respect, the phosphorylase of the fetal liver is different from the phosphorylase of the adult liver as has been demonstrated by isoelectric focussing experiments (K. Sato, H.P. Morris, S. Weinhouse, *Science* 178 (1972) 879). It is not known when the change-over takes place and neither is it known whether such an isoenzyme difference does occur in the human.

Gaull: The rat at birth is a far more immature animal than the human. It seems to me that it becomes of extreme importance to see how these things differ in the human and the rat with regard to relative maturity, especially in the brain.

Hommes: I agree.

Jonxis: Which was the composition of the food of the animals you used? In the human newborn the fatty acid composition of the subcutaneous fat is to a certain extent dependent on the diet of the mother; it could be that the same is true in your experimental animals.

Hommes: The animals received the normal laboratory diet. There is however another problem, the transfer of fatty acids across the placenta. This transfer is rather limited (J. Davis, V. Jansen, H.J. Kayden, H. Schneider, M. Levits, *Pediat. Res.* 7 (1973) 192). So, I doubt whether an effect of dietary changes to the mother will have similar profound effects on the fatty acid composition of membranes of the fetus *in utero* as compared to adipose tissue in the newborn.

Jonxis: We know that in the human newborn the amount of polyunsaturated fatty acids of subcutaneous fat differs from case to case. A certain quantity of polyunsaturated fatty acids, because they are essential fatty acids, must therefore have crossed the placental membrane.

Greengard: Coming back to the difference between rats and humans, with respect to glucose metabolism the difference is not that striking, because it has been known for a very long time that the onset of neonatal hypoglycemia is very similar in humans, apes, rabbits, guinea pigs and rats, irrespective of their maturity in other areas at birth. Similarly, the development of a key enzyme in this respect, glucose-6-phosphatase, again is very similar in all these species: apes, rabbits, guinea pigs, rats and mice. It shows the same pattern in the liver around the time of birth. The differences between these species are much more striking in other areas of metabolism.

Guall: Is that true in the brain as well as in the liver?

Greengard: That is not known.

Snell: I would like to support the suggestion that translocation systems may play a strong role in controlling the development of energy metabolism. I have made some measurements with liver mitochondria from newborn rats and measured pyruvate utilization by these mitochondria between the time of birth and one hour after birth (*Int. J. Biochem.* 5 (1974) 403). There was a rapid increase in pyruvate utilization, which was not wholly accounted for by increases in the number of mitochondria. This may mean that the transport system for pyruvate is being induced in the immediately newborn period. Have you demonstrated that your fetal hepatocytes are sensitive to hormones? In particular, is the glycogenolytic system sensitive to activation by glucagon or by adrenaline? And secondly, when you incubate your fetal hepatocytes, is there a natural decline in glycogen concentration in the hepatocytes or do they maintain their glycogen?

Hommes: The glycogen goes down upon incubation of the hepatocytes. It is a real utilization of glycogen. We have never tested the effect of glucagon or adrenaline on these isolated hepatocytes.

Snell: This bears on the question of the natural stimulus which initiates glycogen breakdown in the immediately newborn rats. The evidence that I have for the *in vivo* situation suggests that immediately after birth glucagon is relatively inadequate in promoting glycogenolysis. This capability develops within an hour after birth (*Biochem. J.* 134 (1973) 899). So I was wondering if the fetal hepatocytes lack the sensitivity to glucagon or whether there are some reasons why glucagon may not be effective *in vivo* immediately at birth.

METHIONINE ADENOSYLTRANSFERASE: DEVELOPMENT AND DEFICIENCY IN THE HUMAN

G.E. Gaull

Department of Pediatric Research
New York State Institute for Research in Mental Retardation
Staten Island, New York, USA
and
Department of Pediatrics and Clinical Genetics Center
Mount Sinai Medical School of the City University of New York
New York, New York, USA

1. INTRODUCTION

The subject of this symposium is the production and utilization of energy in the fetus and its relationship to inborn errors of metabolism. Within this context, it is appropriate to consider the physiological significance and development of methionine adenosyltransferase. I shall review our own studies in the human fetus and also describe what we know about a new deficiency of the activity of this enzyme, which we think is probably a new inborn error of metabolism.

Methionine adenosyltransferase catalyzes the formation of *S*-adenosylmethionine from L-methionine and ATP (Fig.1). It utilizes ATP as a cosubstrate in a relatively unusual way in that there is a complete dephosphorylation of ATP by a pyrophosphate split, with transfer of the 5′-deoxyadenosyl group to one free electron pair of the sulfur of methionine to form a sulfonium group. The elucidation of the stoichiometry of this reaction is the classical work of Cantoni and Mudd (cf. Cantoni, 1951, 1965 and Mudd, 1965).

Methionine adenosyltransferase in the first enzyme on the pathway of transsulfuration of the sulfur of methionine to form cysteine. The transsulfuration pathway is not present in all tissues in mature animals. Furthermore, transsulfuration is absent from brain and liver in man and monkey until sometime after birth, as a result of the absence of the last enzyme on the pathway, cystathionase (Gaull *et al.*, 1972; Sturman *et al.*, 1970). At these early times in development, activity of the vitamin B_{12} dependent 5-methyltetrahydrofolate-homocysteine methyltransferase is strikingly increased (Gaull *et al.*, 1973). Therefore, we have taken this as evidence that the primary course of the metabolism of methionine may be the cycle of demethylation and remethylation of methionine (Fig.2),

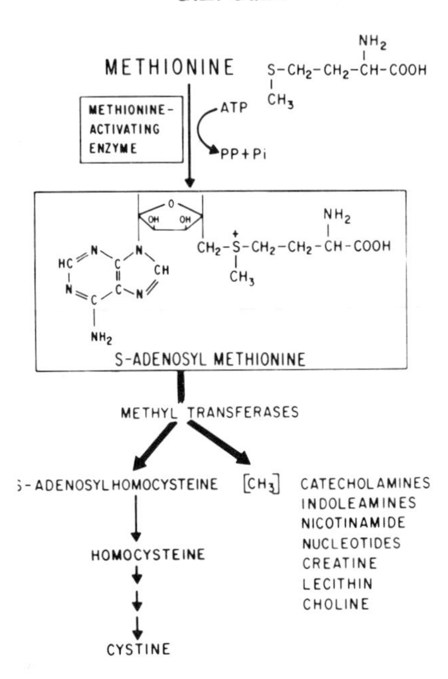

Fig. 1 Methionine adenosyltransferase (previously known as methionine activating enzyme).

rather than transsulfuration. This concept was discussed by me more fully at the last symposium in this series (Gaull, 1973). I should like to make it clear, however, that the cycle concept goes back to the original work on transsulfuration and transmethylation by du Vigneaud (cf. du Vigneaud and Rochele, 1965) and later was discussed by Finkelstein and Mudd in their enzymatic studies on the methionine-sparing effect of cysteine (Finkelstein and Mudd, 1967). More recently Ordonez and Wurtman (1973) have drawn attention to the importance of this cycle in brain. Thus, I would like to underscore the role of methionine adenosyltransferase as the first enzyme in the methionine cycle.

S-adenosylmethionine, the product of this enzyme, has three major functions:

1) *Transmethylation Reactions.* It is a methyl donor in a variety of biological methylation reactions (cf. Shapiro and Schlenk, 1965). Until recently it had been thought to be the only donor important in methyltransferase reactions, but it has been shown that 5-methyltetrahydrofolate is the preferred methyl donor for some methyltransferases (Snyder *et al.*, 1974). It has become clear that in addition to the methylation of small molecules, methylation of a variety of macromolecules, such as DNA, RNA, proteins, polysaccharides, sterols, and the fatty acid chains of phospholipids, also is important (cf. Salvatore *et al.*, 1965). The utilization of *S*-adenosylmethionine for methylation may be considerable, but when 5-methyltetrahydrofolate-homocysteine methyltransferase activity is increased, as it is in fetal tissues, the regeneration of methionine from homocysteine and 5-methyltetrahydrofolate would be facilitated. In regard to the conservation of methionine in the developing fetus, it is noteworthy that folates as well as methionine are concentrated against a gradient in the fetus by the placenta (Szabo and Grimaldi, 1970).

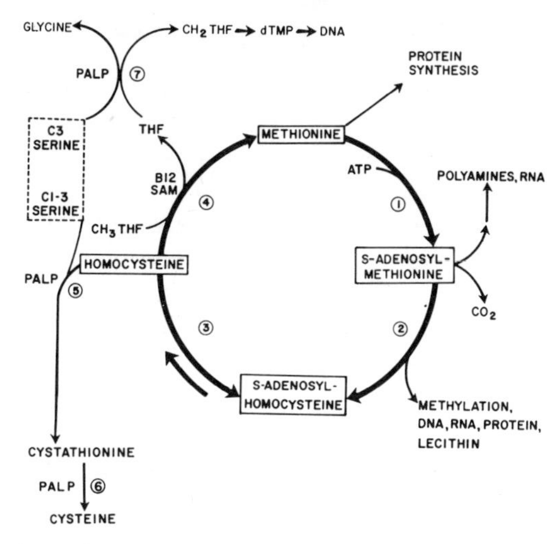

Fig.2 The methionine cycle and related biosynthetic pathways. Abbreviations: CH_3 THF, N^5-methyltetrahydrofolate; THF, tetrahydrofolate; CH_2 THF, $N^{5,10}$-methylenetetrahydrofolate; dTMP, thymidylate; B_{12}, vitamin B_{12} in cofactor form; PALP, pyridoxal phosphate, SAM, *S*-adenosylmethionine; 1) methionine adenosyltransferase; 2) various specific methyltransferases; 3) *S*-adenosylhomocysteine hydrolase; 4) N^5-methyltetrahydrofolate homocysteine methyltransferase; 5) cystathionine synthase; 6) cystathionase; 7) serine-hydroxmethyltransferase.

2) Biosynthesis of polyamines. The studies of the Tabors (1964) and of Williams-Ashman's group (Williams-Ashman *et al.*, 1969) established the mechanism of synthesis of the polyamines and the role of *S*-adenosylmethionine in the transfer of the propylamino group of decarboxylated *S*-adenosylmethionine to putrescine to synthesize spermidine and then spermine (Fig.3). These compounds are involved in control of RNA metabolism and growth in a way as yet incompletely defined. We have demonstrated the high enzymatic capacity and high concentration of these compounds in human fetal liver and brain (Sturman and Gaull, 1974).

3) Enzyme regulation. *S*-adenosylmethionine also is important in modifying the catalytic activity of some enzymatic reactions without donation of either a methyl or a propylamino group, e.g. *S*-adenosylmethionine is required for activity of B_{12}-dependent 5-methyltetrahydrofolate-homocysteine methyltransferase (Taylor and Weissbach, 1969 and 1969a).

The occurrence of metabolic sequences initiated by a common catalytic step suggests that methionine adenosyltransferase might be the site of regulatory control. This idea is supported by reports (Finkelstein and Mudd, 1967; Cantoni, 1951; Gaull *et al.*, 1969) of inhibition of this enzyme by cyst(e)ine, the end product of the transsulfuration pathway. It is noteworthy that in the human fetus accumulation of cyst(e)ine is minimized both by the absence of cystathionase, the enzyme producing it from cystathionine, and by a special mechanism of placental transfer (Gaull *et al.*, 1973). Some kinetic properties suggesting allosterism had been noted earlier (Mudd, 1965; Lombardini and Talalay, 1971). Dr. Harris Tallan in our laboratory has done more detailed studies (Tallan *et al.*,

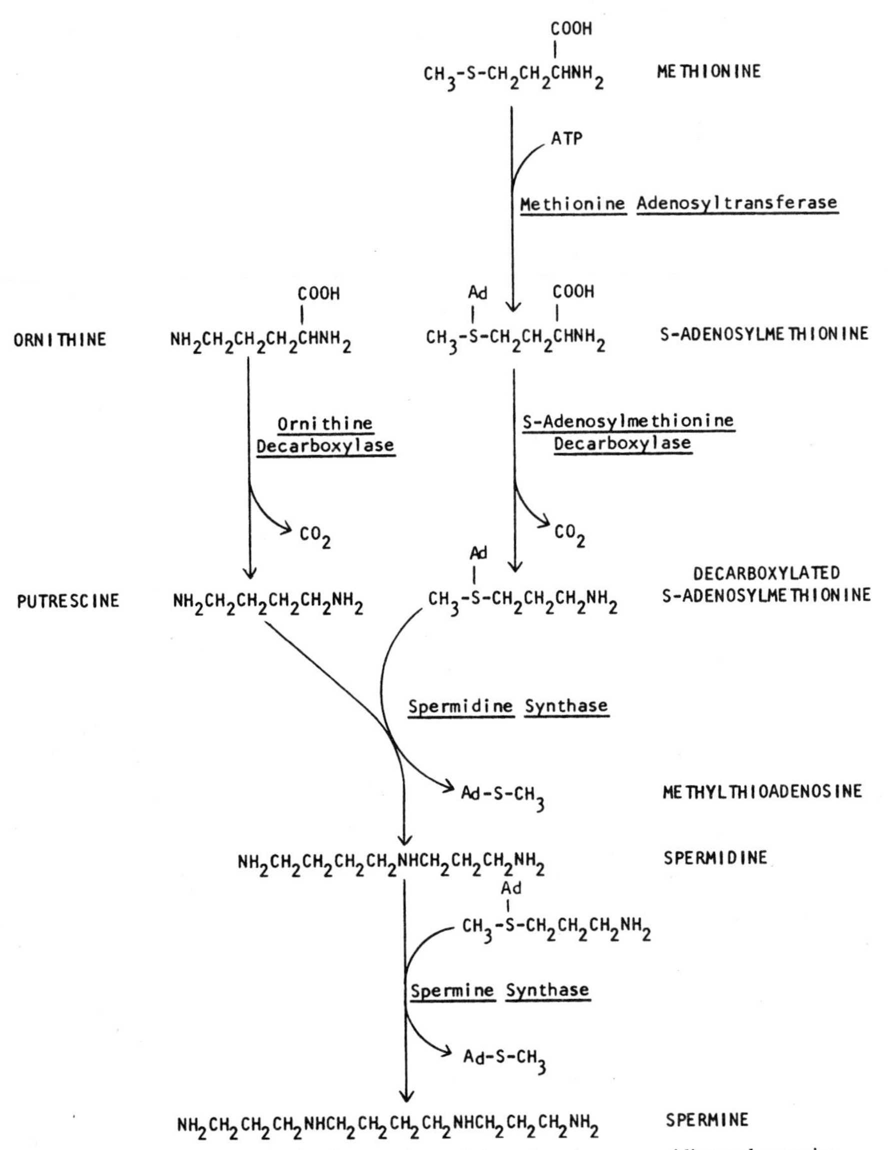

Fig.3 Pathway for biosynthesis of putrescine and the polyamines, spermidine and spermine. Ad: adenosyl.

1973). He clearly established that in rat liver the kinetic behavior of this enzyme is particularly dependent upon the concentration of the previously known activating ions Mg^{2+} and K^{+}. In the absence of both activators, there is a sigmoidal activity curve in response to change in concentration of each substrate, the other substrate being saturating. Mg^{2+} abolishes this apparent cooperative interaction exhibited both by methionine and by $MgATP^{2-}$; the affinity for each substrate is increased, but there is little change in V_{max}. K^+ also increases the affinity of the enzyme for each substrate without changing V_{max} significantly, but has a weaker effect in abolishing the apparent cooperative interaction shown by $MgATP^{2-}$. In the presence of both Mg^{2+} and K^+, however, the major effect is on V_{max}, which is doubled. The total intracellular concentration of Mg^{2+} is very low (about 10 mM); however, the concentration of K^+ is high (100-150 mM) and would be sufficient to increase drastically the affinity of the enzyme for its substrates. Until the enzyme is highly purified an exact kinetic model cannot be constructed, but the diversity of ligand interactions suggests that the enzyme is capable of assuming numerous conformational states in response to the binding of various combinations of ligands. Knowledge of the changes in concentration of K^+, Mg^{2+}, and ATP in fetal tissue is lacking but would be of considerable interest in this regard.

Human liver does not show sigmoidal substrate-activity curves (unpublished observation, H.H. Tallan); however, a number of preparations of human liver have shown some evidence of one type of negative cooperativity (i.e. a "bumpy" $MgATP^{2-}$ curve). What is most significant, however, is that the interactions of human hepatic methionine adenosyltransferase with Mg^{2+} and K^+ and with the substrates are the same as for rat hepatic methionine adenosyltransferase: each metal ion alone increases the affinity of the enzyme for the substrates, and both together increase the V_{max}.

The distribution of methionine adenosyltransferase activity in various tissues and various species has been examined by Finkelstein and Mudd (1967) as well as by our group (Sturman *et al.*, 1970). The enzyme is ubiquitous, although most active in liver. Except for kidney and pancreas, most other tissues have only a small percentage of the activity present in liver. Natori (1965) showed that the methionine adenosyltransferase activity of female rat liver was about twice that of males and that the activity is increased in castrate males. However, control of this activity is also genetic, at least in part: Hancock (1966) showed that the activities in liver of female mice and of male mice varied considerably in various inbred strains; in the rabbit no sex difference was found. Pan and coworkers (1968) have shown that administration of glucocorticosteroids increase methionine adenosyltransferase activity almost 2-fold in the liver of the rat, but not in kidney, pancreas or brain, and that this increase probably involved *de novo* synthesis of protein.

The effect of diet is less clear. Pan and Tarver (1967) found that activity of hepatic methionine adenosyltransferase is low in rats fed a diet low in protein and increases with an increase in protein in the diet, whereas Finkelstein (1967) found an increase in this activity in rats fed a low-protein diet. Finkelstein and Mudd (1967) found that cystine, *in vitro*, inhibited activity of methionine adenosyltransferase and cystathionine synthesis. The activity of both enzymes was decreased on a diet low in methionine but supplemented with cysteine and this decrease could be alleviated by the simultaneous injection of methionine and homocysteine. They pointed out that such changes would tend to alter the distribution of methionine available for synthesis of protein or *S*-adenosylmethionine.

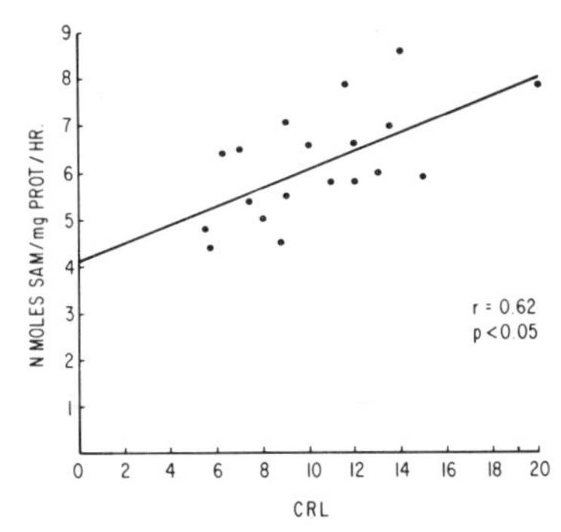

Fig.4 Development of methionine adenosyltransferase specific activity in human fetal brain.

2. DEVELOPMENT OF METHIONINE ADENOSYLTRANSFERASE ACTIVITY

Knowledge about the development of methionine adenosyltransferase activity is somewhat fragmentary but raises some interesting questions. It seemed likely from the work of Hancock (1966) on the liver of the mouse and rabbit and of Sheid and Bilik (1968) on the liver of the rat that the specific activity of methionine adenosyltransferase increases with age. However, their studies were incomplete and used a rather insensitive assay. Finkelstein (1967) used a highly sensitive assay, albeit one rate-limited for substrate, and found a higher activity of rat hepatic methionine adenosyltransferase in the suckling and weanling rat than in the adult rat. Chase and coworkers (1968), using the same assay as Finkelstein, did a more complete study. They found that the specific activity of hepatic methionine adenosyltransferase is low in the rat fetus at 18 days of gestation and increases steadily until birth, at 22 days of gestation. There follows a rapid rise during the first 48 hours after birth. The specific activity then remains constant for at least 30 days, they report, with a slight decrease in the adult rat. However, the overlap in the data for adult liver and neonatal liver makes it unlikely that this decrease is significant. It is also of interest that the total activity per organ increases steadily to maturity. In contrast, the specific activity of methionine adenosyltransferase in brain and in kidney remains constant during the same period, whereas that for intestine increases. Again, total activity per organ for the last three organs increases steadily. The increase in activity in the intestine takes place at the time of weaning, when the methionine intake increases, and the increase in activity in the liver takes place just prior to the period of the animal's most rapid rate of growth.

However, Baldessarini and Kopin (1966) showed that the concentration of *S*-adenosylmethionine in liver and brain decreases during the period from birth to maturity in both the brain and the liver of the rat. Thus, the steady-stage concentration of *S*-adenosylmethionine is decreasing in both organs at a time when the total and specific activity of the enzyme that synthesizes it is increasing.

Let us now look at the development of methionine adenosyltransferase activity in the human. In the liver and kidney of the second trimester human fetus (Gaull *et al.*, 1972, 1973) and of the Rhesus monkey fetus (Sturman *et al.*, 1973) late in gestation (145 days of a 165 day gestation), methionine adenosyltransferase

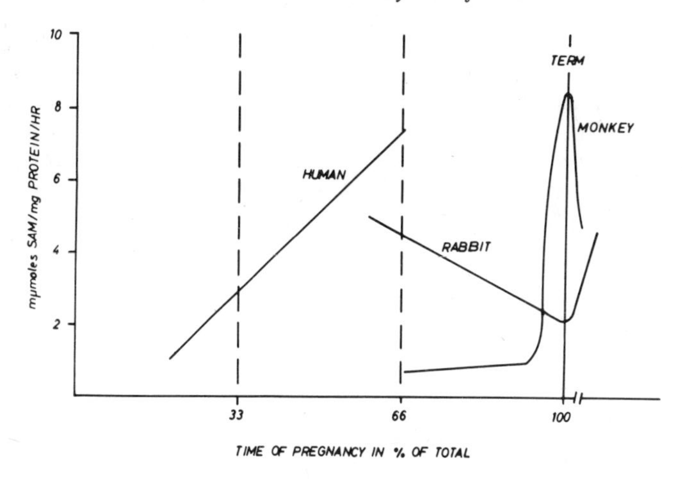

Fig.5 Comparison of development of specific activity of methionine-activating enzyme in lung from human, monkey and rabbit with regard to the time of pregnancy.

activity was about 1/3 of that of the mature liver and kidney. This reduced activity also was found in premature and full-term infants (Gaull *et al.*, 1972), but these specimens were obtained post-mortem, and it is not clear whether such low activity is a post-mortem artifact or represents the actual activity. The situation in the brain is more complicated. In the second trimester human fetus, methionine adenosyltransferase activity in brain is increasing toward values found in mature human brain (Fig.4). In monkey fetuses late in gestation, the activity of brain methionine adenosyltransferase was not statistically different from that of their mothers (Sturman *et al.*, 1973), but only 4 fetuses were examined.

We also have examined the methionine adenosyltransferase activity in fetal lung (Sternowsky *et al.*, 1973). We found that in man, monkey and rabbit, the specific activity of methionine adenosyltransferase in lung increases at the time that it is possible to breathe (Fig.5). This increase in specific activity of methionine adenosyltransferase in the lung may be a response to an increased requirement for methylation associated either with biosynthesis of surface-active agents required to keep the lungs inflated or with detoxification of the blood which perfuses the lung in the extra-uterine situation. It also may represent a compensation for a decrease in tissue concentration of methionine. These possibilities are not mutually exclusive.

Thus, in the human and monkey, we have evidence of a developmental pattern for methionine adenosyltransferase which raises tantalizing teleological possibilities. As yet, however, we have no measurements in the developing human or monkey of the product, *S*-adenosylmethionine. The results from the developing rodent, discussed above, suggest that such data, albeit interesting, cannot give definitive answers, since the problem may be as much a function of the utilization of *S*-adenosylmethionine as of its synthesis.

It is possible that the factor normally limiting the rate of synthesis of *S*-adenosylmethionine by methionine adenosyltransferase is not the concentration of the catalytic protein, but the tissue concentration of methionine. The apparent

$S_{0.5}$ for methionine of rat hepatic methionine adenosyltransferase is about 10^{-3} M. Our unpublished determinations of methionine concentrations of human fetal tissues suggest that the tissue concentration of methionine is probably about 1 order of magnitude lower, so that the enzyme would be functioning at a fraction of its V_{max}. It is known that the concentrations of amino acids in fetal human liver are higher than that in adult human liver (Ryna and Carver, 1966), but their results for methionine are probably spuriously high because cystathionine and methionine were probably not separated. Most amino acids in fetal plasma, including methionine, but with the notable exception of cystine, are concentrated by the placenta against a gradient (Szabo and Grimaldi, 1970). In addition, the lower activity of 5-methyltetrahydrofolate-homocysteine methyltransferase and the lower concentration of its substrate 5-methyltetrahydrofolate in the mature liver and brain would tend to reduce the amount of methionine produced by recycling. Following this line of argument, it is possible that the sudden increase in methionine adenosyltransferase activity at birth is a compensation for the lower tissue concentrations of methionine after birth. Indeed, Baldessarini and Kopin (1966) have pointed out that the availability of methyl acceptors and the rate of utilization of *S*-adenosylmethionine may be the major determinants of the concentration of *S*-adenosylmethionine. Consistent with this hypothesis is the report that the administration of *L*-DOPA to rats results in a decrease in *S*-adenosylmethionine concentrations in brain and adrenal but not in liver (Wurtman *et al.*, 1970), since *L*-DOPA and its decarboxylation product L-DOPamine both undergo rapid 0-methylation at the expense of *S*-adenosylmethionine. The turnover of *S*-adenosylmethionine in liver is rapid, $t_{1/2}$ = 10 min (Lombardini and Talalay, 1971); that of brain has not been calculated, but changes in its specific radioactivity suggest it is very much slower (Baldessarini and Kopin, 1966).

It should also be mentioned that the rate of cellular proliferation *per se* is not correlated with activity of methionine adenosyltransferase. Hancock (1966) showed that a variety of mouse tumors, including a mouse hepatoma, had low activity of methionine adenosyltransferase. Sheid and Bilik (1968) found that Morris hepatomas had about the same amount of methionine adenosyltransferase activity as the host livers and that there was little if any activity in either the solid or the ascites Novikoff tumor. Grossman and coworkers (1974) recently reported methionine adenosyltransferase activity in six different rat hepatomas and found that the specific activity of methionine adenosyltransferase ranged from 2-99% of the mean value for control liver and that the activity did not correlate with rate of growth of the tumor. Lombardini and Talalay (1971) have found that the concentration of *S*-adenosylmethionine in Walker 256 and Lewis lung tumor were comparable to the concentrations of *S*-adenosylmethionine in liver. The concentration of *S*-adenosylmethionine in the blood of normal children is not different from the blood of children with solid tumors or even leukemia; however, all children had higher blood concentrations than those found in normal adults (Bohuon and Caillard, 1971). Thus, a high rate of proliferative growth, *per se*, does not result in a high concentration of *S*-adenosylmethionine or in a high activity of methionine adenosyltransferase. The rate of formation as well as of utilization of *S*-adenosylmethionine are factors in the control of concentrations of *S*-adenosylmethionine in the tissue and, in turn, these factors may relate to specific biosynthetic functions of methyl and propylamino transfer.

Table I. Enzymatic activities in the liver.

	Methionine adenosyl-transferase	Cystathionine β-synthase	Cystathionase	Methyltetrahydrofolate-homocysteine methyltransferase
Patient	5.0; 10.0	118.2	310	2.5
Fetus	26 ± 15 (24)	21 ± 20 (24)	0 (24)	4.7 ± 1.1 (31)
Mature	86 ± 48 (9)	98 ± 57 (9)	126 ± 36 (9)	1.3 ± 0.67 (17)

Activities are expressed as nanomoles of product formed per milligram of soluble protein per hour ± standard deviation. The numbers in parentheses represent the number of separate determinations.

3. DEFICIENCY OF METHIONINE ADENOSYLTRANSFERASE

We recently demonstrated deficiency of hepatic methionine adenosyltransferase (6-12% of mean control activity), unaccompanied by deficiency in activity of cystathionine β-synthase, cystathionase or 5-methyltetrahydrofolate-homocysteine methyltransferase (Table I), in an infant whose plasma concentration of methionine varied between 80 and 128 μmoles/100 ml (20-30 times normal). Even when the infant received a diet low in methionine, the plasma concentration of methionine could not be reduced to less than 10 times normal. No homocystinemia, hypocystinemia, cystathioninuria or tyrosinemia was present. A battery of "liver function" tests and serological tests for hepatogenic infectious agents failed to reveal any general hepatatic disorder which would account for the hypermethioninemia. At the time of percutaneous liver biopsy (6 months of age) there was no cirrhosis; however, electron microscopy by Dr. Fenton Schaffner (unpublished observations) demonstrated breaks in the outer membranes of the mitochondria and hyperplasia of the smooth endoplasmic reticulum. The infant was otherwise fit. Indeed, the sole reason she came under investigation was that hypermethioninemia was discovered in a mass-screening program. A preliminary report of the enzymatic findings has been published (Gaull and Tallan, 1974) and a complete report of the clinical and pathological findings will be published separately.

It is possible that this enzymatic deficiency is the result of a genetic mutation rather than the result of a delay in enzymatic maturation or of an acquired disease. First, the activity of methionine adenosyltransferase on 2 separate biopsies was well below that found in the liver in the second trimester fetuses we examined (Gaull *et al.*, 1972) or in various hepatic cirrhoses (Gaull *et al.*, 1970), including so-called hereditary tyrosinemia. Second, the metabolic abnormality persists to 1½ years of age, well beyond the age that might be considered a delay in maturation. Finally, the deficiency of methionine adenosyltransferase was demonstrated to be specific. There was no accompanying deficiency of cystathionine β-synthase, as in cirrhosis. Hug and coworkers (1968, 1968a) have briefly described an infant with hypermethioninemia and deficiency of methionine adenosyltransferase, but no other enzymes of sulfur metabolism were examined; thus it is not clear that their patients had a specific deficiency. Furthermore, the ultrastructural changes in liver and the clinical picture in Hug's patient were entirely different from those found in our patient. Genetic and tissue culture studies of our patient and her parents will be performed and may help establish this specific enzymatic deficiency as a heritable disease.

The crucial, and as yet unanswered, question is whether the amount of *S*-adenosylmethionine synthesized by the 6-12% of mean control methionine adenosyltransferase activity present in the liver is sufficient to cope with its important functions. If it is inadequate, then replacement therapy with the stabilized *S*-adenosylmethionine now available (cf. Salvatore *et al.*, 1975) should be attempted. However, it is conceivable that the rate of synthesis of *S*-adenosylmethionine by this reduced activity is adequate. The increased concentration of methionine that builds up may allow the reduced amount of functional enzyme to operate closer to its V_{max}, perhaps aided by other regulatory compensations involving K^+ and Mg^{2+}, as discussed above. If the buildup of methionine indeed is compensatory, then perhaps a reduction in dietary methionine is deleterious. It is possible also that the role of 5-methyltetrahydrofolate as a methyl donor for methyltransferase is wider than presently thought. Perhaps it can substitute for *S*-adenosylmethionine in the event of a diminished supply of the latter compound. However, it is also possible that there is insufficient *S*-adenosylmethionine for large excesses of catecholamines, even though routine physiological function is adequate.

Finally, the paucity of overt physical findings in our patient unfortunately gives no assurance that this is a benign disease. Patients with cystathionine synthase deficiency may show few, if any, obvious physical abnormalities until 3-5 years of age and our previous experience with that disease allowed us to identify the morphological abnormality of the hepatocytes in methionine adenosyltransferase deficiency.

ACKNOWLEDGEMENTS

This work was supported by the New York State Department of Mental Hygiene, NIH Clinical Genetics Center Grant GM-19443, NIH Clinical Research Center Grant FR-71 and the Labor Foundation.

REFERENCES

Baldessarini, R.J. and Kopin, I.J. (1966). *J. Neurochem.* **13**, 769-777.
Bohuon, C. and Caillard, L. (1971). *Clinica chim. Acta* **33**, 256.
Cantoni, G.L. (1951). *J. biol. Chem.* **189**, 745-753.
Cantoni, G.L. (1965). *In* "Transmethylation and Methionine Biosynthesis" (S.K. Shapiro and F. Schlenk, eds) pp.21-32. The University of Chicago Press.
Chase, H.P., Volpe, J.J. and Laster, L. (1968). *J. clin. Invest.* **47**, 20-99-2108.
du Vigneaud, V. and Rachele, J.R. (1965). *In* "Transmethylation and Methionine Biosynthesis" (S.K. Shapiro and F. Schlenk, eds) pp.21-32. The University of Chicago Press.
Finkelstein, J.D. (1967). *Arch. biochem. Biophys.* **122**, 583-590.
Finkelstein, J.D. and Mudd, S.H. (1967). *J. biol. Chem.* **242**, 873-880.
Gaull, G.E. (1973). *In* "Symposium on Developmental Biochemistry - Inborn Errors of Metabolism" (F. Hommes and C.J. von den Berg, eds) pp.131-141. Academic Press.
Gaull, G.E., Raiha, N.C.R., Saarikoski, S. and Sturman, J.A. (1973). *Pediat. Res.* **7**, 908-913.
Gaull, G.E., Rassin, D.K., Solomon, G.E., Harris, R.C. and Sturman, J.A. (1970). *Pediat. Res.* **4**, 337.
Gaull, G.E., Sturman, J.A. and Raiha, N.C.R. (1972). *Pediat. Res.* **6**, 538.
Gaull, G.E., Sturman, J.A. and Rassin, D.K. (1969). *Neuropadiatrie* **1**, 199-226.
Gaull, G.E. and Tallan, H.H. (1974). *Science* **184**.
Gaull, G.E., von Berg, W., Raiha, N.C.R. and Sturman, J.A. (1973). *Pediat. Res.* **12**, 527.
Grossman, M.R., Finkelstein, J.D., Kyle, W.E. and Morris, H.P. (1974). *Cancer Res.* **34**, 794-800.
Hancock, R.L. (1966). *Cancer Res.* **26**, 2425-2430.
Hug, G., Cussen, L.J., Schubert, W.K. and Chuck, G. (1968). *J. clin. Invest.* **47**, 49a-50a.
Hug, G., Cussen, L.J., Schubert, W.K. and Chuck, G. (1968). Proceedings Electron Microscopy Society of America **26**, 182-183.
Lombardini, J.B. and Talalay, P. (1971). *Advan. Enzyme Reg.* **9**, 349-384.
Mudd, S.H. (1965). *In* "Transmethylation and Methionine Biosynthesis" (S.K. Shapiro and F. Schlenk, eds) pp.21-32. The University of Chicago Press.

Natori, Y. (1963). *J. biol. Chem.* **238**, 2075-2080.
Ordonez, L.A. and Wurtman, R.J. (1973). *J. Neurochem.* **21**, 1447-1455.
Pan, F., Chang, G., Lee, S. and Tang, M. (1968). Proceedings of Society of Experimental Biology in Medicine **128**, 611.
Pan, F. and Tarver, H. (1967). *J. Nutr.* **92**, 274-280.
Ryan, W.J. and Carver, M.J. (1966). *Nature* **212**, 292.
Salvatore, F., Zappin, V. and Borek, E. (1975). "The Biochemistry of Adenosylmethionine". Columbia University Press, New York.
Shapiro, S.K. and Schlenk, F. (1965). "Transmethylation and Methionine Biosynthesis". University of Chicago Press, Chicago and London.
Sheid, B. and Bilik, E. (1968). *Cancer Res.* **28**, 2512-2515.
Snyder, S.H., Banerjee, S.P., Yamamura, H.I. and Greenberg, D. (1974). *Science* **184**, 1243-1253.
Sternowsky, H.J., Raiha, N.C.R. and Gaull, G. (1973). *Pediat. Res.* **7**, 316.
Sturman, J.A. and Gaull, G.E. (1974). *Pediat. Res.* **8**, 231-237.
Sturman, J.A., Gaull, G. and Raiha, N.C.R. (1970). *Science* **169**, 74-75.
Sturman, J.A., Niemann, W.H. and Gaull, G.E. (1973). *Biol. Neonate* **21**, 16.
Sturman, J.A., Rassin, D.K. and Gaull, G.E. (1970). *Internat. J. Biochem.* **1**, 251.
Szabo, J. and Grimaldi, D. (1970). *Adv. Metab. Dis.* **5**, 185-228.
Tabor, H. and Tabor, C.W. (1964). *Pharmacol. Rev.* **16**, 245-300.
Tallan, H.H., Cohen, P.A. and Gaull, G.E. (1973). *Fedn Proc.* **31**, 576 (**Abs.**).
Taylor, R.T. and Weissbach, H. (1969). *Arch. biochem. Biophys.* **129**, 728-744.
Taylor, R.T. and Weissbach, H. (1969). *Arch. biochem. Biophys.* **129**, 745-766.
Williams-Ashman, H.G., Pegg, A.E. and Lockwood, D.H. (1969). *Advan. Enzyme Reg.* **7**, 291-323.
Wurtman, R.J., Rose, C.M., Matthysse, S., Stephenson, J. and Baldessarini, R. (1970). *Science* **169**, 395-397.

DISCUSSION

Wick: Did you see any clinical symptoms or signs when this child was on a methionine-restricted diet?

Gaull: No apparent clinical differences were seen when she was placed on the low methionine diet by Dr. Derek Lonsdale at the Cleveland Clinic, but the time on that diet was short. However, the child appeared normal from the start. She would not have been discovered had they not been doing routine screening in Ohio. We would not have gone ahead in trying to identify the enzymatic deficiency, if we had not known that patients with cystathionine synthase deficiency may do perfectly well for a number of years. We felt justified both from an ethical as well as from a scientific point of view in going ahead with these studies and trying to define the defect. I do not know what the clinical effect of long-term changes in dietary methionine would be.

Gitzelmann: Did you examine the parents and what did you find as far as methionine in the blood is concerned?

Gaull: We have investigated methionine concentrations in the blood of the parents and have carried out methionine loading tests. By our standards, there is no abnormality. I would point out, in this regard, that loading tests, especially in cystathionine synthase deficiency, are not helpful in identifying heterozygotes. There is no deficiency of methionine adenosyltransferase in skin fibroblasts, not even in the patient. We are looking next at long-term lympoid cell lines, and, if necessary, we shall do liver biopsies in the parents.

Blass: Are there any clinical abnormalities in the parents or family history of neurological or psychiatric disease?

Snell: I have measured the activity of cystothionase, which is one of the enzymes in the transsulfuration pathway in the rat. And here I would suggest that there may be very great differences between the rat and the human in the development of these enzymes, because in the rat the methionine activating enzyme did not rise substantially until after birth. There was activity present in the fetal liver which rapidly increased after birth. The cystathionase activity paralleled this increase very closely, suggesting that the methionine activating enzyme functions together with the cystathionine system. The whole transsulfuration pathway therefore operates in the postnatal rat. I want to turn now to the question of where the methyl donor is coming from in the postnatal rat. Some measurements by my colleague, Dr. Rowsell, showed that there were very high levels on the serine to glycine interconversion system in the neonatal rat, which is very much increased over the adult level and indeed the fetal level, suggesting that this system may play a role in methyl transfer in the postnatal rat.

Gaull: I am familiar, of course, with your work on cystathionase and I very much agree with what has been reported by Heinonen and Raiha (*Biochem. J.* 135 (1973) 1011) for the rat. I also agree that there is a postnatal increase. However, there is still considerable cystathionase activity in fetal liver of every animal that we have been able to examine. The only one that has virtually no cystathionase activity, so far, is the human. We have demonstrated immunochemically that there is no more than a trace of cystathionase. The parallelism you mentioned between the methionine-activating enzyme and cystathionase is an interesting one. Did you actually measure the serine hydroxymethyltransferase?

Snell: No, we did not: neither did I measure the methionine activating enzyme at the same time. The parallelism I suggested was merely a correlation between the results of different groups of workers. The activity measured by Dr. Rowsell was the interconversion of glycine and serine by radiochemical methods involving the liberation of $^{14}CO_2$. So what was really measured was overall glycine oxidation. But this has a component from the serine-hydroxymethyl transferase and other components as well. The real question is: what is the serine-hydroxymethal transferase activity during development in the rat liver?

Gaull: Serine hydroxymethyl transferase activity decreases strikingly in the human brain. Postnatally, serine hydroxymethyl transferase activity in fetal human liver is not different from that in mature human liver (Gaull *et al.*, *Pediat. Res.* 7 (1973) 527).

Snell: The clinical aspect of the question is: do you have any idea about how active the serine to glycine system is in the child you discussed? Because it appears to me that there was an increase in the methyltetrahydrofolate transferase system, which suggests that there was some adaptation to methyl transferase in that direction. Is the methyl donor in this girl, in fact, due to an increase in activity of the serine-glycine system, as the methionine system is not really required. The child would be adapted to the use of serine rather than methionine as a methyl donor.

Gaull: The methyltetrahydrofolate-homocysteine methyl transferase activity of this child was rather high. What I did mention about this child in my talk, is that the other abnormality which we have been able to identify was the failure to clear folates in the usual way. The blood folate concentration should be back to normal within two or three weeks following folic acid administration. This child, however, still had enormously high blood folate concentrations months after she had been given folic acid. So there is probably some abnormality in those two interlocking cycles or methionine regeneration and tetrahydrofolate regeneration. The apparent defect in folate metabolism is interesting, because patients with cystathionine synthase deficiency tend to run low total blood folates as measured by *L. casei.* Methyltetrahydrofolate is the major folate of blood and liver, so you have put your finger on something interesting which we want to investigate in the future. The beta-carbon of serine is used in *de novo* synthesis of thymidelate. We do not know how fast that enters the general one-carbon pool, but it is likely that there must be a separate pool for the methyl carbon of methionine that cannot be made up for by simply giving serine.

Van den Berghe: Have you checked the possibility that there might be something abnormal with the kinetics of the other substrate, ATP?

Gaull: The two biopsies which were taken were about 20 mg wet weight each and therefore we were not able to study the kinetics of the enzyme. We were hoping to do this in fibroblasts, but the fibroblasts did not show the enzyme deficiency. If the long term lympoid cellines go into culture, we will study the kinetics in more detail.

Tyson Tildon: Are there multiple molecular forms of this enzyme? Because you have in the fibroblasts a full complement of activity and only a low value, but not one absent in the liver cells.

Gaull: We think that multiple forms do, indeed, exist. They do occur in yeast (Dr. Stanely Shapiro, personal communication).

BIOGENESIS OF GALACTOSE, A POSSIBLE MECHANISM OF SELF-INTOXICATION IN GALACTOSEMIA

R. Gitzelmann*, R.G. Hansen+ and B. Steinmann*

*Division of Metabolism, Department of Pediatrics,
University of Zürich, Switzerland
+Department of Chemistry and Biochemistry
Utah State University, Logan, Utah, USA

GALACTOSE METABOLISM

Galactose which is consumed in excess of developmental needs is metabolized for energy.

In humans the liver appears to be the primary site for the metabolism of galactose; however other tissues including the red cell have this capacity and hence can conveniently reflect the metabolic capacity of the individual. Fundamental studies using microorganisms and animals together with analysis of the metabolic problems in the human genetic disorders have clarified the pathways of galactose metabolism in man.

The primary metabolic reactions of galactose are illustrated in Fig.1.

1. Non-specific reduction of the aldehyde at carbon-1 occurs which leads to the product galactitol (Hers, 1960; Hayman and Kinoshita, 1965). This is especially significant in some genetically defective humans since the lens of the eye forms galactitol from excess galactose, and with no further capacity to metabolize galactitol it accumulates giving rise to cataracts (van Heyningen, 1971).
2. Dehydrogenation at carbon-1 leading to the formation of galactonic acid occurs in animals and man (Cuatrecasas and Segal, 1966). In man, whether galactonic acid arises by the action of a specific enzyme or of a general aldehyde dehydrogenase or whether by some phosphorylated intermediates has not been solved, but galactonic acid is excreted in galactosemia and in galactokinase deficiency when galactose is consumed (Bergren *et al.*, 1972; Gitzelmann *et al.*, 1974).
3. Microorganisms contain an enzyme which catalyzes an oxidation of galactose at carbon-6 to galactose hemialdehyde (Avigad *et al.*, 1962). While this reaction is of no known significance in humans it does serve as the basis for the quantitative estimation of galactose and some of its derivatives.
4. Direct phosphorylation of galactose at carbon-6 is of questionable significance in man. Some evidence for the occurrence of gal-6-P has been presented

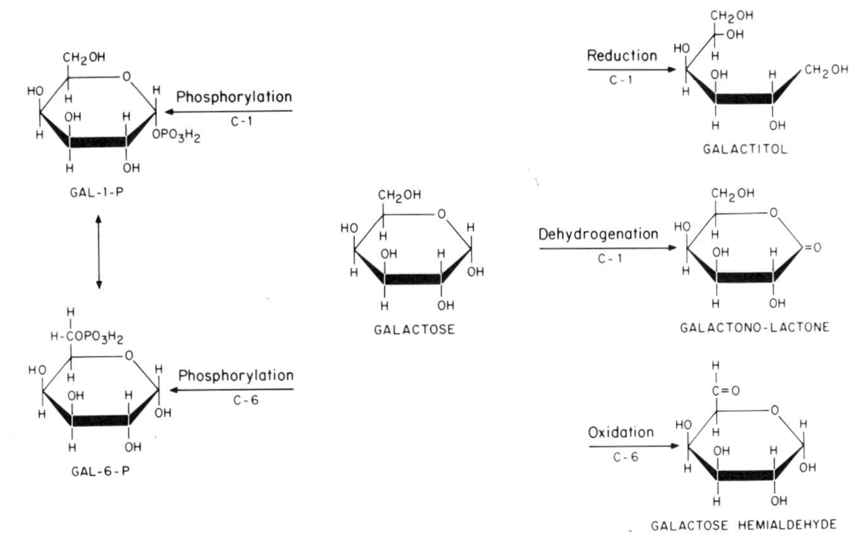

Fig. 1 Initial steps of galactose metabolism.

(Inouye *et al.*, 1962), but as an alternative to direct phosphorylation it could arise from gal-1-P with a mutase type reaction (Posternak and Rosselet, 1954).

5. Phosphorylation catalyzed by a kinase initiates the Leloir pathway of galactose metabolism in man. The resulting product gal-1-P was identified in 1943 in rabbit liver following ingestion of galactose (Kosterlitz, 1943). Normally human liver metabolizes most of the ingested galactose through the Leloir pathway which in addition to the kinase requires two more enzymes, namely a transferase and an epimerase (Leloir, 1951):

a) kinase
galactose + ATP → gal-1-P + ADP

b) transferase
gal-1-P + UDP-glc ⇌ UDP-gal + glc-1-P

c) epimerase
UDP-gal ⇌ UDP-glc
Sum of b + c:
gal-1-P ⇌ glc-1-P

In addition to the function of UDP-glc as a cofactor in the Leloir pathway for galactose utilization it serves broader needs of the cell (Caputto *et al.*, 1967; Ginsburg and Neufeld, 1969). The synthesis of glycogen requires UDP-glc. Many chemical compounds needed for growth and for the differentiation of structures such as nervous tissue, bone, antigenic determinants on cell surfaces, etc., contain a variety of saccharides. UDP-glc is the primary source of glycosyl residues for the biosynthesis of many of these saccharides. It is not surprising therefore that the enzyme for catalyzing the synthesis of UDP-glc is ubiquitous and abundant:

d) pyrophosphorylase
glc-1-P + UTP ⇌ UDP-glc + PPi

In aqueous extracts of liver more than 0.5% of the protein is this synthetase for

UDP-glc (Albrecht *et al.*, 1966; Levine *et al.*, 1969). For the present paper it is relevant that the reaction is reversible; the equilibrium *in vitro* favors reactants as written which fact gives rise to the enzyme being commonly designated as a pyrophosphorylase.

In addition to the occurrence in abundant quantities the pyrophosphorylase is not highly specific. It will catalyze the reaction with UDP-gal at about 5% of the rate with UDP-glc (Turnquist *et al., in press*).

d) gal-1-P + UTP $\rightleftharpoons$ UDP-gal + PPi

Isselbacher (1958) has proposed a pyrophosphorylase pathway of galactose utilization which is significant in humans:

d) pyrophosphorylase
gal-1-P + UTP $\rightleftharpoons$ UDP-gal + PPi

c) epimerase
UDP-gal $\rightleftharpoons$ UDP-glc

d′) pyrophosphorylase
UDP-glc + PPi $\rightleftharpoons$ UTP + glc-1-P

Sum d, c and d′:
gal-1-P $\rightleftharpoons$ glc-1-P

Due to the low specificity of the UDP-glc pyrophosphorylase it will catalyze both reactions d and d′ (although claims have been made that a separate enzyme (Abraham and Howell, 1969; Chacko *et al.*, 1972) exists in human liver for each of the two reactions).

The epimerase, which is required together with the pyrophosphorylase seems to be ubiquitous and occurs early in human development (Gitzelmann and Steinmann, 1973). To produce gal-1-P from UDP-gal in the cell obviously PPi is required. It is now clear that PPi is a by-product of most biosynthetic reactions (Lehninger, 1965) producing polysaccharides, proteins and lipids, and furthermore PPi occurs in sufficient quantities to play a regulatory function in metabolism of plant cells (Schwenn *et al.*, 1973; Levine and Bassham, 1974) and to form crystals with calcium, notably in some human joint disorders (Russell *et al.*, 1970).

GALACTOSE-1-PHOSPHATE IN GALACTOSEMIA

Galactosemia is an inborn error of metabolism (Segal, 1972) caused by the inherited deficiency of gal-1-P uridyltransferase (Fig.10) which normally catalyzes the second step in the conversion of galactose to glucose. Disease manifestations are cataracts, liver and kidney dysfunction and disturbed mental development due to organic brain damage. Symptoms arise when galactose is ingested and gal-1-P, i.e. the product of the kinase reaction accumulates. This phosphate ester is the incriminated noxious agent causing most of the pathology. This can be inferred from the comparison of disease manifestations in transferase deficiency and those in galactokinase deficiency (Table I).

Cataracts are the only common sign; they are caused by osmotic swelling and disruption of lens fibers due to the accumulation of galactitol (van Heyningen, 1971). Brain, liver and kidney pathology is not observed in kinase deficiency and since gal-1-P cannot be formed in this disorder, one can conclude that it must play the deciding pathogenetic role in transferase deficiency.

Transferase deficient infants and children are treated with a galactose-exclusion diet. The stringency of the diet can be monitored by the estimation of gal-1-P in erythrocytes (Gitzelmann, 1970), a level of $\leqslant 4$ mg/100 ml being considered suf-

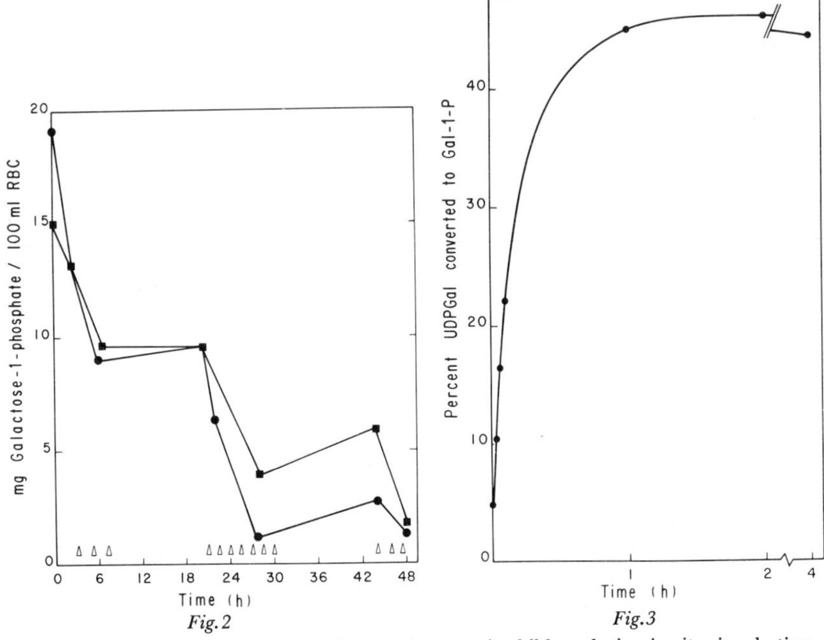

Fig. 2 Gal-1-P levels in erythrocytes of two galactosemic children during *in vitro* incubation with glucose as the sole carbohydrate. Points indicate changes of medium (Gitzelmann, 1969).

Fig. 3 Formation of gal-1-P from UDP-gal by an erythrocyte lysate lacking gal-1-P uridyltransferase (Gitzelmann, 1969).

Table I. Affected organs in untreated galactokinase and gal-1-P uridyltransferase deficiencies.

	Kinase Deficiency	Transferase Deficiency
Lens	+	+
Brain	—	+
Liver	—	+
Kidney	—	+

ficiently low. Galactose ingestion is followed by an immediate rise of red cell gal-1-P. We were therefore alarmed in 1966 when we discovered that in a galactosemic infant, diagnosed at birth and treated adequately, gal-1-P had risen from 8 mg/100 ml in cord blood to 22 mg/100 ml on her tenth day and stayed high for weeks (Gitzelmann, 1969). Her red cells, maintained *in vitro* on a galactose-free medium, were not only capable of disposing rapidly of gal-1-P but, similarly to those of her elder sister, seemed able to synthesize it as well (Fig. 2).

If red cells can form gal-1-P from glucose, other tissues must have the same capability. Evidently, such biosynthesis of gal-1-P constituted a mechanism of potential self-intoxication in well-treated galactosemic infants (Gitzelmann

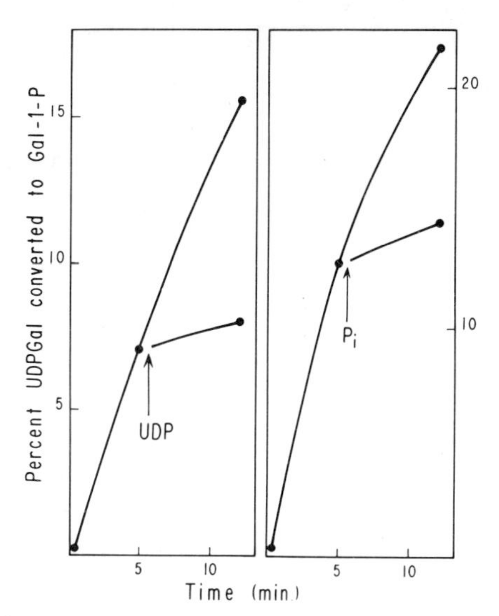

Fig.4 Inhibition of gal-1-P formation from UDP-gal by UDP and Pi in erythrocytes lacking gal-1-P uridyltransferase (Gitzelmann, 1969).

and Hansen, 1974). Therefore, we started to investigate which biochemical route was being used. We have obtained convincing evidence that it is the pyrophosphorylase pathway (Gitzelmann and Hansen, *in press*).

GALACTOSE-1-PHOSPHATE FROM UDP-GALACTOSE

That the pyrophosphorylase pathway is a means of galactose metabolism in the human erythrocyte stems from the following observations: epimerase and UDP-glc are normal red cell constituents. Hence UDP-gal is also available as a substrate for other reactions. Incubation of hemolysates lacking transferase (Gitzelmann, 1969) with UDP-gal produced gal-1-P in a reaction which had an absolute dependence on inorganic pyrophosphate and was stimulated by magnesium (Fig.3); the production of gal-1-P was inhibited by DUP-glc, by UDP and Pi (Fig.4). Under conditions which are identical, the crystalline enzyme from human liver catalyzes the formation of gal-1-P from UDP-gal (Knop and Hansen, 1970; Turnquist *et al., in press*).

UDP-glucose pyrophosphorylase of both calf and rabbit have also been purified and crystallized (Albrecht *et al.*, 1966; Knop and Hansen, 1970; Turnquist *et al., in press*). Biochemical evidence is convincing that one protein catalyzes reactions with both glucose and galactose derivatives. Throughout purification and crystallization the ratio of activity of the enzyme towards the various substrates remains constant (Ting and Hansen, 1968; Knop and Hansen, 1970; Turnquist *et al., in press*). UDP-gal is bound to the purified enzyme as a function of the number of protomer sub-units of pyrophosphorylase (Levine *et al.*, 1969; Turnquist *et al.*, 1974). This bound UDP-gal may then be stoichiometrically replaced by UDP-glc (Fig.5: Turnquist *et al.*, 1974). UDP-glc also limits the synthesis of gal-1-P from UDP-gal by hemolysates (Gitzelmann, 1969). It is concluded therefore that although less effectively, UDP-gal competes for the same site on the enzyme.

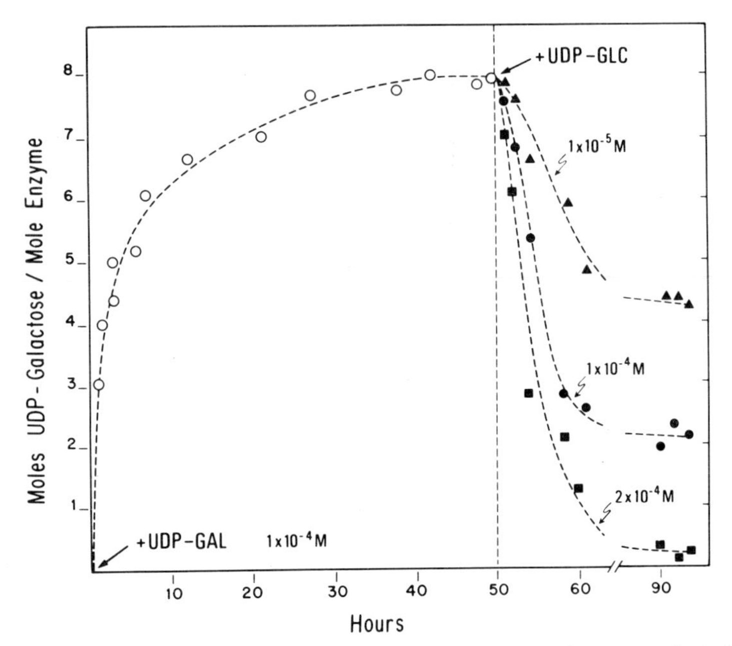

Fig.5 Substrate binding by UDP-hexose pyrophosphorylase. To estimate quantitatively substrate binding by crystallized pyrophosphorylase from human liver, dialysis cells were constructed using silicone rubber sheeting. Two mg of enzyme were added to one cell. Both cells were then filled with 1×10^{-4} M UDP-gal, 2 nM UDP-gal (^{14}C-uridine), 2 mM Mg^{++}, and .01 molar tricine buffer (pH 8.5) to 0.5 ml volume in each cell. At the time intervals indicated the equilibrium was monitored by placing 5 µl aliquots in 15 ml of Aquasol^{R} and measuring radioactivity in a liquid scintillation counter. To displace UDP-gal after 50 hours, UDP-glc was added in different concentrations (Turnquist *et al.*, 1974).

Immunological evidence has also been obtained which supports this conclusion (Gitzelmann and Hansen, *in press*).

Guinea pigs were immunized with crystalline UDP-glc pyrophosphorylase from human, calf, and rabbit liver. Double diffusion analyses showed cross-reactivity between all three enzymes. Red cell lysates and extracts of kidney, liver and brain of humans reacted with antisera against the human and calf liver enzyme, and so did erythrolysates of galactosemic patients. Both the human liver and calf liver enzymes were precipitated as well as inhibited by antisera against rabbit, human and calf liver enzyme. Immunotitration experiments are shown in Fig.6. As equal amounts of the two enzymes were reacted with increasing volumes of antiserum, the activities which remained in solution decreased in the same proportion. In all samples, the relative activities with UDP-gal compared to that with UDP-glc remained constant within the experimental error: i.e., 3-5 percent. This fact strongly suggested the presence of a single enzyme protein in both the human and calf liver preparations. Much less probably the three enzymes used for the stimulation of antibody shared common antigenic determinants on a similar subunit (Gitzelmann and Hansen, *in press*).

Similar immunotitration experiments were performed on red cell lysates from

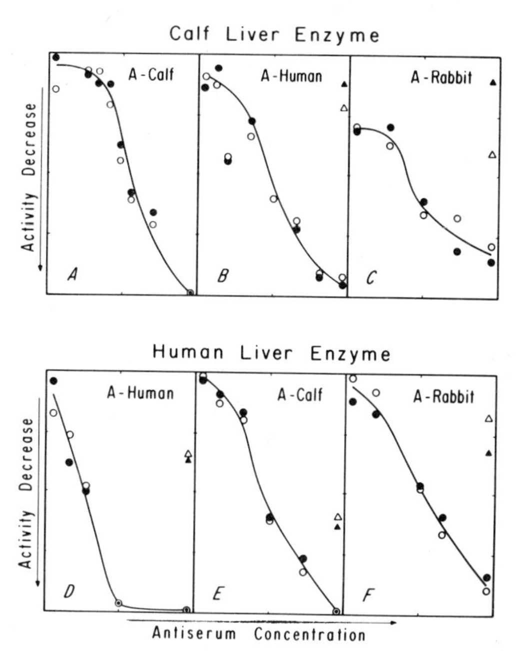

Fig. 6 Immunoprecipitation of calf liver and human liver UDP-hexose pyrophosphorylase with serum from guinea pigs immunized with the crystalline enzyme from either calf, man or rabbit. For immunoprecipitation of calf enzyme (A, B and C), aliquots of dialyzed enzyme containing approximately 40 μg of protein and 6 U of activity were treated with varying volumes of decomplemented antiserum, in 0.01 M glycine buffer pH 8.0 and 0.02 M mercaptoethanol in the presence of 160 μg of bovine serum albumin free of pyrophosphorylase activity, in a volume of 0.2 ml, at 2° overnight. For precipitation of one half of the antigen, 14 μl of anti-calf, 60 μl of anti-human and 75 μl of anti-rabbit serum was needed. For immunoprecipitation of human enzyme (D, E and F), aliquots of 5 μg of 0.5 U of dialyzed enzyme were incubated with increasing volumes of antiserum, but for better enzyme stability the volume of serum was kept constant in all tubes by supplementing antiserum with non-specific serum. Bovine serum albumin (120 μg in saline) was added only to the control tube. For precipitation of one half of the antigen, 5 μl of anti-human, 17 μl of anti-calf and 17 μl of anti-rabbit serum was needed. Samples were centrifuged at 27,000 g for 20 min, and pyrophosphorylase activity remaining in the supernatants was estimated. UDP-glc pyrophosphorylase was assayed (Albrecht *et al.*, 1966) in the presence of 1 mM UDP-glc, 2 mM PPi and magnesium; UDP-gal pyrophosphorylase (Verachtert *et al.*, 1965) with 10 mM UDP-galactose and 2 mM PPi and magnesium. The antisera are abbreviated A - Human, etc.. Activity with saline replacing immune serum is shown as △, ▲, for UDP-glc pyrophosphorylase ○, △, and for UDP-galactose pyrophosphorylase ●, ▲. For purposes of illustration, the scale for UDP-gal pyrophosphorylase is expanded 20 to 40 fold. In all but one (C) titrations, precipitation was complete or nearly complete as shown in the graph. Precipitation of calf liver enzyme by the anti-rabbit enzyme remained incomplete, e.g. in C: two-thirds of enzyme remained in solution and active in presence of 160 μl of serum, (Gitzelmann and Hansen, *in press*).

galactosemic patients with complete absence of transferase. As is shown on the titration curve (Fig.7), the activity rapidly disappeared from the supernatants; it was initially recovered in part from the precipitates, but as antiserum concentrations increased, the activity became progressively inhibited. Thus, the lysates

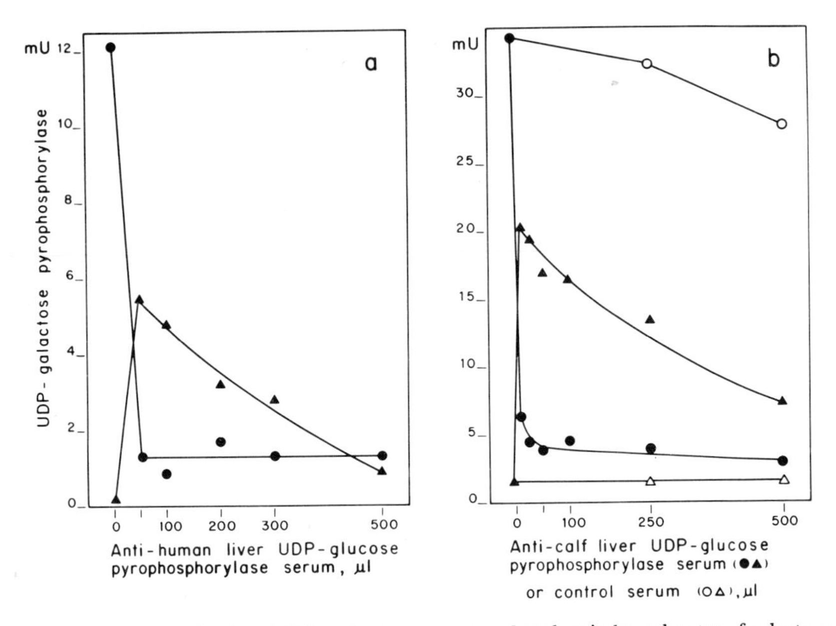

Fig. 7 Immunoprecipitation of UDP-galactose pyrophosphorylase in hemolysates of galactosemic patients with guinea pig antisera against (a) human and (b) calf liver UDP-glucose pyrophosphorylases. Red cells of galactosemic children were washed, frozen and thawed, cleared from ghosts by centrifugation and dialyzed against water, in the cold. Lysates were pre-incubated at 37° for 10 min, cooled and diluted with one part of 0.06 M Tris-HCl pH 8.0. Equal portions of lysate (1.5 ml) were reacted with increasing volumes of antiserum (0.01 to 0.05 ml) or control serum which had been decomplemented and dialyzed against 0.9% NaCl, at 20° for 20 min, and then at 2° overnight. Samples were centrifuged at 113,000 g for 60 min, supernatants were aspirated, and the precipitates were resuspended in 2 ml of 0.08 M Tris-HCl pH 8.0 using motor driven Teflon pestles. Supernatants and precipitates were assayed for UDP-gal pyrophosphorylase activity at 37°. The reaction mixtures contained UDP-gal (1 mM), sodium pyrophosphate (40 mM), $MgCl_2$ (10 mM), in 0.06 M Tris buffer pH 8.0, and UDP-galactose-^{14}C (galactose µl). Samples were deproteinized at 100° for 90 sec, cooled on ice, centrifuged, and 10 µl amounts of the supernatants were analyzed by ascending chromatography, radio-chromatography scanning and liquid scintillation counting as described (Gitzelmann, 1969). Activity was calculated as the percentage of UDP-gal converted to gal-1-P and is expressed in mU per immunoreaction tube, for supernatants (●) and precipitates (▲) separately. One experiment with non-immune serum is shown with open symbols (○, △). Evidently, the enzyme was precipitated completely at low antiserum concentrations; inhibition (see precipitates) progressed as serum concentrations rose (Gitzelmann and Hansen, *in press*).

lost their capability of converting UDP-gal to gal-1-P through precipitation/inhibition by the antibody to UDP-glc pyrophosphorylase of human liver. Not unexpectedly, the antiserum to the calf liver enzyme had a similar inhibitory effect (Gitzelmann and Hansen, *in press*).

BIOSYNTHESIS OF GALACTOSE FROM GLUCOSE

Today there is much additional evidence, both clinical and experimental, documenting biosynthesis of galactose from glucose in man. Mayes and Miller (1973) have grown transferase deficient skin fibroblasts in a medium devoid of galactose and observed the formation of gal-1-P from glucose, to concentrations (approx. 2 mg/g protein) similar to those attained by red cells *in vitro* (Fig.2; approx. 3mg per g non-Hb protein).

Galactokinase deficient children grow and develop normally when maintained on a diet free of glactose. Pregnant women who are heterozygotes for transferase deficiency and have previously had transferase deficient offspring are subjected to a galactose exclusion diet; in spite of galactose deprivation, their fetuses develop normally. One completely transferase deficient woman who followed a lactose-free regimen throughout pregnancy delivered a healthy infant and, presumably still on the diet, produced lactose (2.8 g/100m l) in her colostrum on the second post partum day (Roe *et al.*, 1971). In galactokinase deficient twin infants on diet, small amounts of galactose remained detectable in plasma and urine (Olambiwonnu *et al.*, 1974). Both, galactokinase and transferase deficient infants though on galactose exclusion diets excreted some galactitol in the urine (Olambiwonnu *et al.*, 1974; Roe *et al.*, 1973). High gal-1-P in cord blood of transferase deficient newborns born of mothers on galactose-restricted diets is not uncommon (Donnell *et al.*, 1969).

The need for galactose-containing polymers for functional and structural integrity of the cells and tissues is met by biosynthetic reactions which have been detailed. Hence there is a substantial capacity to synthesize galactose and its derivatives at all stages of human development. For instance, we have seen one transferase deficient newborn (B.H.) in whom diagnosis was suspected because of an older sibling with the deficiency. We established the diagnosis by demonstrating complete lack of transferase in erythrocytes. At age 5 hrs, his gal-1-P was 17 mg/ 100 ml of RBC. At the end of his first day, gal-1-P had risen to 26 mg/100 ml (Fig.8) although he had been given intravenous glucose only and had taken no food (Gitzelmann and Hansen, *in press*). Clearly his red blood cells had synthesized gal-1-P, either from endogenous galactose stores but more likely from glucose.

We have observed what appears to be uncontrolled biosynthesis of gal-1-P in transferase deficient infants (Gitzelmann and Hansen, *in press*). We followed two newborn galactosemics, R.B. and D.A., who had been detected through mass screening. After the initial exposure to milk and after diagnosis was made, both received "galactose-free" formula (Table II) and galactose exclusion diets. When the children left the hospital at approximately one month of age both sets of parents were thoroughly instructed and appeared to follow schedules rigidly. Very surprisingly, upon initiation of diet at age 7 and 6 days respectively, red cell gal-1-P levels dropped rather slowly and potentially toxic levels were maintained for months (Fig.9). At age 27 days, infant R.B. was fed isocaloric amounts of dextrimaltose in water for 36 hrs; however, gal-1-P did not drop but rose slightly. Female infant D.A. had breast swellings until her 25th day of age. At age 19 days, her breast secretions, collected and compared to that of 3 other newborns, contained what appeared to be normal concentrations of lactose (Table III), obviously synthesized from glucose. It is interesting that in this patient after an initial drop, red cell gal-1-P peaked again on her 15th day, possibly because of endogenous lactose production in her breasts. Such biogenesis of galactose could take place in tissues other than the red cells and the breast glands, e.g. in the liver and the central nervous system. It is remarkable in this connection that in the few existing long-term follow-up studies (Komrower and Lee, 1970; Fishler *et al.*, 1972), some of the galactosemia patients who had been diagnosed at birth and

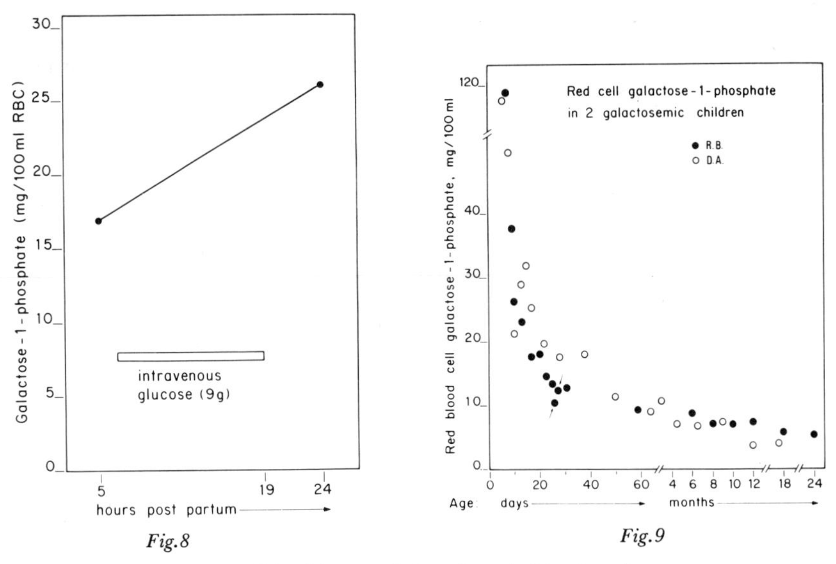

Fig.8 Red blood cell gal-1-P level in an unfed newborn lacking transferase.

Fig.9 Red blood cell gal-1-P levels in two children lacking transferase. Infant R.B. was fed dextrimaltose for 36 h (arrows). Infant D.A. had breast swellings until her 25th day (Gitzelmann *et al., in preparation*).

Table II. Lactose and galactose content* of commercial infant formulae.

		Lactose (g/100 ml)	Galactose (g/100 ml)
Bebenago®	(soya, dialyzed), 15% in water, as fed to infants R.B. & A.D.	<0.001	0.007
Nutramigen®	(casein hydrolysate), 15% in water, for comparison	0.011	<0.001

* Enzymatic analysis (Kurz and Wallenfels, 1970).
From: R. Gitzelmann *et al., in preparation.*

treated since, showed signs of organic brain damage at school age, evidenced by difficulties in visual perception. Whether self-intoxication with gal-1-P through prolonged maintenance of high levels during infant life was indeed the cause of the damage, can now be speculated.

To recapitulate, the important pathways of galactose metabolism in man are illustrated in the integrative Fig.10. The *Leloir pathway* consists of galactokinase (step 1), the gal-1-P uridyltransferase (step 2), the UDP-hexose 4-epimerase (step 3). The *pyrophosphorylase pathway* which is ancillary in normal man becomes

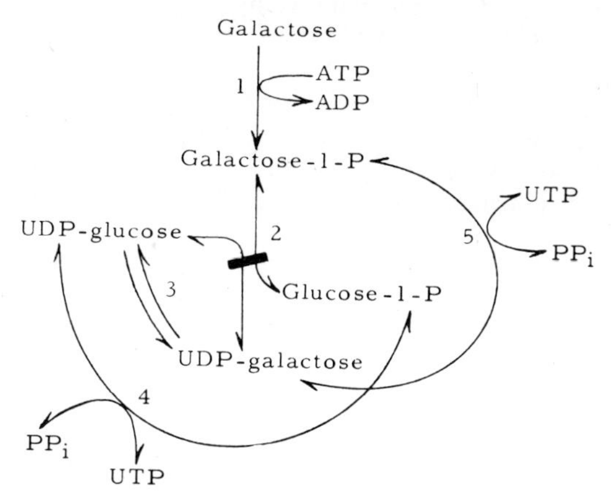

Fig.10 Galactose metabolism in man. For explanation, see text (Gitzelmann, 1969).

Table III. Lactose and galactose in breast secretions of newborns*.

		Lactose (g/100 ml)	Galactose (g/100 ml)
D.A.	Galactosemia, 19 days (transferase activity: 0.0 U/g Hb)	2.9	0.027
		2.5	0.016
M.J.	Compound heterozygote, 22 days (transferase activity: 0.3 U/g Hb)	3.6	0.013
		2.5	0.013
E.M.	Control, 11 days	5.5	0.005
		6.1	0.005
A.K.	Control, 13 days	4.6	0.022
		5.1	0.016

* R. Gitzelmann *et al.*, *in preparation.*

a primary pathway of galactose metabolism when transferase (step 2) is blocked in the genetic defect. This pathway involves, in order, pyrophosphorylase (step 5), the same epimerase (step 3) and pyrophosphorylase (step 4). Hence only *one* additional enzyme is required for the pyrophosphorylase pathway. Since all reactions except that of the kinase are freely reversible under physiological conditions both pathways can be a means of converting gal-1-P into glc-1-P. The reverse is also operative: gal-1-P is formed from glc-1-P depending upon the substrate pressures and the needs of the cell for intermediates.

The regulation of this process to limit the amount of UDP-gal formed to that required for biosynthetic reactions and thereby prevent the formation of toxic quantities of gal-1-P is the central problem of *some* transferase deficient children.

SUMMARY AND CONCLUSIONS

1. Galactosemic infants lacking transferase totally are now suspected not only of degrading but also of synthesizing galactose-1-phosphate. UDP-hexose pyrophosphorylase and epimerase are the two enzymes for this metabolism in man:

d) pyrophosphorylase
glc-1-P + UTP $\rightleftharpoons$ UDP-glc + PPi

c) epimerase
UDP-glc $\rightleftharpoons$ UDP-gal

d) pyrophosphorylase
UDP-gal + PPi $\rightleftharpoons$ gal-1-P + UTP
Sum: glc-1-P $\rightleftharpoons$ gal-1-P

In galactosemic infants, galactose-1-phosphate could be derived from UDP-galactose in toxic amounts.

2. UDP-hexose pyrophosphorylase is indeed the catalyst for the formation of galactose-1-phosphate from UDP-galactose in hemolysates of galactosemia patients.

3. UDP-glucose pyrophosphorylase and UDP-galactose pyrophosphorylase are identical, structurally and immunologically.

4. There is close structural relationship between the liver UDP-hexose pyrophosphorylase of man, calf and also rabbit.

REFERENCES

Abraham, D.H. and Howell, R.R. (1969). *J. biol. Chem.* **244**, 545-550.
Albrecht, G.J., Bass, S.T., Seifert, L.L. and Hansen, R.G. (1966). *J. biol. Chem.* **241**, 2968-2975.
Avigad, G., Amaral, D., Asensio, C. and Horecker, B.L. (1962). *J. biol. Chem.* **237**, 2736-2743.
Bergren, W.R., Ng, W.G., Donnell, G.N. and Markey, S.P. (1972). *Science* **176**, 683-684.
Caputto, R., Barra, H.S. and Cumar, F.A. (1967). *Ann. Rev. Biochem.* **36**, 211-246.
Chacko, C.M., McCrone, L. and Nadler, H.L. (1972). *Biochim. biophys. Acta* **268**, 113-120.
Cuatrecasas, P. and Segal, S. (1966). *Science* **153**, 549-551.
Donnell, G.N., Koch, R. and Bergren, W.R. (1969). *In* "Galactosemia" (D.Y.-Y. Hsia, ed) pp.247-268. Thomas Springfield.
Fishler, K., Donnell, G.N., Bergren, W.R. and Koch, R. (1972). *Pediatrics* **50**, 412-419.
Ginsburg, V. and Neufeld, E.F. (1969). *Ann. Rev. Biochem.* **38**, 371-388.
Gitzelmann, R. (1969). *Pediat. Res.* **3**, 279-286.
Gitzelmann, R. (1970). *In* "Methoden der enzymatischen Analyse" (H.U. Bergmeyer, ed) pp.1253-1257. Chemie Weinheim.
Gitzelmann, R. and Steinmann, B. (1973). *Helv. paed. Acta* **28**, 497-510.
Gitzelman, R. and Hansen, R.G. (1974). Abstr. Europ. Soc. Ped. Res., Ann. Meeting Sevilla 1973. *Pediat. Res.* **8**, 137.
Gitzelmann, R. and Hansen, R.G. *Biochim. biophys. Acta (in press).*
Gitzelmann, R., Wells, H.J. and Segal, S. (1974). *Europ. J. clin. Invest.* **4**, 79-84.
Hayman, S. and Kinoshita, J.H. (1965). *J. biol. Chem.* **240**, 877-882.
Hers, H.G. (1960). *Biochim. biophys. Acta* **37**, 120-126.
Inouye, T., Tannenbaum, M. and Hsia, D.Y.-Y. (1962). *Nature* **193**, 67-68.
Isselbacher, K.J. (1958). *J. biol. Chem.* **232**, 429-444.
Knop, J.K. and Hansen, R.G. (1970). *J. biol. Chem.* **245**, 2499-2504.
Komrower, G.M. and Lee, D.H. (1970). *Arch. Dis. Child.* **45**, 367-373.
Kosterlitz, H.W. (1943). *Biochem. J.* **37**, 318-321.
Kurz, G. and Wallenfels, K. (1970). *In* "Methoden der enzymatischen Analyse" (H.U. Bergmeyer, ed) pp.1147-1151. Chemie Weinheim.
Lehninger, A.L. (1965). "Bioenergetics", p.183. Benjamin, New York.
Leloir, L.F. (1951). *Arch. Biochem. Biophys.* **33**, 186-190.

Levine, G. and Bassham, J.A. (1974). *Biochim. biophys. Acta* **333**, 136-140.
Levine, S., Gillett, T.A., Hageman, E. and Hansen, R.G. (1969). *J. biol. Chem.* **244**, 5729-5734.
Mayes, J.S. and Miller, L.R. (1973). *Biochim. biophys. Acta* **313**, 9-16.
Olambiwonnu, N.O., McVie, R., Ng, W.G., Frasier, S.D. and Donnell, G.N. (1974). *Pediatrics* **53**, 314-318.
Posternak, T. and Rosselet, J.P. (1954). *Helv. chim. Acta* **37**, 246-250.
Roe, T.F., Hallatt, J.G., Donnell, G.N. and Ng, W.G. (1971). *J. Pediat.* **78**, 1026-1030.
Roe, T.F., Ng, W.G., Bergren, W.R. and Donnell, G.N. (1973). *Biochem. Med.* **7**, 266-273.
Russell, R.G.G., Bisaz, S., Fleisch, H., Currey, H.L.F., Rubinstein, H.M., Dietz, A.A., Bousinna, J., Micheli, A. and Fallet, G. (1970). *Lancet* **II**, 899-902.
Schwenn, J.D., Lilley, R.M. and Walker, D.A. (1973). *Biochim. biophys. Acta* **325**, 586-595.
Segal, S. (1972). *In* "The Metabolic Basis of Inherited Disease" (J.B. Stanbury, J.B. Wyngaarden and D.S. Fredrickson, eds) pp.174-195. McGraw-Hill, New York.
Ting, W.K. and Hansen, R.G. (1968). *Proc. Soc. exp. Biol. Med.* **127**, 960-962.
Turnquist, R.L., Gillett, T.A. and Hansen, R.G. *J. biol. Chem. (in press).*
Turnquist, R.L., Turnquist, M.M., Bachmann, R.C. and Hansen, R.G. (1974). *Biochim. biophys. Acta* **364**, 59-67.
Van Heyningen, R. (1971). *Exp. Eye Res.* **11**, 415-428.
Verachtert, H., Bass, S.T., Seifert, L.L. and Hansen, R.G. (1965). *Anal. Biochem.* **13**, 259-264.

DISCUSSION

Wick: I just wonder if there are not certain clinical implications of your work. Do you think that it is possible to decrease this galactose biosynthesis by giving a diet low in carbohydrate?

Gitzelmann: It is difficult to predict what effects diets high or low in glucose would exert on galactosemic patients. On the one hand, by giving a diet rich in glucose one might expect that levels of glucose-1-phosphate would rise; glucose-1-phosphate might then occupy the enzyme so as to prevent it from interacting with galactose intermediates, in which case no galactose-1-phosphate could be formed from UDP-galactose. On the other hand, red cells *in vitro* during glucose starvation hold their galactose-1-phosphate level or even raise it.

Van den Berghe: The enzymes involving the formation of pyrophosphate are considered irreversible because of the very active pyrophosphates that degrade pyrophosphate. Have you considered in this case that one of the metabolites of galactose might inhibit this pyrophosphatase and shift the equilibrium towards the formation of galactose-1-phosphate?

Gitzelmann: We have not done any measurements on this. The role of pyrophosphatases and the actual levels of pyrophosphate at the site of the reaction are not known presently.

Tager: The formation of precipitating antibodies is a function of the ratio between the concentration of the antigen and antiserum. If you titrate the formation of precipitating antibody-antigen complexes, which in your case you are doing by measuring the activity remaining in the supernatant, as a function of antiserum concentration, you get first a decrease of the activity in the supernatant, and then, if you go sufficiently high with your antiserum concentration, you get an increase again. There is an optimum ratio of antigen to antiserum. That did not seem to occur in your experiment.

Gitzelmann: In our precipitation experiments we have refrained from working in antibody excess in order to economize antiserum. Therefore, in our figures only one part of the entire precipitation curve is represented, i.e. that of antigen excess.

Gaull: You demonstrated two lines of identity, in your Ouchterlony plates. The inner line had a spur on it. I am wondering whether the inner line represents mutual contaminants or whether you consider it a monospecific antibody.

Gitzelmann: I would suspect that in most double diffusion precipitation reactions where a complex antigen is involved more than one precipitation line will occur. We have experienced this with human fructosediphosphate aldolase B and, as you have just seen, also with crystallized human pyrophosphorylase. One must remember that the enzyme, an octomer, also occurs in units of 16 and 32 subunits. Thus there is a certain molecular diversity within the pure antigen. The spurs could easily be explained by different isoenzymes occurring in different tissues. Although at present we do not know much about pyrophosphorylase isoenzymes, we do think that this may be due to isoenzyme differences between red cells and the liver.

Gaull: Did you make antibody to the red cell enzyme?

Gitzelmann: No, the red cell enzyme has not been isolated in the pure form. It has never been crystallized from red cells, but it has been crystallized from human liver.

Blass: It does in no way detract from the very great interest in this work to mention that the dogma in the literature is that well-treated galactosemic patients do well. A recent study (Donnell *et al., in* "Galactosemia", D.Y.Y. Wrialed, ed., C.C. Thomas, Springfield, 1969, p.247) reported some visual perceptual difficulties; however the mean I.Q. was very close to 100.

Gitzelmann: I.Q. measurements are very popular, because you arrive at a figure. However in addition to the visual perception difficulties (which later may fade and which we do not yet know) some patients had rather serious problems of social adjustment due to character deficiencies. We ought to pay more attention to these aspects when we evaluate normal versus abnormal mental development in galactosemic children.

Blass: I agree with you.

Koster: Is the binding of UDP-glucose to the enzyme independent of the pyrophosphate concentration?

Hansen: Yes, it is. UDP-glucose has a higher affinity for the enzyme than has UDP-galactose. UDP-glucose will in fact replace UDP-galactose from the enzyme.

Koster: Then UDP-glucose will occupy all the sites on the enzyme. How do you get galactose-1-phosphate?

Hansen: This depends on the relative concentrations of the two substrates and the relative binding constants. The *in vivo* concentrations of UDP-glucose and UDP-galactose are not too far apart; I can only give it to you in terms of reaction rates: it is something between 5 and 10% of the same rate of reaction at equimolar concentrations, and operatively that has more meaning.

THE INFLUENCE OF FOOD ON THE BODY COMPOSITION OF THE LOW BIRTH WEIGHT INFANT

J.H.P. Jonxis

Department of Pediatrics, University of Groningen, Oostersingel 59
Groningen, The Netherlands

The relation between the doctor in the medical school on the one hand and the biochemist and physiologist on the other hand is a complicated one. The medical specialist in the University Hospital should be a man of science. However, practical medicine involves a certain and, in emergency cases, a large amount of routine, especially when a group of doctors has to work in co-operation. Routine, however, seldom leads to independent observation and discovery. During his busy hours, the doctor may envy the pure scientist for his freedom to study his problems at leisure. The scientist may envy the busy doctors, particularly at a moment when he cannot solve his scientific problem.

That the medical profession needs the biochemist and the physiologist is an established fact. The medical man may from his side discover by simple observation and a few measurements phenomena in his patients which may form new stimuli for the biochemist.

The inborn errors of metabolism and, in my opinion also, the prematurely or dysmaturely born child demonstrate that the information that these groups of patients give might be important for the development of our biochemical knowledge. On the other hand, the doctor should be able to follow the new discoveries in biochemistry closely enough to enable him to apply them in practical medicine.

As a paediatrician I should like to call your attention to a problem concerning the low birth weight infant and the influence both quantitatively and qualitatively that food intake has on his body composition. Compared with other mammals, the human infant born at term is a very slow grower. His weight gain seldom exceeds 0.6% daily. Growth *in utero* is much quicker: in the last months of pregnancy it still amounts to 1.5-2%.

In the low birth weight infant, whether it is prematurely born or underfed *in utero*, we nowadays try with different measures to achieve growth rates which equal the normal intrauterine growth rate. To effect this, a high caloric intake is necessary together with conditions in which the energy requirements to keep the body temperature constantly at 37° are minimal.

I should like to discuss the aminoacid metabolism and fat metabolism of such infants. The caloric intake of a normal full-term infant (birth weight 3700 g) is 100-120 cal/kg bodyweight after the first weeks of life, his growth rate being 0.6% daily. In the prematurely born infant (birth weight 1500 g), a food intake of a caloric value of about 180 cal is needed to obtain a growth rate, equal to the intrauterine growth rate, of 1.7%. The caloric values of the new tissues that are daily deposited is difficult to estimate, because of the wide varieties of the amount of fat present in the adipose tissue. But we may assume that the fat percentage of the prematurely born infant is about 7% and that of the full-term infant about 15%. This makes the difference in caloric value of the tissue that is daily deposited per kg bodyweight far less than one might expect from the growth rate of these two groups of children.

A factor that we may take into account is the limited fat absorption in premature children, which means that a part of the caloric intake is lost with the faeces. In the premature infant about 11 calories of the 180 calories he is taking are lost in this way, whereas in the full-term infant taking 100 calories the loss is only about 2.8 calories (Widdowson, 1965).

Our knowledge about the minimal amount of protein that these two groups of children need to achieve their optimal growth rates is still limited. We know, however, that on the basis of human milk protein 2 g per kg body weight or even a little less are enough to achieve a normal growth rate, provided the caloric requirement in the full-term infant is met. For in the premature infant 2.7 g of cow's milk protein are enough to meet the protein requirements to achieve a growth rate equal to the intrauterine growth rate, provided that the high caloric requirements are met (Jonxis, 1974).

How do the plasma aminoacid levels react on different protein and caloric intakes in the low birth weight infant? We know that the plasma aminoacid levels in cord blood are higher than those in maternal blood. For some aminoacids, like lysine, arginine and ornithine, levels up to three times higher than those of maternal blood have been reported. For many aminoacids the concentrations in cord blood are 1.5-2 times higher than those found in maternal blood (Young, 1971). In infants of an age less than one month the levels of lysine, serine, proline, glycine, valine, leucine and tyrosine are still higher than those found in the blood of somewhat older infants and adults (Snyderman *et al.*, 1968). The differences, however, are not very large. Levine and Marples (1939) found that a high protein intake combined with a low vitamin C intake caused tyrosinuria in the young, and particularly in the prematurely born infant. The increased aminoacid excretion in the young infant was found to be caused by a limited tubular reabsorption of aminoacids on the one hand and increased plasma levels on the other hand.

What is the influence of protein load and growth rate on the plasma aminoacid levels in low birth weight infants? To study this problem, we gave alternatively formulas that contained 100 calories and 2.7 g of protein, 180 calories and 2.7 g of protein, and 180 calories and 4.7 g of protein per kg body weight to 10 infants weighing 1300-1800 g during their 3rd to 7th week of life. The weight gain of the children who got 180 calories and 4.7 g of protein amounted to 16.9 g daily; for those who received 180 calories and 2.7 g of protein it was 14.2 g. In those who received 180 calories and 2.7 g of protein weight gain was 7.6 g daily. The differences in the growth rates at 180 calories and 2.7 g and 4.7 g respectively are not large and may be due entirely to the higher mineral content of the formula, rich in protein.

The values just mentioned suggest that a formula with a protein content of 2.7 g per kg body weight is sufficient to obtain an optimal growth rate, provided that the caloric intake amounts to 180 calories per kg body weight daily.

What are the plasma aminoacid levels of the babies fed on these formulas? We have to compare them with the cord blood aminoacid levels and with the aminoacid levels of full-term infants at the age of two weeks, who are fed on a formula of 100 calories and 2.7 g of protein per kg bodyweight daily. The free aminoacids were determined on heparinized plasma. Details are given in Table I. Plasma samples were precipitated immediately with picric acid and prepared for analysis by the method of Stein and Moore (1958).

Table I. Plasma aminoacid levels.

	Prematures, birth weight 1300-1800 g			
Aminoacid μmol/ml	100 cal., 2.7 g of protein	180 cal., 4.7 g of protein	180 cal., 2.7 g of protein	Full-term babies
Ornithine	8	9	7	6
Lysine	16	23	14	15
Histidine	8	11	10	7
Tryptofane	2	3	3	4
Arginine	8	7	6	5
Threonine	27	33	21	12
Serine	17	22	22	14
Glutamic asparagine	50	55	60	44
Proline	33	42	31	17
Glutamic acid	17	16	17	18
Citrulline	3	3	3	2
Glycine	30	32	30	24
Alanine	30	43	41	25
Valine	22	30	19	18
Methionine	4	7	6	3
Isoleucine	8	9	7	5
Leucine	13	14	7	8
Tyrosine	27	36	25	6
Phenylalanine	8	9	8	4

Compared with the values found in the full-term infants the plasma values for threonine, proline, tyrosine and phenylalanine tend to be higher in the premature infant on the same diet. The plasma aminoacid values in protein deficient infants are characterized by low valine, leucine, isoleucine and tyrosine values. Such are not found in our infants, not even in those with a rather low protein intake and a high growth rate. An increased protein load of 4.7 g in the high caloric diet caused a marked rise in the plasma values of many aminoacids, especially of those of lysine, histidine, threonine, proline, alanine, valine, methionine, isoleucine and tyrosine. The increase, however, is never more than 50% over the value found when the diet contained only 2.7 g of protein. Among these aminoacids are threonine, proline and tyrosine, which are already found to be high on a diet with lower protein and caloric values, but not phenylalanine, however. Threonine, proline and tyrosine also are increased in normal full-term infants of 2 weeks, but to a lesser extent.

When we compare the values during the period in which the children had respectively a high or a normal caloric intake, both with 2.7 g of protein, we find that in the period of the high caloric intake in which the growth rate equals the intrauterine growth rate, the levels of the essential aminoacids, leucine, valine and threonine, tend to be somewhat lower in the infant with a high weight gain than might be expected.

When we compare our values with those mentioned in the literature for cord blood we find that, in general, the values found in our children during the period of a high caloric and high protein intake are about the same as those found in cord blood. The values of threonine, proline and tyrosine, however, markedly exceed those found in cord blood.

Low birth weight infants on the whole tend to have somewhat higher plasma aminoacid levels than full-term infants of the same age (Jonxis, 1974). In our formula-fed infants the values for threonine, proline and tyrosine are higher and rise still further when the amount of protein in the formula is increased.

What are the causes of the abnormal behaviour of the aminoacids just mentioned? We do not know. Is there any harm to be expected from the high values, particularly when a formula rich in protein is administered? Mental retardation in some inborn errors in aminoacid metabolism, in which high plasma aminoacid values occur, has made us suspicious. On the other hand, we know that the high plasma tyrosine values in infants with transitory tyrosinaemia do not cause brain damage (Martin, 1974). Still, until more is known we should avoid giving more than 3 g of protein per kg of bodyweight daily to low birth weight infants.

The fat percentage of the foetus rises quickly during the last months of intrauterine life. There are, however, wide variations. On the average it rises from 7% in the child of 1500 g to 16% in the full-term infant of 3700 g. About 550 g of fat are thus stored in the last 50 days of intrauterine life. The presence of some linoleic acid (essential fatty acid) in the fat of the foetus and the newborn indicates that some transport of fat or fatty acids through the placental membrane takes place. To what extent the body fat of the foetus is directly derived from fat or fatty acids in the maternal blood is uncertain. Recent investigations have made it likely that fat transport through the placental membrane is more important than was assumed previously. During the last year we have made determinations of the composition of the body fat of infants, mainly low birth weight ones, from birth up to an age of 8 months. All babies received the same formula, Almiron, in which the fat is maize-oil with a linoleic acid percentage of 57. We may suppose that the same important increase in fat percentage that takes place during the last months of intrauterine growth also happens in the low birth weight baby who rapidly gains weight during his first months of life.

Specially owing to the rapid accumulation of fat, it is likely that the composition of his body fat resembles, to a certain extent, the fatty acid composition of the fat in his food. The reason why we have chosen a formula with a high content

Table II. Fatty acid composition of the body fat of babies at birth. g/100 g total fatty acids. Mean ± S.D.

C14:0	3.3 ± 0.44
C16:0	45.8 ± 1.6
C18:0	3.8 ± 0.37
C16:1	15.2 ± 1.2
C18:1	29.0 ± 1.8
C18:2	2.9 ± 0.71

Table III. Fatty acid composition of body fat of babies. g/100 g total fatty acids. Mean ± S.D.

	0-4 weeks N 20		4-16 weeks N 19		Droese
C14:0	3.15	(2.0-3.8)	2.54	(1.0-4.4)	6
C16:0	39	(29-49	37.1	(15.5-39.2)	29
C18:0	3.4	(1.8-4.5)	2.2	(1.5-2.8)	
C16:1	13.8	(8.8-17.4)	10.9	(8.2-14.5)	8
C18:1	27.7	(22.8-31.8)	27.9	(24.1-32.2)	36
C18:2	11.4	(3.5-27)	28.8	(20.0-46.7)	7

Table IV. Fatty acid composition of body fat of babies, 6-12 months old, having a mixed diet. g/100 g total fatty acids. Mean ± S.D.

C14:0	4.8 ± 1.2
C16:0	29.7 ± 4.6
C18:0	2.9 ± 0.72
C16:1	13.8 ± 2.7
C18:1	40.8 ± 3.3
C18:2	8.0 ± 4.1

of linoleic acid (maize-oil) is that this fat gives a much better fat absorption in the young infant than the fat of most other formulas, and in this aspect nearly equals human milk fat, although the percentage of linoleic acid in human milk is much lower. The determinations were performed in subcutaneous fat obtained by puncture. For the gas chromatographic determinations of the fatty acids we used the method as described by Lipsky *et al.* (1959) and of Metcalfe *et al.* (1966).

Our results found in newborn babies are collected in Table II, those found in infants between 0 and 4 weeks and in infants between 4 and 16 weeks in Table III. Some older babies were already on a mixed diet (Table IV).

The data shown confirm the supposition that the composition of the body fat of the baby reflects, to a large extent, the composition of the fat in the formula. Few comparable data are available up till now from children fed on formulas with another fatty acid composition. Droese and coworkers (1974) recently published some figures from Germany. They did not mention the fatty acid composition of the milk which those babies received. For children under 6 months, probably none premature, they found a linoleic acid percentage of 11 (± 2.65). For the somewhat older children on a mixed diet they found a value of 7.8% (± 2.60), for children 12 years old 10.7% (± 2.83).

We do not yet know what all this means. The Dutch as well as the German children were in a good nutritional condition. We may assume that the fat and calcium absorption was particularly good in the Dutch children. The plasma cholesterol values found in our children are low compared with values found in children who got other formulas (Darmady *et al.*, 1972). Our total plasma cholesterol values were, on an average, 106 mg/100 ml (variation 70-158). Literature gives an average value of 170 mg/100 ml for this age group. Once more, what this means is uncertain. But to our present knowledge, it is more likely to be an advantage than a disadvantage. About the influence of the fatty acid composition of the food on the membrane composition we know very little. Preliminary results regarding investigations on the fatty acid composition of the membrane of the erythrocyte make it likely that a certain influence is there, but that it is not a considerable one. The fact remains that the fatty acid composition of the young infant's food has an important influence on the composition of his body fat.

REFERENCES

Darmady, J. *et al.* (1972). *Brit. Med. J.* **2**, 685.
Droese *et al.* (1974). *Zeitschr. Kinderheilk* **116**, 269.
Jonxis, J.H.P. (1974). *Verslag Vergadering Afdeling Natuurkunde, Koninklijke Nederlandse Akademie van Wetenschappen* **83**, 35.
Levine, S.Z. *et al.* (1939). *Science* **90**, 620.
Lipsky, S.R., Landown, R.A. and Godet, M.R. (1959). *Biochim. biophys. Acta* **31**, 336.
Martin, H.P. *et al.* (1974). *Pediatrics* **53**, 212.
Metcalfe, L.D. , Smitz, A.A. and Pelka, J.R.P. (1966). *Anal. Chem.* **38**, 514.
Moore, S., Spackman, D.H. and Stein, W.H. (1958). *Anal. Chem.* **30**, 1185.
Snyderman, E. *et al.* (1968). *Ped. Res.* **2**, 131.
Widdowson (1965). *Lancet* **II**, 1099.
Young, M. (1971). *In:* "Metaholic Processes in the Foetus and Newborn Infant" (J.H.P. Jonxis, H.K.A. Visser and J.A. Troelstra). H.E. Stenfert Kroese N.V., Leiden, p.97.

DISCUSSION

Gaull: What was the quality of the protein you gave the infants: how much whey protein versus caseine?

Jonxis: We took normal cow's milk protein, with a normal ratio of caseine, lactalbumine and globulins. The milk was relatively low in cysteine as compared to human milk.

Gaull: Dr. Räihä and I have recently completed a large-scale study of low birth weight infants, with emphasis on the quality and quantity of protein. We fed four different types of protein against pooled human breast milk. I do not want to go into all the details, but certainly the general conclusions that we came to about the amino acids, especially threonine, the essential amino acids and tyrosine, are entirely in keeping with your study. We would also raise the question as to whether this is necessarily a good thing to do.

Jonxis: I have no definite answer. We agree that we should be careful in using diets with too high amounts of protein.

Clark: You have commented on the eating habits of the British family and those of the Dutch people. Have you any evidence that the cholesterol content in British babies is higher than in Dutch babies?

Jonxis: These values have been taken from a few publications about some children fed with butterfat. I have no idea what is happening in England.

Stern: In neonatal mass screening we often find that infants with blood methionine levels in the range 4-6 mg/100 ml turn out to be of low birth weight and on a high protein diet.

Jonxis: Yes, this agrees with our data with regard to prematurity.

Gaull: Was your whey protein derived from bovine milk?

Jonxis: Yes, it was.

Gaull: Because the whey protein from bovine milk is not the same as the whey protein of human milk. The methionine concentration in blood of low birth weight infants in our experience is even lower on human milk than it is on so-called humanized milk. We would consider your values as high values. Human milk is almost unique in animal proteins: the ratio of cysteine to methionine is very high; most animal proteins are high in methionine relative to cysteine. Human milk is peculiar in regard to having a very low concentration of the phenolic amino acids. We have been two million years evolving a diet which contains about 1% protein, which has an amino acid composition which is quite different from all other animal proteins. For instance the rabbit, which is a very rapidly growing animal, has a 15% milk protein and ours is about 0.8 to 1%. I certainly would agree with Professor Jonxis that before we start meddling with human diets we ought to know a little bit more about what we are doing.

Jonxis: We should first of all not forget that the human being is a very slow grower. Moreover, the human, except for some farm animals, is the only mammal whose diet has been changed during the past 10,000 years. Before the fire was invented for cooking food, he was not able to eat potatoes or grain. The original diet of the human being must have been very different. He had to live on some fruit, perhaps one or two eggs, and molluscs. He was a very poor hunter as long as he had no dogs; so the primitive diet must have been quite low in animal protein. In our intestinal tract we still have an enzyme, trehalase, which is only present in lower animals such as insects and molluscs. I believe human milk reflects more or less the limitations of the diet of the female in very old times, because genetically the human has not changed much in the last 100,000 years.

Sternowsky: One question concerns carbohydrates. If you feed your energy intravenously, you are forced to feed glucose. Is it really good for the children to be fed fructose?

Jonxis: As a practical clinician I know that feeding these small-for-date infants intravenously you run into all kinds of difficulties. They utilize glucose rather poorly and often lose glucose with the urine. Fructose, on the other hand, easily makes your infants acidotic.

Blass: It is becoming increasingly clear that although fructose is converted to glucose metabolically, it is not an equivalent foodstuff. Some of the data is experimental, related to the effect of fructose on activation of the pyruvate dehydrogenase complex and alternations in metabolic pathways in liver, *in vivo*, *in vitro* and in cell-free preparations. The second is epidemiologic. There is evidence that a number of degenerative diseases, particularly of vessels, have become more common as the proportion of refined sugar in the diet has increased in a number of populations in the world.

MODULATIONS OF DIFFERENTIATIVE COMPETENCE DURING EMBRYOLOGICAL DEVELOPMENT

W.J. Rutter, R. Pictet and L. Rall

Department of Biochemistry and Biophysics, University of California
San Francisco, California, USA

The pancreas is a useful paradigm for morphogenesis and cytodifferentiation. This organ is comprised of endocrine (A,B,D), exocrine (acinar), and duct cells. The several cells have well-defined physiological functions and the pertinent cell-specific molecules are well studied. The development of the pancreas, like other organ systems, is dependent on an interaction between epithelial and mesenchymal tissues. Intact pancreatic rudiments (pancreatic epithelia plus surrounding mesenchyme) or pancreatic epithelia cultivated in the presence of mesenchymal tissue, develop *in vitro* into tissues having normal pancreatic morphology and approximately normal content of the endocrine proteins, insulin and glucagon, as well as the exocrine digestive enzymes. My colleagues and I have previously described pancreatic morphogenesis and the patterns of synthesis and accumulation of the specific cellular proteins in both *in vitro* and *in vivo* systems (Rutter *et al.*, 1968; Kemp *et al.*, 1972; Pictet and Rutter, 1972; Pictet *et al.*, 1972; Rall *et al.*, 1973). We have also identified a factor derived from mesenchymal tissue which is required for the growth and normal differentiation of pancreatic epithelial tissue (Rutter *et al.*, 1964; Ronzio and Rutter, 1972; Levine *et al.*, 1973).

The cellular composition and synthetic capabilities of the differentiated tissue depend significantly on 1) the presence of the specific mesenchymal factor, 2) the nutritional quality of the medium and 3) glucocorticoids.

MESENCHYMAL FACTOR IS REQUIRED FOR EPITHELIAL CELL PROLIFERATION

Mesenchymal tissue is required for the *in vitro* growth of the pancreatic epithelium (Grobstein, 1967). An extract from whole embryo can be substituted for this mesenchyme (Rutter *et al.*, 1964; Ronzio and Rutter, 1972). This effect of the mesenchymal factor has been used as an assay for the isolation of the factor. It has been purified (~1000-fold) from the membranous fractions of the chicken embryo at a stage when it is relatively rich in mesenchymal tissues.

Table I. Mesenchymal factor affects growth and differentiation of rat embryonic pancreas epithelium.

Days in Culture		DNA/ Epithelium	Amylase (ng Maltose/ min/μg DNA)	Insulin (ng/μg DNA)
0	-	0.03	0.005	0.006
7	MF	0.4	180.	35.
7	Inactivated MF	0.03	30.	41.

Table II. Amino acid levels in the nutrient medium affect growth and differentiation of the embryonic rat pancreas*

	Low Amino Acids (BME)	High Amino Acids (6 X BME)
μg protein/pancreas	50	150
μg DNA/pancreas	3	6
protein/DNA ratio	17	25
Exocrine cells		
% total protein		
Chymotrypsin	13	25
Amylase	6	12
Endocrine B cells		
Insulin		
ng hormone/μg DNA	28	74

* Day 13 embryonic pancreases, plus 7 days in culture.

The mesenchymal factor may be a glycoprotein since it is sensitive to the action of trypsin and periodate, but it is unaffected by RNase, DNase, etc.. It has been shown to affect growth and differentiation by acting directly with the cell membrane (Levine *et al.*, 1973). Presumably, its effect on gene expression is therefore mediated via an intermediate pleiotypic messenger. Experiments carried out with Dr. Silvana Filosa suggest that the mesenchymal factor activity depends partly upon cAMP. In fact, inactive preparations of MF are fully activated when tested in the presence of cAMP analogues which penetrate cells, such as dibutyryl cAMP or 8-hydroxy-cAMP (but not cAMP itself). The inactive factor is still required for this effect, and is not replaceable by dbcGMP or 8-BrcGMP. These results suggest the possibility that mesenchymal factor preparations contain multiple active components.

Pancreatic epithelia dissociated from their surrounding mesenchyme by mechanical agitation and trypsin treatment carry out little DNA synthesis as measured by thymidine uptake into DNA. In the presence of the mesenchymal factor there is rapid uptake into DNA and increased growth as measured both by the increased DNA and protein content (Ronzio and Rutter, 1972; Levine *et al.*, 1973; Pictet *et al.*, 1974). In the absence of such factor, the cells of the pancreatic rudiment undergo little if any cell division and the cultivated rudiment contains few acinar cells and, instead, a large proportion of endocrine cells. In the presence of mesenchymal factor, the rudiment contains the normal high proportion (~80%) of acinar cells and low proportion (~5%) of endocrine cells. These changes are reflected in the accumulation of the exocrine enzymes and insulin as seen in Table I.

THE NUTRITIONAL ENVIRONMENT

Pancreatic rudiments develop normally in a simplified culture medium consisting of the usual basic salts, vitamins and amino acids. Unlike many other systems, the development of this organ does not require embryo extract or other crude biological fluids containing an ill-defined array of proteins, hormones and nutrients. Pancreatic epithelia, however, require proteins such as albumin in the medium or the cells slough off from the tissue and eventually perish. The cultivated pancreatic rudiment grows considerably more slowly than the *in vivo* rudiment even under optimal nutritional conditions. Both the rate of growth of the pancreas and the proportion of specific exocrine and endocrine proteins are a function of the culture medium. As shown in Table II and Figs 1 and 2, cultivations in Eagle's minimal medium results in a significantly lower growth rate, a lower cell size, and a different proportion of exocrine and endocrine proteins, as compared to cultivation in the same medium supplemented with 6 times the essential amino acids. Measurements of the protein accumulation in the rudiment during development indicate that at the time of the maximum protein synthetic rate approximately half of the amino acids in the minimal medium are utilized for protein synthesis during the course of the usual 24 hour incubation between medium changes. The increased protein synthesis in the rudiments cultivated in the medium containing 6x amino acids would have resulted in the utilization of nearly all of the amino pool in the BME medium. The nutritionally deficient medium not only decreases the rate of growth and accumulation of protein, but alters the specific activity of the specific proteins. A 13 day pancreas cultured for a period of 7 days using lower amino acid conditions has approximately 17 μg of protein per μg DNA; a similar rudiment cultured in enriched medium has 25 μg of protein per μg DNA; a difference of 8 μg of protein per μg of DNA. In the first case, amylase and chymotrypsinogen, the two major proteins synthesized by the pancreas, account for approximately 20% of the cellular protein or 3 μg of enzyme per μg of DNA. In the second case, the two activities together account for 41% of the cellular protein or 10 μg of enzyme per μg of DNA; an increase of 7 μg of protein per μg of DNA. Thus, in the presence of high levels of amino acids, the majority of the increase in cellular protein is due to synthesis of the cell specific proteins. These enzymes are the dominant entities, other exocrine activities account for less than 5% of the protein accumulated in the late embryonic pancreas (Kemp *et al.*, 1972). The synthetic processes for cell specific proteins are more sensitive to the nutritional environment than those involved in the synthesis of other components of the cells. Furthermore, there appears to be a selective change in the proportion of exocrine and endocrine cells under the two nutritional states. Although quantitative morphometry has not been carried out, the number of B cells in the amino acid poor medium is considerably lower than those in the enriched medium. This observation correlates with the higher insulin content in the enriched medium.

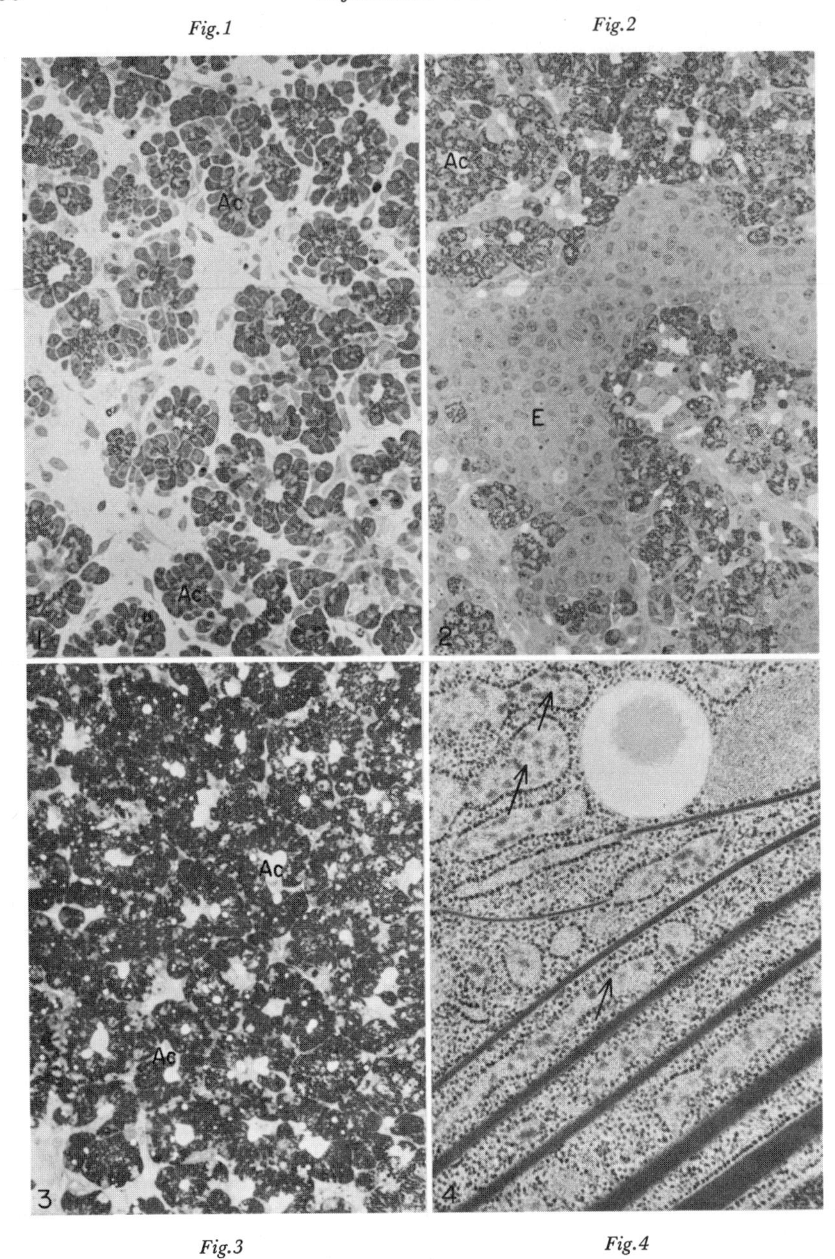

Figs 1-4. Pancreatic rudiments were explanted at day 13 and cultured for 7 days. Figures 1 and 2 show pancreases cultured in BME (Fig.1), and amino acids enriched BME (Fig.2). The density of the acini (Ac) is greater in Fig.2 than in Fig.1. In addition the enriched medium results in a greater development of the endocrine tissue consisting mostly of B cells as seen with

These results do not necessarily imply that the endocrine cells are themselves more sensitive to the nutritional environment. The effect here may be a result of sequential development of the cell types. Other studies have shown that division and differentiation of B cells occur prior to acinar cells. Thus, the effect can be explained on the basis that the nutritional effects are selective temporally. This possibility may be a model for nutritional effects on the development of the entire embryo. The development and differentiation of various tissues proceed sequentially and thus, nutritional modifications may produce significantly altered ratios of particular cell types. These changes may be reversible and trivial or alternatively may prove long lasting due to qualitative effects on the organism. If the organisms have been evolutionarily selected in order to minimize the dangers of nutritional deprivation it seems reasonable that the general proteins associated with cell functions will be maintained when specific functions are lost. Furthermore, the regulatory tissues may be less affected than those associated with digestive and metabolic functions.

MODULATION OF SPECIFIC SYNTHESIS BY GLUCOCORTICOIDS

Glucocorticoids have been shown to induce the synthesis of a number of specific proteins and even to affect the premature differentiation of a number of embryonic tissues such as the retina, the gut, etc.. Kulka and his colleagues (Cohen and Kulka, 1974; Yalowsky *et al.*, 1973) have demonstrated that glucocorticoids induce differentiation of the chicken pancreas. Unfortunately results obtained with this system are complicated by the large endogenous secretory rate. Thus, the effects of the glucocorticoids may be on specific accumulation or secretion of the pancreatic proteins. We have investigated the effects of dexamethasone on the developing rat pancreas where the secretion of both insulin and acinar proteins in the absence of secretagogues is insignificant. The experiments show that glucocorticoids in the physiological ranges exert a selective stimulatory effect on the synthesis and accumulation of specific proteins. In the absence of glucocorticoids, the day 20 rudiment cultivated in enriched medium contains approximately 40% specific exocrine proteins. In the presence of glucocorticoids, more than 70% of the proteins are exocrine. This is illustrated in Table III and Fig.3. The high magnification electron microscopic observation of the tissues treated with dexamethasone show the endoplasmic reticulum is distended with material having the same electron density as that found in the zymogen granules (Fig.4); and furthermore, there are crystalloid structures found frequently within the rough endoplasmic reticulum as if the proteins are concentrated enough to crystallize within these structures. This has never been observed under *in vivo* circumstances. The predominant effect of glucocorticoids is due to an increase in amylase levels although there is a larger proportionate increase in carboxypeptidase B. Other proteins are affected to a smaller degree or not at all. The physiological basis (if there is one) for a selective effect of glucocorticoids on some but not all of the acinar proteins is not known. In fact, the relative levels of enzymes produced in the

---

Figs 1-4 continued

the electron microscope. When dexamethasone is present (Fig.3) acinar cells accumulate more zymogen granules. Figure 4 shows the appearance of zymogen-like material which accumulates in the presence of dexamethasone in the rough endoplasmic reticulum of some acinar cells. This material is probably the same as is seen in dilated part of the rough endoplasmic reticulum (arrows). Figs 1-3, × 220; Fig.4, × 33,000.

Table III. Glucocorticoids modulate the levels of exocrine enzymes and insulin in the embryonic pancreas.

	% Total Proteins, Day 20		
	In vitro		*In vivo*
	− Dex	+ Dex	
Chymotrypsin	25	25	30
Amylase	12	43	13
Procarboxypeptidase A	2	2	3
Procarboxypeptidase B	0.1	0.7	0.8
Lipase	0.1	0.05	0.1
Insulin	0.3	0.06	0.05
Cell-specific proteins	40	71	47
Protein+DNA ratio	25	50	70

presence of glucocorticoids are somewhat altered from those found in the differentiated embyro. Thus, there are other as yet undefined influences on the relative secretory rates of the specific exocrine proteins.

The effects of glucocorticoids are not only to induce selectively synthesis of the specific proteins but also to decrease the rate of proliferation. Thus, the rudiments have somewhat less DNA but the cells are larger due to the accumulation of increased quantities of the exocrine proteins. Glucocorticoids are not required for acinar differentiation nor do they produce precocious development of acinar cells. Cultivation of the rudiments in the complete absence of glucocorticoids develop normally.

Desnuelle and colleagues (1963) have demonstrated that the relative levels of the pancreatic acinar proteins vary according to the diet. Thus, with a high carbohydrate diet there is a relative increase in amylase levels, and with a high protein diet, a relative increase in chymotrypsin. These effects are examples of dietary adaptations, and may have been due to enzyme induction by specific dietary components. We have cultivated pancreatic rudiments *in vitro* in the presence of high glucose and/or high amino acid levels, without altering significantly the specific levels of amylase or chymotrypsin in the differentiated tissue. The substantial effects of glucocorticoids observed here suggest that the effects observed by Desnuelle and his colleagues may have been due in part to an intermediate change in the level of glucocorticoids.

The present studies emphasize that the competence for macromolecular synthesis in embryonic tissues is dependent not only on the number of cells associated with the particular differentiated function but also on the rate of synthesis of the proteins in those cells. The mesenchymal factor is specifically required for cell proliferation in the pancreatic epithelium; the rate of cell proliferation can also be modulated both by glucocorticoids and the nutritional state. The rate of specific cellular protein synthesis is sharply dependent upon the amino acid concentration. At lower amino acid levels, the non-specific proteins are selectively synthesized;

in addition, glucocorticoids enhance the relative proportion of cell specific proteins synthesized. These experimental results illustrate the dynamic modulation of differentiation during development.

ACKNOWLEDGEMENTS

This work was supported by a grant from the National Science Foundation No.3-5256 and Genetics Center grant GM-19527.

REFERENCES

Cohen, A. and Kulka, R.G. (1974). *J. biol. Chem.* **249**, 4522-4527.
Desnuelle, P. (1963). *Rev. Franc. Etudes Clin. et Biol.* **8**, 494-500.
Grobstein, C. (1967). *Natl Cancer Inst. Monogr.* **26**, 279-285.
Kemp, J.D., Walther, B.T. and Rutter, W.J. (1972). *J. biol. Chem.* **247**, 3941-3946.
Levine, S., Pictet, R. and Rutter, W.J. (1973). *Nature New Biol.* **246**, 49-52.
Pictet, R.L. and Rutter, W.J. (1972). *In:* "Handbook of Physiology", Section 7 (D.F. Steiner and N. Freinkel, eds) Vol.1, pp.25-66.
Pictet, R.L., Clark, W.R., Williams, R.H. and Rutter, W.J. (1972). *Develop. Biol.* **29**, 436-467.
Pictet, R.L., Filosa, S., Levine, S., Phelps, P. and Rutter, W.J. (1974). *In:* "Extracellular Matrix Influences on Gene Expression" *2nd Internatl Santa Catalina Island Colloq., Sept.18-24, 1974 (in press).*
Rall, L., Pictet, R. and Rutter, W.J. (1973). *Proc. natn. Acad. Sci. U.S.A.* **70**, 3478-3482.
Ronzio, R.A. and Rutter, W.J. (1973). *Develop. Biol.* **30**, 307-320.
Rutter, W.J., Wessells, N.K. and Grobstein, C. (1964). *J. Natl Cancer Inst. Monogr.* **13**, 51-70.
Rutter, W.J., Kemp, J.D., Bradshaw, W.S., Clark, W.R., Ronzio, R.A. and Sanders, T.G. (1968). *J. Cell Physiol.* 72 **(Suppl.1)**, 1-18.
Yalovsky, U., Heller, H. and Kulka, R.G. (1973). *Exptl Cell Res.* **80**, 322-328.

DISCUSSION

Tager: You showed some data on the binding of the mesenchyme factor to sepharose beads. This would seem to indicate that the factor is acting externally, that it does not have to be internalized. Have you tried the effect of other glycoproteins which might have the same carbohydrate structure and which might also bind to the cell surface in the same way?

Rutter: Yes, we tried a number of other growth factors which have glycoprotein structures. Some of those bind, but none of those produce this specific effect. We do not know the specific carbohydrate structure. So we cannot answer the question whether the carbohydrate structure is sufficient in itself.

Charles: Is there a critical period in the development of the epithelial cells in which they respond to the mesenchymal factor and, if so, do the epithelial cells then lose the activity to bind the mesenchymal factor?

Rutter: The epithelial cells have the competence to react with the factor once the reticulum is formed. The original gut cells do not. Subsequent to that, the experiment can only be performed during a period of a two- or three-day interval because after that the mesenchyme and the epithelium are so intermxied *in vivo* that you cannot carry out the experiment.

HORMONAL REGULATION OF ENZYME SYNTHESIS DIFFERENTIATING MAMMALIAN TISSUES

O. Greengard

Department of Biological Chemistry, Harvard Medical School
and
Cancer Research Institute, New England Deaconess Hospital
Boston, Massachusetts, USA

INTRODUCTION

Our studies in the last few years have been aimed at establishing correlations between endocrine changes and the maturation of enzyme patterns in the liver of fetal and young postnatal rats. So far, three hormones, glucocorticoid, thyroxine and glucagon, have been identified which are responsible for promoting the synthesis of specific groups of enzymes at well-defined stages of development. The present chapter describes the general approach employed in such studies, discusses the implication of the results to mechanisms involved in selective gene expression during differentiation and illustrates some difficulties which arise in investigating developmental enzyme formation in organs other than liver.

RESULTS AND DISCUSSION

Ample evidence indicates that the beginning of glucocorticoid and thyroid hormone secretion in the normal rat fetus is followed by the emergence of a cluster of enzymes in the liver and that another cluster of enzymes appears upon the secretion of glucagon in the neonate. Evidence for causal connections between the endocrine and enzymic events was obtained by showing that the concentration of some enzymes of the late fetal cluster could be raised prematurely by the administration of cortisol, while others were increased by the administration of thyroxine; the evocation of enzymes of the neonatal cluster required the injection of glucagon to the fetus (Greengard, 1969; Greengard, 1971).

It is well known that in the rat there is an ebb of corticosterone and thyroid secretion after birth and a resurgence after the 10th postnatal day (Taylor and Howard, 1971; Hommes *et al.*, 1969). The late suckling cluster of enzymes in liver emerges shortly after this time. Experiments, in which some enzymes of this cluster were prematurely evoked by cortisol and others by thyroxine administration, proved that the normal, postnatal changes in these pituitary-related activi-

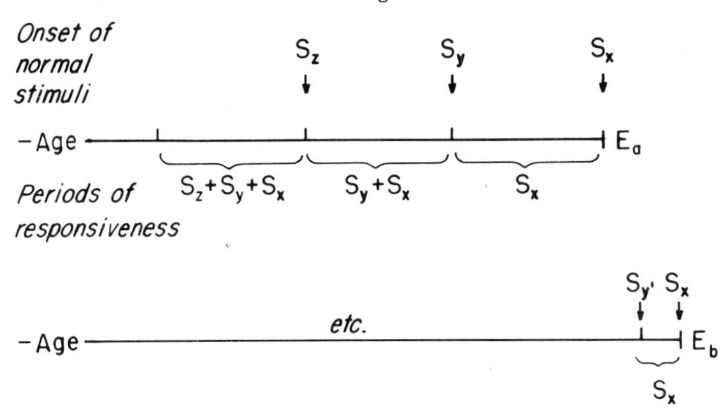

Fig. 1

ties also play important roles in hepatic biochemical differentiation (Greengard, 1971; Greengard and Jamdar, 1971). However, several additional observations indicated that the impact of these hormones does not provide a sufficient explanation for the scheduled appearance of enzyme clusters. For example, in the late suckling cluster, the rise of tryptophan oxygenase (Greengard and Dewey, 1971) or glucokinase (Jamdar and Greengard, 1970) was inhibited by adrenalectomy (performed on day 11); the premature evocation of these enzymes required sequential treatment with cortisol as well as substrate (tryptophan and glucose, respectively). Thus, the natural emergence of these enzymes is dependent on both endocrine and metabolic changes in the developing animal. Even those enzymes of the late suckling cluster which could be evoked prematurely (7-10 days before the scheduled time) by cortisol alone, or by thyroxine alone, present a problem: Why did they not appear in the late fetal liver in response to the first impact of these hormones?

The most likely explanation is that the expression of a given gene in terms of a significant amount of the finished product requires the sequential action of a series of stimuli (see arrows above the upper horizontal line in Fig.1). Let S_x be the last stimulus, the final natural trigger, which results in the immediate formation of a given enzyme (E_a). The system is competent to respond to S_x a little before the normal time of its impact and this is why (if we know the identity of S_x) we can evoke premature enzyme formation during the last of the intervals (bracketed S_x beneath horizontal line). During the preceding interval (bracketed $S_y + S_x$), i.e., before the natural impact of stimulus S_y, we cannot cause the premature appearance of E_a unless we are lucky enough to know the identity of S_y and can thus administer both S_y and S_x in the appropriate sequence. Determination of the time points at which stimuli S_x or $S_y + S_x$ become effective identifies the stage in development when positive steps towards the full competence of the expression of a given gene (or a set of genes) have occurred. Consideration of the physiological state at these critical times may provide clues to the identity of the natural stimulus (S_y or S_z). To illustrate the practical advantages of this approach, let E_a (upper line) and E_b (lower line) stand for those enzymes of the late fetal and late suckling clusters, respectively, whose synthesis requires glucocorticoid. Although their final trigger is identical

(S_x = corticosterone), only E_a appears prenatally in response to the beginning of corticosterone secretion. The reason why E_b appears later, i.e., awaits the postnatal rise in circulating corticosterone, is that the competence of the liver to synthesize this enzyme necessitates an additional stimulus. This stimulus (S_y') must have acted around term, because administered cortisol can evoke enzymes such as E_b as early as one or two days after birth. Thus, consideration of endocrine or metabolic events which occur in association with the transition to extrauterine existence should provide the clue to the identity of stimuli such as S_y'.

Glucagon, secreted by the neonate, might be such a stimulus, but only in one special case were we able to demonstrate that glucagon injected into the fetus could facilitate responsiveness to subsequently injected cortisol. We showed some time ago that tyrosine aminotransferase (an enzyme of the neonatal cluster) could be evoked in fetal liver by glucagon (Greengard and Dewey, 1967). In view of the fact that immediately after birth (and throughout postnatal life) (Franz and Knox, 1967) cortisol is a very effective inducer of tyrosine aminotransferase, it was surprising to find that in fetal rats cortisol was without effect whether

Table I. Glucagon treatment as a prerequisite for the cortisol induction of tyrosine aminotransferase (TAT) in fetal liver.

		TAT activity (μmoles/h/g liver)	
Age	Pretreatment	Saline	Cortisol, 5 h
4 days	-	34	200
Fetal	-	< 5	< 5
Fetal	Glucaton, 5 h	28	-
Fetal	Glucagon, 29 h	< 5	30

Cortisol was injected *i.p.* into 4-day-old rats (2.5 mg hydrocortisone acetate per 100 g body weight) or to fetal rats (0.125 mg per fetus, 20th day of gestation) 5 h before assay. Fetal rats, injected with glucagon (0.25 mg per fetus), were either assayed 5 h later (penultimate line) or were given cortisol 24 h later and assayed 5 h after that (last line). For additional details and reproducibility of measurements see Greengard (1974).

administered alone or simultaneously with glucagon. More recent experiments revealed that cortisol can induce before birth but only if fetuses were pretreated with glucagon. After a single injection of glucagon tyrosine aminotransferase reached maximal levels rapidly (in about 5 h); if cortisol was given to the same fetuses 24 h later (when the glucagon-induced enzyme was no longer there) considerable induction was obtained (Table I). From these observations one may conclude that the secretion of glucagon, which occurs normally immediately after birth, is responsible not only for initiating the synthesis of tyrosine aminotransferase but also for permitting its further increase in response to the other hormone, cortisol.

In rats, the emergence of the late suckling cluster of enzymes in liver coincides with the accumulation of several enzymes in kidney and brain (Greengard, 1971). However, attempts to induce these enzymes in brain prematurely with thyroxine or cortisol were generally unsuccessful. One renal enzyme, aspartate aminotrans-

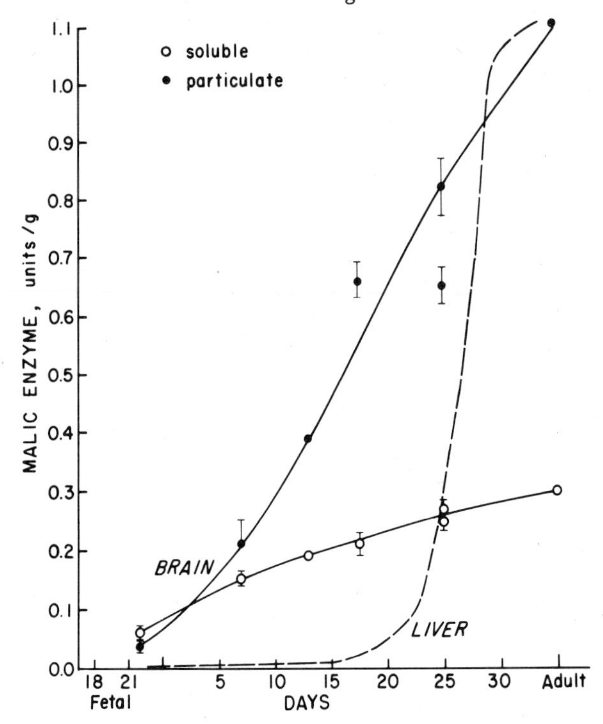

Fig.2 The development of malate-NADP dehydrogenase (malic enzyme) in brain and liver. Homogenates of cerebral hemispheres (in 9 volumes of 0.25 M sucrose) were centrifuged for 50 min at 100,000 × g to separate the soluble (○) and particulate (●) fractions. The points are means of results with 2 rats (no brackets) or 3-5 rats (brackets indicate SD). The experimental details for the development of the enzyme in the soluble fraction of liver (———) and its premature rise (△) as a result of throxine treatment are given by Greengard and Jamdar (1971).

ferase (Greengard, *in press*), which did rise prematurely after cortisol administration to the neonate, required a dose high enough to interfere seriously with growth. Even such excessive doses of cortisol (or thyroxine) did not influence the accumulation of aspartate aminotransferase or of other enzymes in brain. A closer scrutiny of the "developmental history" of frequently studied cerebral enzymes which accumulate rapidly during the third postnatal week should have predicted these negative results. The developmental increase of brain enzymes (e.g., glutamate decarboxylase, glutamate dehydrogenase, hexokinase, etc. (MacDonnell and Greengard, 1974)), like that of renal aspartate aminotransferase (Herzfeld and Greengard, 1971), actually begins at term or shortly thereafter, thus preceding the postnatal increase in thyroxine (or corticosterone) secretion. Recent studies of malate-NADP dehydrogenase, for example, have shown that it accumulates in both liver and brain during postnatal life (Fig.2). In liver, malate-NADP dehydrogenase is an entirely soluble enzyme; in brain, however, while some of the activity is found in the soluble portion of the cells, a considerable portion of the enzyme can be sedimented by centrifugation and is probably concentrated in synaptosomes (MacDonnell and Greengard, 1974). More important for the present discussion is the difference in timing: in liver this enzyme appears after the 15th postnatal day (i.e., soon after the normal rise in thyroid activity (Hommes *et al.*, 1969)), whereas in brain the activities in both portions of the cell start rising at birth.While thyroxine treatment (△, Fig.2) can evoke the hepatic enzyme a week or more before the physiologically scheduled time (Greengard and Jamdar, 1971), it has no effect on the cerebral enzyme. Thus, the normal rise of thyroid activity in the second postnatal week which triggers the synthesis of the hepatic malate-NADP dehydrogenase does not seem to influence the same enzyme in brain. The accumulation of enzymes such as malate-NADP dehydrogenase in postnatal brain and kidney may well be a delayed consequence of prenatal programming, but no evidence is available as yet which would indicate that the beginning of thyroid or adrenocortical activity in the late fetus contributes to this programming.

ACKNOWLEDGEMENTS

This investigation was supported by USPHS CA-08676 from the National Cancer Institute, and the Atomic Energy Commission Contract AT(11-1)-3085 with the New England Deaconess Hospital

REFERENCES

Franz, J.M. and Knox, W.E. (1967). *Biochemistry* **6**, 3464.
Greengard, O. (1969). *Science* **163**, 891.
Greengard, O. (1971). *Essays Biochem.* **7**, 159.
Greengard, O. (1974). *In:* "Problems and Priorities in Perinatal Pharmacology" (J. Dancis and J.C. Hwang, eds). Raven Press, New York.
Greengard, O. *J. Steroid Biochem. (in press).*
Greengard, O. and Dewey, H.K. (1967). *J. biol. Chem.* **242**, 2986.
Greengard, O. and Dewey, H.K. (1971). *Proc. natn. Acad. Sci. U.S.A.* **68**, 1698.
Greengard, O. and Jamdar, S.C. (1971). *Biochim. biophys. Acta* **237**, 476.
Herzfeld, A. and Greengard, O. (1971). *Biochim. biophys. Acta* **237**, 88.
Hommes, F.A., Wilmink, C.W. and Richters, A. (1969). *Biol. Neonatorum* **14**, 69.
Jamdar, S.C. and Greengard, O. (1970). *J. biol. Chem.* **245**, 2779.
MacDonnell, P.C. and Greengard, O. (1974). *Arch. Biochem. Biophys.* **163**, 644.
Taylor, M.H. and Howard, E. (1971). *Endocrinology* **88**, 1190.

DISCUSSION

Gaull: I have been interested in another complication in the induction of enzymes that you did not discuss. Dr. Snell and Dr. Räihä, who have both been studying cystathionase in fetal rat liver, found that the administration of corticosteroids could not prematurely induce this enzyme. But when Dr. Räihä took the liver out of the fetus and grew it in organ culture he then was able to stimulate the cystathionase activity. I wonder whether we are talking about a balance between something which is inhibiting and something which is stimulating.

Greengard: I think that there is no evidence for the existence of such an inhibitor. It is true, of course, what you are saying about the organ culture, but I think it shows that the cultured liver is not like the liver *in vivo*.Under the usual conditions, while we can induce an enzyme like tyrosine transaminase the explants lose, for example, something like 80% of the soluble protein content. This indicates that the culture conditions are rather unhealthy.

Rutter: I was interested in another thing that you did not talk about, concerning the receptors for the hormones themselves. When do the receptors for glucocorticoid and for glucagon arise in these systems? At early times one might not have an effect because of the receptors.

Greengard: This specificity with respect to age and enzyme cannot be explained by the absence or presence of the receptor, unless you assume a separate receptor for each enzyme or each group of enzymes. The glucocorticoid receptor is certainly present in fetal liver.

Rutter: Ballard, I know, measured the glucocorticoid receptor.

Greengard: Yes, in the lung. He did not find the receptor to be limiting in this organ either.

Rutter: I wonder if there might be specific receptors for each group of enzymes. If you just had two receptors, would that suffice?

Greengard: No, because I did not mention some further complications, namely that responsiveness continues to change with age. You can take three enzymes and find that one would respond to the given hormone only between the age of 8 and 10 days, the other only in the fetus, and the third one only after the 20th day. There are too many variations to explain by two receptors.

De Groot: I have a general question. In your fascinating paper you have described a critical period of responsiveness of enzymes for endocrine stimuli. Do you think that the critical period is limited to a small period, and that an enzyme will not respond at all to the endocrine stimulus when it does not get its endocrine stimulus at a certain period?

Greengard: This is quite possible, although we have not done experiments in a way to prove it.

Van den Berg: Are there long-lasting effects of premature induction of an enzyme? Is it still evident at a much later stage?

Greengard: We have not found this. We treated neonatal animals, with cortisol for example, and looked for permanent effects, but did not find any on the ensymes we studied. This does not mean that there are no such effects.

Sternowsky: Do you think that inducing an enzyme by injecting dexamethasone into the mother makes a completely different picture from that by injecting the hormone into the fetus, when it is done *in vitro*? I refer with this question to the well-known efforts to induce lecithin synthesis in the lung. It works if you inject the hormone into the mother, but it does not if you inject it into the fetus.

Greengard: I do not think that cortisol has ever been injected into the human fetus for that purpose.

Sternowsky: No, but in the newborn and the prematurely born with respiration difficulties.

Greengard: That is a different question. The organism was no longer *in utero*.

Sternowksy: When you inject the hormone the moment the baby is born, or a minute or so afterwards, I cannot see why it should make any difference between injecting the hormone into the mother or into the newborn, which is still practically a fetus.

Greengard: You may inject it within a minute after birth, but the consequences develop during the subsequent day or days. The conditions during that time are not identical to the conditions *in utero*.

Sternowsky: That is correct.

Tyson Tildon: I just wondered when you said, "No inhibitors." The effect which you saw with cortisol, where you get the reduction, is that you interpreted as inhibitory to the PPH enzyme?

Greengard: No, I think that either the degradation of the enzyme was enhanced or its synthesis was decreased.

Tyson Tildon: One other question. It is possible that some effects of hormones are mediated through the changes in concentration of metabolites, rather than being a direct effect of the hormones?

Greengard: Yes, we never know whether the hormone action is direct or mediated through a change in metabolites. Enzyme amounts could certainly be altered by either.

AMMONIA AND ENERGY METABOLISM IN ISOLATED MITOCHONDRIA AND INTACT LIVER CELLS

J.M. Tager[1], T.P.M. Akerboom[1], J.B. Hoek[2]*, A.J. Meijer[1,3]
W. Vaartjes[3]†, L. Ernster[2] and J.R. Williamson[3]

[1]*Laboratory of Biochemistry, University of Amsterdam*
Plantage Muidergracht 12, Amsterdam, The Netherlands
[2]*Biochemical Institute, University of Stockholm*
Box 6409, S-11 382 Stockholm, Sweden
[3]*Johnson Research Foundation, University of Pennsylvania*
Philadelphia, Pennsylvania, USA

Glutamate dehydrogenase occupies a central position in the catabolic processes leading to the synthesis of urea (Chamalaun and Tager, 1970). By deamination of glutamate, it provides the bulk of the ammonia required for carbamoyl phosphate synthesis. On the other hand, it can remove ammonia by the reductive amination of α-oxoglutarate to glutamate. The activity of this enzyme, which is localized in the mitochondria, is extremely high in the liver, and also in certain other tissues like brain. The enzyme reacts with NAD and NADP. Thus the components of the glutamate dehydrogenase reaction can be used to estimate the redox state of mitochondrial nicotinamide nucleotides in tissues in which the activity of the enzyme is high (Williamson *et al.*, 1967; Krebs and Veech, 1969). Conversely, if one of the components of the reaction is present in excess, it can influence the mitochondrial redox state. Thus accumulation of ammonia leads to a more oxidized state of the mitochondrial nicotinamide nucleotides (Klingenberg and Slenczka, 1959; Klingenberg, 1961; Hoek and Tager, 1973).

Ammonia is the toxic factor in the congenital hyperammonemias (Shih and Efron, 1972), and also plays an important role in the pathogenesis of hepatic coma (Zieve, 1966). It has been suggested that ammonia exerts its toxic action by interfering with energy metabolism in the cell (see Hindfelt and Siesjö (1971a, 1971b) for a review).

One possibility is that removal of α-oxoglutarate in the glutamate dehydrogenase reaction may lead to depletion of Krebs-cycle intermediates and thus to decreased oxidative activity (Recknagel and Potter, 1951; Bessman and Bessman,

* Present address: Department of Biochemistry, University of Nairobi, Nairobi, Kenya.
† Present address: Laboratory of Veterinary Biochemistry, University of Utrecht, Biltstraat 172, Utrecht, The Netherlands.

1955; Clark and Eiseman, 1958). Ammonia can inhibit oxygen uptake with pyruvate as substrate in isolated mitochondria, for instance those from liver (Worcel and Erecinska, 1962). However, there is no change in the amount of α-oxoglutarate in the brain when rats are fed with ammonia (Shorey *et al.*, 1967; Hindfelt and Siesjö, 1971b). This finding suggests that depletion of Krebs-cycle intermediates may not be the primary cause of ammonia toxicity. Indeed, in experimental animals a *stimulation* of oxidative activity can be observed upon ammonia intoxication (Hawkins *et al.*, 1973).

A second possibility is that incorporation of ammonia into glutamine, an ATP-requiring process, may impose an increased energy drain on the tissue (Weil-Malherbe, 1962). Although the glutamine content of the brain is increased when ammonia is administered to rats, Hawkins *et al.* (1973) have found that the release of the amide to the circulation is too slow to account for the stimulation by ammonia of oxidative metabolism.

A third possibility, proposed by Hawkins *et al.* (1973) is that an increased level of ammonium ion in the plasma may lead to a decrease in the resting potential across the plasma membrane in brain and a stimulation of the Na^+-K^+ ATPase.

Several workers have measured the effect of ammonia intoxication on the adenine nucleotide and creatine phosphate content of brain. Two significant points emerge. Firstly, when rapid quenching methods are used, no effect of ammonia on total ATP and ADP content of the tissue can be observed, in spite of the fact that oxidative metabolism is stimulated (Hindfelt and Siesjö, 1971b; Hawkins *et al.*, 1973). Secondly, ammonia intoxication leads to a small, but significant decrease in creatine phosphate, as observed by Hindfelt and Siesjö (1971b) and by Hawkins *et al.* (1973).

The question arises of what causes the stimulation by ammonia of oxidative metabolism. Is it, as Hawkins *et al.* (1973) suggest, due mainly to a decrease in the membrane potential across the plasma membrane and a stimulation of the Na^+-K^+ ATPase? Does ammonia impose an energy demand in other ways? Is the increased energy demand the cause of ammonia toxicity?

We have investigated this problem by studying the effect of ammonia on metabolic processes in the liver, a tissue in which considerable progress has been made in elucidating metabolic pathways and control mechanisms. In these studies suspensions of isolated rat-liver cells were used. The cells were prepared according to the procedure of Berry and Friend (1969), in which the liver is perfused with collagenase and hyaluronidase in order to digest intercellular material. The washed hepatocytes obtained by this procedure are exactly comparable to the perfused liver with regard to rates of ureogenesis and gluconeogenesis, and other metabolic and biochemical parameters (Krebs *et al.*, 1974).

Table I shows the effect of ammonia on urea synthesis and on the ATP content of isolated rat-liver cells. When ammonia plus ornithine are added, the rate of urea synthesis is about 33 μmoles/g weight per h. Concomitantly, the ATP content decreases from 2.28 to 1.53 μmoles/g wet weight. When oleate is added as well as ammonia plus ornithine, the rate of urea synthesis is almost doubled, but the ATP content actually increases. This effect of ammonia on the steady-state level of ATP cannot be due solely to an increased energy demand for urea synthesis. When aminooxyacetate, a transaminase inhibitor (Hopper and Segal, 1962) was present as well as ammonia, there was an inhibition of urea synthesis, as expected

Table I. Effect of ammonia on ATP level in rat-liver cells.

Additions	No. of expts.	ATP (μmoles/g wet weight at 30 min)	Δ Urea (μmoles/g w.w. per h)
None	12	2.28 ± 0.12	—
Orn + NH_3	7	1.53 ± 0.07	32.8 ± 5.6
Orn + NH_3 + oleate	7	1.86 ± 0.07	59.0 ± 6.4

Rat-liver cells (3-6 mg dry weight of cells/ml incubation mixture) were isolated by the method of Berry and Friend (1969) and incubated at 37° C in a final volume of 4 ml with Krebs-Ringer bicarbonate (pH 7.4), 4% (w/v) defatted bovine serum albumin and (where indicated) 3 mM ornithine, 8 mM NH_4Cl and 1 mM oleate. The gas phase was 95% O_2-5% CO_2. The values are the means ± S.E.M.

Table II. Effect of ammonia ± aminooxyacetate on ATP level in rat-liver cells.

Additions	ATP (μmoles/g wet weight at 30 min)	Δ Urea (μmoles/g wet weight per h)	Δ Glucose (μmoles/g wet weight per h)
Pyruvate (10 mM)	2.48	16	32
Pyruvate + AOA (1 mM)	2.83	15	32
Pyruvate + NH_3	1.86	162	24
Pyruvate + NH_3 + AOA	1.88	57	12

For conditions, see Table I. Abbreviation: AOA, aminooxyacetate.

Table III. Effect of ammonia and aminooxyacetate on β-hydroxybutyrate oxidation and ATP level in rat-liver cells.

Additions	ATP (μmoles/g wet weight at 30 min)	Δ Urea (μmoles/g wet weight per h)	Δ Acetoac. (μmoles/g wet weight per h)
None	2.48	15	111
AOA (1 mM)	2.44	10	99
NH_3	2.24	32	149
NH_3 + AOA	2.24	9	122

The reaction mixture contained 3 mM ornithine. For other conditions, see Table I. Abbreviations: Acetoac., acetoacetate; AOA, aminooxyacetate.

(Table II). Yet even in the presence of aminooxyacetate, the steady state level of ATP was decreased (Table II). In the experiment shown in Table III, cells were incubated with β-hydroxybutyrate. The oxidation of β-hydroxybutyrate to acetoacetate was stimulated 34% by the addition of ammonia. However, in the presence of aminooxyacetate, which inhibited urea synthesis, there was still a 23% stimulation of β-hydroxybutyrate oxidation. It is clear that ammonia stimulates oxidative metabolism not only because extra ATP is required for urea synthesis, but also for some other reason or reasons as well.

ATP is synthesized during the passage of electrons along the respiratory chain, for instance from NADH to oxygen. There is an obligatory coupling of ATP synthesis to oxidation, so that an increased demand for ATP leads to an increase in the rate of oxidation, and *vice versa.* Three molecules of ATP are synthesized for each molecule of NADH oxidized by oxygen.

$$NADH + H^+ + \tfrac{1}{2}O_2 + 3ADP + 3P_i \rightleftharpoons NAD^+ + 3ATP + 4H_2O \qquad (1)$$

In the span NADH to cytochrome c, two molecules of ATP are synthesized for each molecule of NADH oxidized.

$$NADH + 2\text{ cytochrome } c^{3+} + 2ADP + 2P_i \rightleftharpoons NAD^+ + 2\text{ cytochrome } c^{2+} + 2ATP \qquad (2)$$

The equilibrium constant K of reaction 2 is:

$$K = \frac{[NAD^+]}{[NADH]} \times \frac{[\text{cytochrome } c^{2+}]^2}{[\text{cytochrome } c^{3+}]^3} \times \frac{[ATP]^2}{[ADP]^2\,[P_i]^2}$$

The value of K can be calculated from the half-reduction potential of NAD, the half-reduction potential of cytochrome c, and the standard free energy of hydrolysis of ATP.

Wilson *et al.* (1974) have recently investigated the relationship between redox reactions and the phosphorylation potential in suspensions of intact rat-liver cells. They calculated the redox state of mitochondrial NAD from the ratio β-hydroxybutyrate: acetoacetate, and measured the redox state of cytochrome c spectrophotometrically. To estimate the phosphorylation potential, total cellular ATP, ADP and phosphate were used.

Wilson *et al.* (1974) found that the measured value of K differed from the calculated value by a factor of 15-100. They considered this value to be sufficiently low to be able to conclude that the mitochondrial respiratory chain is in near-equilibrium with the phosphorylation state calculated from the total content of adenine nucleotides and phosphate in the cell.

Although the validity of the conclusion may be questioned*, it occurred to us that these relationships might be used to compare different metabolic conditions, and, in particular, to investigate the effect of ammonia on energetic relationships in the liver cell.

---

* Equilibrium relations between the oxidation-reduction reactions of the mitochondrial respiratory chain and ATP synthesis have been extensively investigated in isolated mitochondria (see, e.g., Chance and Hollunger (1961); Klingenberg and Schollmeyer (1963); Cockrell *et al.* (1966); Muraoka and Slater (1969); Slater (1969); Slater *et al.* (1973); and Rosing (1974); cf. also Erecinska *et al.* (1974)). Muraoka and Slater (1969) have concluded that near thermodynamic equilibrium between the redox reactions of the respiratory chain and phosphorylation reactions is reached when mitochondria are in State 4 (cf. Klingenberg and Schollmeyer, 1963). However, in the metabolizing liver cell, the mitochondria are most probably not in State 4, so that the relations between the oxidation-reactions and ATP synthesis measured by Wilson *et al.* (1974) and by ourselves (Tables V and VII) should more correctly be referred to as steady-state relationships.

Table IV. Effect of uncoupler or ammonia on redox state and level of ATP and ADP in isolated rat-liver cells.

Reactant or reactant ratio	Concentration or ratio		
	No additions	5 mM FCCP	10 mM NH_4Cl
BOH/acetoacetate	0.194	0.113	0.095
Cyt^{2+}/cyt^{3+}	0.075	0.044	0.043
ATP*	1.79	1.33	1.48
ADP*	0.70	1.62	1.28
Urea*	0.91	0.88	3.47

Rat-liver cells (18 mg dry weight/ml) were incubated at 25°C in a final volume of 5 ml with Krebs-Ringer bicarbonate (pH 7.4) and 2% dialysed bovine serum albumin (Sigma Fraction V). After 15 min, the reaction was stopped in 2 ml of the incubation mixture by adding $HClO_4$. β-hydroxybutyrate, acetoacetate, urea, ATP and ADP were assayed in the neutralized acid extract. In another portion of the incubation mixture, oxidized and reduced cytochrome c (cyt^{3+} and cyt^{2+}) were assayed as described by Wilson *et al.* (1974).

* μmoles/g wet weight found at 15 min.

When the uncoupler of oxidative phosphorylation carbonylcyanide p-trifluoromethoxyphenylhydrazone (FCCP) was added to a suspension of rat-liver cells, the β-hydroxybutyrate:acetoacetate ratio declined, indicating that there was an oxidation of mitochondrial NAD. Concomitantly, cytochrome c became more oxidized, ATP decreased, and ADP increased (Table IV). Similar changes were induced by ammonia (Table IV).

In the experiment of Table IV, the measured values of K for the reaction

$$NADH + 2\ \text{cytochrome}\ c^{3+} + 2ADP + 2P_i \rightleftharpoons NAD^+ + 2\ \text{cytochrome}\ c^{2+} + 2ATP \qquad (2)$$

could be obtained from the measured metabolite contents. These values are shown in the 4th line of Table V. In the last line, these values are compared with that calculated from the half-reduction potentials of NAD and cytochrome c, and the $\Delta G'_0$ for ATP hydrolysis. With no additions, this value differed from the calculated value by a factor of 88, which is within the range reported by Wilson *et al.* (1974).

When the uncoupler was added the difference between the measured and calculated values of K increased from 88 to 1740, indicating that reaction 2 became more displaced from equilibrium (Table V). A similar increase in displacement from equilibrium was obtained when ammonia was added (Table V).

In order to ascertain if this increase in the displacement from equilibrium was due solely to the increased energy demand for urea synthesis, an experiment was carried out in which the effect of ammonia was tested both in the presence and in the absence of CO_2. As Haussinger *et al.* (1975) have shown, the synthesis of urea is completely inhibited when CO_2 is absent. Nevertheless, similar changes were observed on adding ammonia to rat-liver cells in the absence of CO_2 as in its presence: a decrease in the β-hydroxybutyrate:acetoacetate ratio, an oxidation of cytochrome c, a decrease in ATP, and an increase in ADP (Table VI).

The results of three separate experiments on the effect of ammonia on the energetic relationships in rat-liver cells are summarized in Table VII. Thus the addition

Table V. Redox and phosphorylation states in isolated rat-liver cells .

$$NADH + 2\ cyt\ c^{3+} + 2ADP + 2P_i \rightleftharpoons NAD^+ + 2\ cyt\ c^{2+} + 2ATP$$

$$K = \frac{[NAD^+]}{[NADH]} \times \frac{[cyt\ c^{2+}]^2}{[cyt\ c^{3+}]^2} \times \frac{[ATP]^2}{[ADP]^2\,[P_i]^2}$$

	No additions	5 μM FCCP	10 mM NH_4Cl
$[NAD^+]/[NADH]$	3.3×10^2	5.7×10^2	6.8×10^2
$([Cyt\ c^{2+}]/[cyt\ c^{3+}])^2$	0.57×10^{-2}	0.20×10^{-2}	0.19×10^{-2}
$([ATP]/[ADP]\,[P_i])^2$	2.8×10^5	0.23×10^5	0.51×10^5
$K_{measured}$ (M^{-2})	5.3×10^5	0.25×10^5	0.68×10^5
$K_{calculated}$*/$K_{measured}$	88	1740	685

The $[P_i]$ was taken to be 4.8 mM, which was the concentration found in the freeze-clamped perfused liver (cf. Wilson *et al.* (1974)).

* From half-reduction potentials of NAD and cytochrome c (see Wilson *et al.*, 1974) and $\Delta G_0'$ for ATP hydrolysis (Rosing and Slater, 1972).

Table VI. Effect of ammonia and CO_2 on redox state and on ATP and ADP levels and urea synthesis in isolated rat-liver cells.

Reactant or reactant ratio	Concentration or ratio			
	CO_2 present		CO_2 absent	
	control	$+NH_4Cl$	control	$+NH_4Cl$
BOH/acetoacetate	1.09	0.38	0.56	0.37
$Cyt\ c^{2+}/cyt\ c^{3+}$	0.125	0.078	0.042	0.019
ATP*	3.17	2.58	2.64	2.11
ADP*	0.30	0.71	0.42	0.84
Urea*	1.5	3.0	<0.2	<0.2

Rat-liver cells (7 mg dry weight/ml) were incubated at 25° C in Krebs-Ringer solution (final volume, 5 ml) containing 3 mM ornithine and either 26.6 mM bicarbonate (gas phase, 95% O_2 - 5% CO_2) or 26.6 mM 5,5'-dimethyloxazolidine-2,4-dione (gas phase, 100% O_2). The pH was 7.4. Further procedure as in Table IV.

* μmoles/g wet weight found at 10 min.

of oleate, which brought about a marked reduction of mitochondrial NAD, had no effect on the measured value of K (cf. Wilson *et al.* (1974)). When ammonia was added, the measured value of K was decreased under all conditions, even when CO_2 was absent. The last two columns of Table VII show that ammonia brought about an increase in the difference between $K_{calculated}$ and $K_{measured}$ indicating that the reaction

Table VII. Energetic relations in rat-liver cells.

$NADH + 2\ cyt\ c^{3+} + 2ADP + 2P_i \rightleftharpoons NAD^+ + 2cyt^{2+} + 2ATP$

$$K = \frac{[NAD^+]}{[NADH]} \times \frac{[cytochrome\ c^{2+}]^2}{[cytochrome\ c^{3+}]^2} \times \frac{[ATP]^2}{[ADP]^2\,[P_i]^2}$$

Expt.	Additions or omission	$K_{measured}$		$K_{calculated}$*/ $K_{measured}$	
		$-NH_4Cl$	$+NH_4Cl$	$-NH_4Cl$	$+NH_4Cl$
1	None	5.3×10^5	0.68×10^5	88	685
2	None	4.9×10^5		97	
	+ Ala	5.4×10^5	0.45×10^5	86	995
	+ Ala, + oleate	4.4×10^5		100	
3	None	29×10^5	9.2×10^5	15	48
	+ Orn	44×10^5	5.1×10^5	10	86
	$- CO_2$	0.52×10^5	0.20×10^5	850	2200
	$- CO_2$, + Orn	3.0×10^5	0.13×10^5	147	3880

Reaction conditions for Expt.1 described in Table IV and for Expt.3 in Table VI. In Expt.2, the reaction mixture contained Krebs-Ringer bicarbonate, 2% albumin, 15 mg/ml dry weight of cells and (where indicated) 10 mM alanine and 1 mM oleate; other conditions as in Table IV.

* From half-reduction potentials of NAD and cytochrome c (see Wilson *et al.*, 1974) and $\Delta G_0'$ for ATP hydrolysis (Rosing and Slater, 1972).

$$NADH + 2\ cytochrome\ c^{3+} + 2ADP + 2P_i \rightleftharpoons NAD^+ + 2\ cytochrome\ c^{2+} + 2ATP \quad (2)$$

becomes more displaced from equilibrium when ammonia is present.

These results suggest that ammonia imposes an increased energy demand on rat-liver cells, over and above that required for urea synthesis.

Possible consequences of the increased energy demand imposed by ammonia are indicated by the results of the experiment presented in Fig.1. Rat-liver cells were incubated with glutamine in the presence of oleate and ethanol, and urea production and ammonia content were measured in the presence of different concentrations of the uncoupler dinitrophenol. In the control experiment, uncoupler at 0.05 and 0.1 mM stimulated urea production slightly. As the concentration of uncoupler was increased, urea synthesis was progressively inhibited. The concentration of ammonia in the reaction mixture was 0.4 mM in the absence of uncoupler, and was increased to 1.2 mM or higher in its presence. When ornithine was present, much higher concentrations of uncoupler had to be added before urea synthesis became inhibited. In the presence of ornithine, which stimulates citrulline synthesis from carbamoyl phosphate, the level of ammonia was low.

We conclude that the increased uncoupler sensitivity of urea synthesis in the absence of ornithine was due to the high concentration of ammonia present, which imposed an extra energy demand on the system. This extra energy demand may be due, in part at least, to an uncoupler-like effect of ammonia.

A possible mechanism for an uncoupling by ammonia is the following. Gluta-

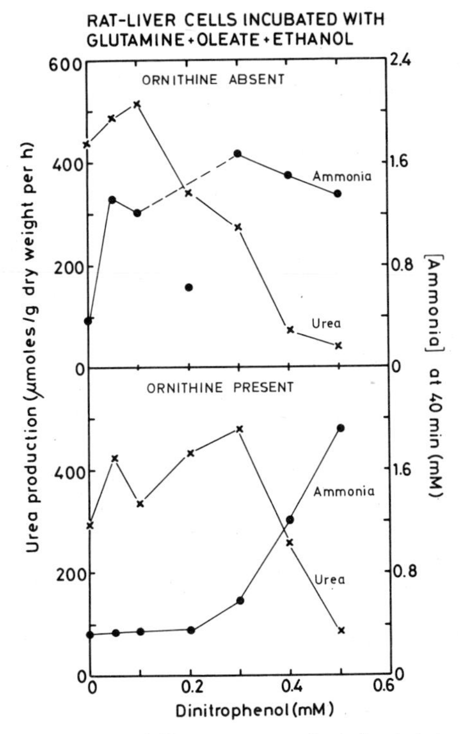

Fig.1 Effect of uncoupler and ornithine on urea synthesis from glutamine and ammonia level in rat-liver cells. Rat-liver cells (6.7 mg dry weight/ml) incubated in a final volume of 4 ml at 37° C with Krebs-Ringer bicarbonate (pH 7.4), 2% defatted bovine serum albumin, 3 mM ornithine (where indicated), 5 mM ethanol, 1 mM oleate and 10 mM glutatmine.

mate dehydrogenase is able to react with both NAD and NADP. In isolated liver mitochondria, NAD is highly oxidized, and NADP is kept highly reduced by the presence of the energy-linked transhydrogenase (Danielson and Ernster, 1963; Klingenberg and Schollmeyer, 1963; Estabrook van Nissley, 1963). Thus kinetic considerations lead one to expect that glutamate will react mainly with NAD^+, and α-oxoglutarate + ammonia with NADPH, simply because NAD^+ and NADPH are present in high concentrations (Hoek *et al.*, 1974). Coupling of these two reactions would lead to an oxidation of NADPH by NAD^+ (reactions 3 and 4).

$$\text{Glutamate} + NAD^+ \rightleftharpoons \alpha\text{-oxoglutarate} + NH_3 + NADH + H^+ \quad (3)$$

$$\alpha\text{-oxoglutarate} + NH_3 + NADHP + H^+ \rightleftharpoons \text{Glutamate} + NADP^+ \quad (4)$$

Furthermore, coupling of these two reactions of glutamate dehydrogenase with the energy-linked transhydrogenase should, in principle, lead to energy dissipation (reactions 3-6).

$$\text{Glutamate} + NAD^+ \rightleftharpoons \alpha\text{-oxoglutarate} + NH_3 + NADH + H^+ \quad (3)$$

$$NADH + NADP^+ + ATP + H_2O \rightarrow NADPH + NAD^+ + ADP + P_i \quad (4)$$

$$NADPH + H^+ + \alpha\text{-oxoglutarate} + NH_3 \rightleftharpoons \text{glutamate} + NADP^+ \quad (5)$$

$$\text{Sum:} \quad ATP + H_2O \rightarrow ADP + P_i$$

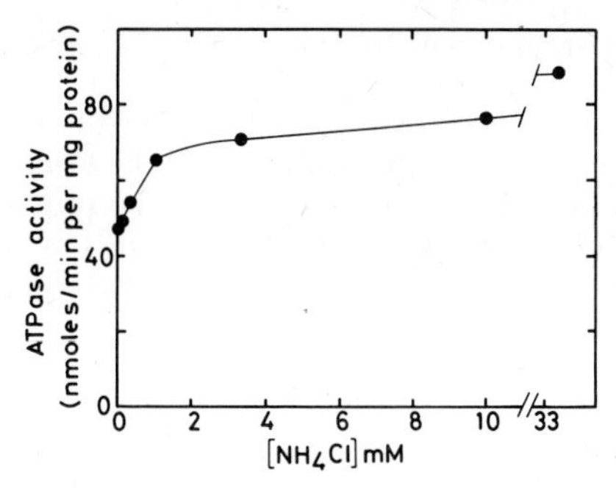

Fig. 2 Glutamate dehydrogenase-induced ATPase activity in rat-liver mitochondria. For experimental conditions, see legend to Table II of Hoek *et al.* (1974). 16 mM glutamate and 0.2 mM α-oxoglutarate were present.

Table VIII. Effect of metabolites of the glutamate dehydrogenase reaction on ATP hydrolysis in rat-liver mitochondria.

Glu (mM)	α-Oxoglutarate (mM)	NH_4Cl (mM)	FCCP (μM)	ATPase activity (nmoles/min.mg Protein)
16	0.2	3.3	0	71
16	0	3.3	0	44
0	0.5	3.3	0	52
0	0	0	1.7	154

For reaction conditions, see legend to Table II of Hoek *et al.* (1974).

The results of the experiment shown in Fig.2 show that hydrolysis of ATP does, in fact, occur when ammonia is added to rat-liver mitochondria in the presence of glutamate and α-oxoglutarate. This ATPase activity is dependent on the presence of all the components of the glutamate dehydrogenase reaction (Table VIII). For comparison, the last line of Table VIII shows the ATPase activity of rat-liver mitochondria in the presence of the uncoupler FCCP at a concentration of 1.7 μM.

In summary, the results of the studies with isolated liver cells reported in this paper indicate that ammonia imposes an energy demand over and above that required for urea synthesis. Part of this energy demand may be due to stimulation of the Na^+-K^+ ATPase, as suggested by Hawkins *et al.* (1973). Part may be due to a coupling of the glutamine synthetase and glutaminase reactions. Finally, our results with isolated mitochondria suggest that part of the energy demand may be due to an energy-dissipating coupling of the energy-linked transhydrogenase with a transhydrogenation brought about by the components of the glutamate dehyrogenase reaction. This increased energy demand may play a role in the pathogenesis of ammonia toxicity.

ACKNOWLEDGEMENTS

This study was supported in part by grants from the Netherlands Foundation for the Advancement of Pure Research (Z.W.O.) under the auspices of the Netherlands Foundation for Chemical Research (S.O.N.), The Swedish Cancer Society, the Swedish National Science Research Council and the United States Public Health Service (Grants AM-15120 and AA-00292). The authors wish to thank Mr. G.M. van Woerkom for his technical assistance. J.B.H. is grateful to the Swedish Science Research Council for the award of a postdoctoral research fellowship

REFERENCES

Berry, M.N. and Friend, D.S. (1969). *J. Cell Biol.* **43**, 506-520.

Bessman, S.P. and Bessman, A.N. (1955). *J. clin. Invest.* **34**, 622-628.

Chamalaun, R.A.F.M. and Tager, J.M. (1970). *Biochim. biophys. Acta* **222**, 119-134.

Chance, B. and Hollunger, G. (1961). *J. biol. Chem.* **236**, 1577-1584.

Clark, G.M. and Eiseman, B. (1958). *New Engl. J. Med.* **259**, 178-180.

Cockrell, R.S., Harris, E.J. and Pressman, B.C. (1966). *Biochemistry* **5**, 2326-2335.

Danielson, L. and Ernster, L. (1963). *Biochem. Z.* **338**, 188-205.

Erecinska, M., Veech, R.L. and Wilson, D.F. (1974). *Arch. Biochem. Biophys.* **160**, 412-421.

Estabrook, R.W. and Nissley, S.P. (1963). *In:* "Functionelle und Morphologische Organisation der Zelle" (P. Karlson, ed) pp.119-131. Springer, Berlin.

Haussinger, D., Weiss, L. and Sies, H. (1975). *Eur. J. Biochem. (in press).*

Hawkins, R.A., Miller, A.L., Nielsen, R.C. and Veech, R.L. (1973). *Biochem. J.* **134**, 1001-1008.

Hindfelt, B. and Siesjö, B.K. (1971a). *Scad. J. Clin. Lab. Invest.* **28**, 353-364.

Hindfelt, B. and Siesjö, B.K. (1971b). *Scand. J. Clin. Lab. Invest.* **28**, 365-374.

Hoek, J.B. and Tager, J.M. (1973). *Biochim. biophys. Acta* **325**, 197-212.

Hoek, J.B., Ernster, L., de Haan, E.J. and Tager, J.M. (1974). *Biochim. biophys. Acta* **333**, 546-559.

Hopper, S. and Segal, H.L. (1962). *J. biol. Chem.* **237**, 3189-3195.

Klingenberg, M. (1961). *In:* "Zur Bedeutung der freien Nukleotiden", 11 Colloquium der Gesellschaft für physiologische Chemie, Mosbach/Baden, pp.82-114. Springer, Berlin.

Klingenberg, M. and Schollmeyer, P. (1963). Symp. Intracellular Respiration: Phosphorylating and Non-phosphorylating Oxidation Reactions; Proc. 5th Int. Congr. Biochem., Moscow, 1961, Vol.5, pp.46-65. Pergamon, London.

Klingenberg, M. and Slenczka, W. (1959). *Biochem. Z.* **331**, 486-517.

Krebs, H.A., Cornell, N.W, Lund, P. and Hems, R. (1974). *In:* "Regulation of Hepatic Metabolism" (F. Lundquist and N. Tygstrup, eds) pp.726-750. Munsgaard, Copenhagen.

Krebs, H.A. and Veech, R.L. (1969). *In:* "The Energy Level and Metabolic Control in Mitochondria" (S. Papa, J.M. Tager, E. Quagliariello and E.C. Slater, eds) pp.329-382. Adriatica Editrice, Bari, Italy.

Muraoka, S. and Slater, E.C. (1969). *Biochem. biophys. Acta* **180**, 221-226.

Recknagel, R.O. and Potter, V.R. (1951). *J. biol. Chem.* **191**, 263-270.

Rosing, J. (1974). "Studies on Thermodynamics and Mechanism of Oxidative Phosphorylation", Ph.D. Thesis, Univ. of Amsterdam, pp.1-101. Gerja, Waarland, Netherlands.

Rosing, J. and Slater, E.C. (1972). *Biochim. biophys. Acta* **267**, 275-290.

Shih, V.E. and Efron, M.L. (1972). *In:* "The Metabolic Basis of Inherited Disease" (J.B. Stanbury, J.B. Wijngaarden and D.S. Fredrickson, eds) pp.370-392. McGraw-Hill, New York.

Shorey, J., McCandless, D.W. and Schenke, S. (1967). *Gastroenterology* **53**, 706-711.

Slater, E.C. (1969). *In:* "The Energy Level and Metabolic Control Mitochondria" (S. Papa, J.M. Tager, E. Quagliariello and E.C. Slater, eds) pp.255-259. Adriatica Editrice, Bari, Italy.

Slater, E.C., Rosing, J. and Mol, A. (1973). *Biochim. biophys. Acta* **292**, 534-553.

Weil-Malherbe, H. (1962). *In:* "Neurochemistry" (K.A.C. Elliott, I.H. Page and J.H. Quastel, eds) pp.321-329. Thomas, Springfield.

Williamson, D.H., Lund, P. and Krebs, H.A. (1967). *Biochem. J.* **103**, 514-526.

Wilson, D.F., Stubbs, M., Veech, R.L., Erecinska, M. (1974). *Biochem. J.* **140**, 57-64.

Worcel, A. and Erecinska, M. (1962). *Biochim. biophys. Acta* **65**, 27-33.

Zieve, L. (1966). *Arch. Intern. Med.* **118**, 211-223.

DISCUSSION

Hommes: I think these are very beautiful studies, especially if you take into account your last sentence relating these studies to inborn errors of the urea cycle. But I have some difficulties there with the interpretation. You add ammonium chloride to the isolated cells to a final concentration which is in the millimolar range. In hyperammonemia due, for instance, to ornithine transcarbamylase deficiency, one will consider a plasma concentration of 150 micromolar as already pathological. That is one order of magnitude lower as compared to what you have added to the cells. My question is: do you see the same effects when you lower the concentration of ammonium chloride?

Tager: This is an extremely important experiment, which we have not done yet. We plan to do this type of experiment with isolated liver cells, and also plan some similar experiments in collaboration with Helmut Sies. Perhaps I may quote one of the experiments carried out by Sies and co-workers. They have found that ammonia stimulates oxygen uptake in the flow-through perfused rat liver (D. Haussinger, L. Weiss and H. Sies, *Europ. J. Biochem., in press).* If you omit CO_2 from the system, this stimulation of oxygen uptake does not occur. However, I believe that oxygen uptake in this particular case may not be a good parameter. Sies has found that the addition of ammonia stimulates pyruvate disappearance, as measured by the oxidation of [1-^{14}C] pyruvate to $^{14}CO_2$. Concomitantly, the percentage of pyruvate dehydrogenase in the active form was increased by ammonia. This stimulation of pyruvate oxidation also occurs when CO_2 is absent, so that it is not dependent on the energy demand for urea synthesis. These experiments of Sies were carried out at much lower concentrations of ammonia than we use, the maximal concentration that he used in most of his experiments being 1.2 mM. He also carried out a titration of the effect of ammonia on pyruvate oxidation. The stimulation is dependent on the concentration of ammonia added; you see a very beautiful stepwise stimulation of pyruvate oxidation on the stepwise increase of the ammonia concentration, both in the presence and in the absence of CO_2.

Wick: Do you see any latency of the ammonia effect? I ask this because clinically it is well known that such a latency of ammonia toxicity exists. I recall one patient who had about 300 μg of ammonia per 100 ml and looked perfectly healthy. It was thought that a wrong estimation had been done, but the patient was dead the following morning (Frenton, personal communication). Thus, there is a latency between ammonia increase and coma of several hours. Do you see any latency of this type?

Tager: In the studies of Sies and co-workers, no latency effect was observed. One may, however, speculate that the effects that ammonia has may start very early, and that a cumulative action may cause a final lethal effect.

Van den Berg: I would like to make one general point. I do not think there is any strict relationship between ammonia levels and clinical signs. One has to be very careful with looking only at ammonia itself to find one common mechanism which will explain everything. You did say that some of your K's calculated differ appreciably from the measured K's. There is always the possibility for another explanation: compartmentation, and therefore the calculation on the basis of compounds can be wrong if components are separated by barriers, and therefore you may have local parts where equilibrium can occur. In most of your experiments a very large flux is going through the system. The relation between a static measurement at a certain time and something going on in a dynamically changing system will only be possible when the activity of an enzyme is in excess over its flux.

Tager: Your comment is based on a misunderstanding. The condition of a near equilibrium was not proposed by me but by Wilson *et al.* (1974). I have stated clearly that our approach in using the value of measured K (I never used the word equilibrium constant for the measured values) is to compare different metabolic conditions.

Van den Berg: I would then prefer to use another term, but that is a minor point. Still, the concept of fluxes going on and measuring at a certain time is a more basic difficulty than the other one.

Tager: I certainly do not think that I am measuring even near equilibrium; I am only measuring steady states. But we are hoping in this way to be able to compare different metabolic conditions. In my introduction I pointed out that there is very little change in total ATP and ADP, but what you do detect is a small change in creatine phosphate (Hindfelt and Siesjö, 1971b; Hawkins *et al.*, 1973). I believe myself, which would support your idea, that this small change in creatine phosphate really does reflect an increased flux of some kind.

Cremer: I would like to ask Dr. Tager a question related to his liver cell preparations, in comparison with some work that has been done quite recently on the whole brain *in situ.* This relates to the possible difference between acute experiments with ammonia, in which you are actually changing the steady-state concentration of glutamine in the tissue, compared to the chronic situation. In a chronic situation of hyperammonemia we found that the glutamine is very high in the brain, but it stays at a new steady-state level. We do not think that the actual rate of oxidation is faster then. There is no evidence for an increased rate through glutamine synthetase. In your liver cells, did you find an initial increase and then reaching a new steady-state glutamine concentration?

Tager: I cannot answer your question since we did not measure glutamine in our experiments.

Wick: I would just like to make a comment about this parallelism of ammonia and clinical coma. I think you can never say that there is no parallelism because the coma is there without ammonia. There are many other types of coma without increased ammonia; but on the other hand when a certain level of ammonia is reached you will inevitably get a coma. The fact that this is not yet so clear is due mainly to the difficulties of ammonia estimations. That makes the theoretical work much more valid; it is really at the root of specific toxicity.

Land: Of interest in relation to Dr. Tager's work is the finding of Drs. Clark and Nicklas, who have shown that ammonium ions uncouple glutamate oxidation by brain mitochondria. Ammonium ions both stimulate and uncouple pyruvate oxidation. This is somewhat difficult to rationalize. One can rationalize stimulation of pyruvate oxidation, and one can understand the uncoupling of the glutamate oxidation by the mechanism you just elucidated so nicely, but one has difficulty in explaining an uncoupling of pyruvate oxidation by ammonia.

Tager: The effect of ammonia in stimulating glutamate and pyruvate oxidation in brain mitochondria is not specific for ammonia; you found it also for potassium and sodium.

Clark: Not quite in the same way. Potassium does not uncouple.

Tager: No, but you did find stimulation. I thought that your interpretation, with which we would agree, is that there is a stimulation of the entry of pyruvate and of glutamate into the mitochondria (F.A.J.T.M. van den Bergh and J.A. Gimpel, unpublished observations).

Clark: Yes, this may be part of the answer.

Tager: As far as the uncoupling effect of ammonia on both pyruvate and glutamate oxidation is concerned, I agree. The uncoupling of pyruvate oxidation by ammonia is more difficult to understand.

Blass: Roche and Reed (1974, *Fed. Proc.* 33, 1427) have recently reported that ammonium ion acts on the system which inactivates and reactivates the pyruvate dehydrogenase complex. It favors activation, by accentuating the inhibitory effect of ADP on the inactivating kinase. Concentrations of 1-5 millimolar are active *in vitro*. As Dr. Clark points out, higher concentrations of potassium ions have a similar effect.

Tager: Reed reported not only on ammonia, but also on other monovalent cations.

Blass: Yes, but specifically ammonia and potassium.

Tager: I want to emphasize that Reed and co-workers found that monovalent cations in general stimulate the phosphatase.

GLUCONEOGENESIS IN THE NEONATAL RAT: THE METABOLISM AND DISPOSITION OF ALANINE DURING POSTNATAL DEVELOPMENT

K. Snell

Department of Biochemistry, University of Surrey
Guildford, Surrey, England

ENERGY CONSIDERATIONS AND GLUCONEOGENESIS

Gluconeogenesis is one of the major metabolic roles of the liver and one of the major sites of energy utilisation in this organ. Theoretically the synthesis of one molecule of glucose from two molecules of pyruvate requires six molecules of ATP, or other high-energy phosphate equivalents. The ATP is required for the conversion of pyruvate to oxaloacetate, for the conversion of oxaloacetate to phosphoenolpyruvate and for the conversion of 3-phosphoglycerate to 1,3-diphosphoglycerate. The pathway of gluconeogenesis is essentially a reversal of the reactions of glycolysis except for the involvement of four unique enzymes which ensure the reversibility of what would otherwise be irreversible steps (Fig.1). These steps are the phosphorylation of glucose by hexokinase which is reversed by glucose 6-phosphatase; the phosphorylation of fructose 6-phosphate by phosphofructokinase which is reversed by fructose diphosphatase; and the transphosphorylation of phosphoenolpyruvate with ADP to form ATP and pyruvate which is reversed by the sequential action of pyruvate carboxylase and phosphoenolpyruvate carboxykinase. Each of these reactions involves the net utilisation of ATP in one direction with no net formation of ATP in the other direction. The simultaneous operation of these opposing enzymes could set up a series of futile cycles in which continual interconversion of the substrates and products occurred with a net consumption of ATP. The extent to which these three substrate cycles occurs during gluconeogenesis will determine the actual amount of energy, in the form of ATP, which is required for the overall process. It has been suggested that such substrate cycling might be significant in terms of the amplification of metabolic signals for the regulation of gluconeogenesis and glycolysis (Newsholme and Start, 1973) or as a factor contributing to nonshivering thermogenesis (Newsholme *et al.*, 1972; Clark *et al.*, 1973).

Recently attempts have been made *in vivo* using suitable isotopic tracer procedures to estimate in quantitative terms the extent to which recycling occurs

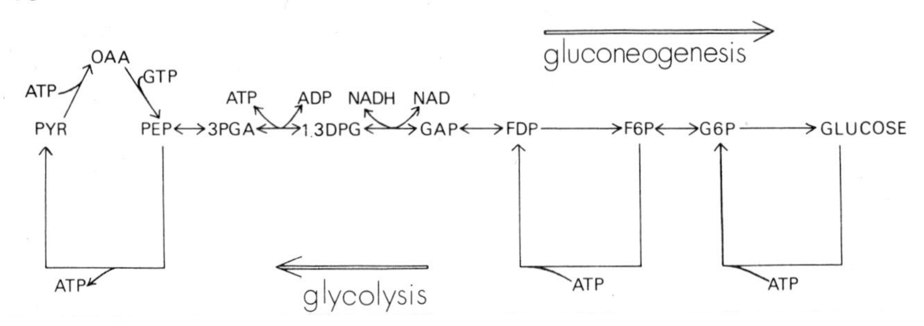

Fig. 1 Substrate cycles of gluconeogenesis. Abbreviations: PYR, pyruvate; OAA, oxaloacetate; PEP, phosphoenolpyruvate; 3PGA, 3-phosphoglycerate; 1,3DPG, 1,3-diphosphoglycerate; GAP, glyceraldehyde 3-phosphate; FDP, fructose 1,6-diphosphate; F6P, fructose 6-phosphate; G6P, glucose-6-phosphate. For the enzymes involved see the text and Fig.2 of Snell and Walker (1973b).

through each of the three glycolytic-gluconeogenic substrate cycles in various physiological circumstances. Such measurements suggest that during gluconeogenesis from lactate in the suckling 2 day-old rat, recycling through the fructose diphosphate-fructose-6-phosphate substrate couple and the glucose-6-phosphate-glucose substrate couple amounts to increases in energy expenditure of 7.5 and 9.8%, respectively (Clark *et al.*, 1974) and that through the pyruvate-phosphenolpyruvate couple amounts to about a 14% increase (Friedmann *et al.*, 1971). Thus, overall the increase in energy expenditure during gluconeogenesis from lactate would be in the order of 30%, i.e. the equivalent of about 8 molecules of ATP per molecule of glucose formed rather than the theoretical value of 6. In addition there is evidence from *in vitro* studies on the isolated perfused liver of adult rats that such energy expenditure is greater when the gluconeogenic substrate is 10 mM-alanine than when it is 10 mM-lactate, so that the nature of the actual glucose precursors *in vivo* might also influence the extra energy expenditure involved in gluconeogenesis.

Since a good deal of the metabolic energy produced in the suckling rat is required for synthetic processes leading to the accumulation of body protein and brain lipids during the neonatal period of active growth, the amount which can be spared for glucose synthesis needs to be carefully controlled in order to avoid unnecessary wastage. To a large extent this control is achieved by the activities of the enzymes which catalyse the rate-limiting steps of gluconeogenesis and also by the activities of these enzymes relative to those which catalyse the opposing reactions of glycolysis. Some crude estimate of these enzymic relations can be obtained by comparing the maximal activites of the enzyme as measured in assays carried out under optimal conditions *in vitro*. Of course the actual activity of these enzymes which obtains *in vivo* will be related to the concentrations not only of substrates for the enzymes but also of intermediary metabolites and other effectors which regulate precisely the expression of enzyme activity. In this context it is pertinent to note that the adenine nucleotides, the ratio of which is a sensitive indicator of the energy status of the cell, are regulatory effectors of, for example, both fructose diphosphatase and phosphofructokinase. AMP acts as a negative effector on fructose diphosphatase and a positive effector on phosphofructokinase and, in addition, the latter enzyme is also negatively controlled by ATP levels. Nevertheless a comparison of maximal activities may still provide some rough guide as to the enzymic capacity of the opposing pathways of gluconeogenesis and glycolysis. It is found that *in vitro* activities of the key enzymes of gluconeogenesis, involved in the unidirectional steps in Fig.1, are increased in the suckling neonatal rat as compared to the young weaned and adult rat (see later, Table III). This suggests that the neonatal rat has the capacity for increased rates of gluconeogenesis relative to the adult. In addition the ratios of activities of glycolytic enzymes to gluconeogenic enzymes are lower in the neonatal rat than in the adult, e.g. pyruvate kinase/phosphoenolpyruvate carboxykinase is about 6.2 in the 10 day-old rat and about 30 in the adult (calculated from data of Vernon and Walker, 1968). This again suggests that the flux in the direction of gluconeogenesis is very much predominant over that in the opposite direction in the neonatal rat. (The fact that even in the neonatal rat the ratio of pyruvate kinase to phosphoenolpyruvate carboxykinase is greater than unity suggests that *in vivo* the activity of pyruvate kinase must be considerably inhibited in order to achieve a net gluconeogenic flux.)

MEASUREMENTS OF GLUCONEOGENESIS IN THE DEVELOPING NEONATAL RAT

In agreement with the enzymic data, measurements of the incorporation of radioactively-labelled precursors into carbohydrate in liver slices from developing rats have shown that the maximal rate of gluconeogenesis is greater in the suckling rat than in the normal, fed adult (Ballard and Oliver, 1963; Vernon *et al.*, 1968). These findings have been confirmed *in vivo* by measuring the incorporation from tracer amounts of labelled *L*-lactate and *L*-alanine into the body glucose pool (Walker and Snell, 1973). Similar *in vivo* studies involving the administration of tracer amounts of isotopically-labelled glucose and measurements of the turnover of the body glucose pool in developing rats have shown an increased quantitative requirement for glucose synthesis during the suckling period (Vernon and Walker, 1972; Snell and Walker, 1973a; Walker and Snell, 1973). Both lactate and alanine have pyruvate as a common intermediate on the carbon pathway to glucose formation, but because isotope incorporation is not direct evidence of the net synthesis of glucose from a precursor these *in vivo* studies do not permit us to identify the actual precursor which is responsible for the observed increase in carbon flux through pyruvate to glucose. In order to gain information on this point it is necessary to measure the net accumulation of glucose in the presence of a precursor. Such measurements have been made using liver slices (Vernon *et al.*, 1968) but these were only of limited value because of the well-known deficiencies of this preparation and also because the rats at the different ages in the study were pre-treated in different ways to deplete the hepatic glycogen content. In the adult rat such studies have been carried out using the isolated perfused liver and, in addition, this preparation has also proved useful for studying aspects of the regulation of gluconeogenesis, particularly with respect to hormonal control. More recently the use of isolated liver cells for such studies has been developed and this technique requires an initial perfusion of the liver with collagenase or a collagenase-hyaluronidase mixture to achieve a separation of the hepatocytes. For all these various reasons it was decided to adapt the method of liver perfusion for use with neonatal rats, and the lowest practicable age was found to be about 10 days *post partum* when the liver mass is about 0.4 g.

The perfusion technique used was essentially that of Hems *et al.* (1966) and

Table I. Gluconeogenesis from various substrates in perfused livers of adult and neonatal rats.

Substrate	Initial concn. (mM)	Rate of gluconeogenesis Neonatal	(μmol/min/g of liver) Adult
None		0.14 (10)	0.08 (6)
L-lactate	11.2	0.90 (6)[a]	0.49 (6)
L-serine	10[b]	0.40 (5)[a]	0.13 (5)
L-alanine	9.02	0.28 (17)	0.25 (10)

Neonatal rats were 10 days *post partum.* All animals were starved for 16-20 h before the experiment. Livers were perfused *in situ* as described by Snell (1974). The first line gives the apparent endogenous rate of glucose synthesis in the absence of added substrate; all other results are corrected for this endogenous carbohydrate formation. [a]Significantly different from adult value ($p < 0.001$). [b]Not determined experimentally. The number of observations is given in parentheses. From Snell (1974).

brief details have been given by Snell (1974a). All rats were starved overnight for 16-20 h to deplete the livers of glycogen and liver perfusion was then carried out *in situ* with 50 ml of bicarbonate-saline, containing dialysed bovine serum albumin and aged human erythrocytes. Livers were perfused for 38 min with this medium before the addition of glucogenic precursors as 0.2 M neutral solutions to give an initial circulating concentration of about 10 mM. Samples of medium were removed at time-intervals into perchloric acid and metabolites were determined on the neutralized, deproteinised extracts.

Table I shows measurements of net glucose synthesis from various precursors in the perfused liver of 10 day-old and adult rats. A higher endogenous rate of glucose formation was observed in the 10 day-old rat compared to the adult. In the neonatal animal this endogenous rate arises almost entirely from gluconeogenesis from the small residual amount of lactate (0.2 - 0.4 mM) present in the red blood cells used in the perfusion medium. The shedding of endogenous glucose units from any small amount of residual glycogen in the liver will be essentially complete during the perfusion period before samples are taken. On the other hand in the adult where residual glycogen levels are higher and where there is over 12-times the mass of tissue for the same constant volume of perfusion medium compared to the neonatal liver, the shedding of these glucose units will continue throughout the experimental period. The rate of glucose formation from the lactate in the medium will be lower in the adult but, in any case, because of the large mass of tissue the lactate will be quickly depleted in the 38 min period before measurements are made. After correcting the rates of glucose formation in the presence of various substrates for the corresponding endogenous rate at the two ages, *L*-lactate and *L*-serine, but not *L*-alanine, were found to give significantly higher gluconeogenic rates in neonatal rat liver than in adult liver. The higher rate of gluconeogenesis from lactate is in agreement with the evidence obtained with labelled lactate *in vivo*. It has been suggested that the higher rate from serine may not reflect the increased gluconeogenic pathway via pyruvate to glucose, but may be due to the enhanced operation of a unique pathway involving hydroxypyruvate and D-glycerate as intermediates rather than pyruvate (Snell, 1974a). The similar rates of gluconeogenesis at the two ages with 10 mM-alanine as substrate is in apparent contradiction to the data using isotopically-labelled alanine *in vivo*, but

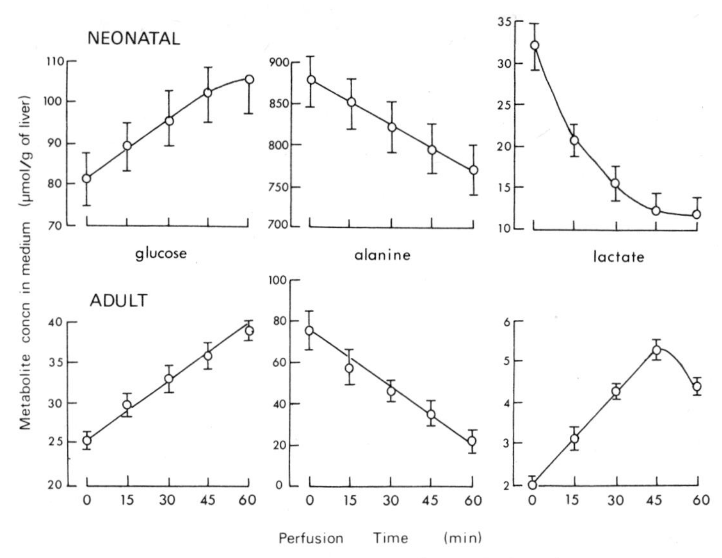

Fig. 2 Time course of the changes in metabolites during perfusion of neonatal and adult rat liver. Livers from 16-20 h-starved rats were perfused with an initial concentration of 10 mM-alanine. Substrate was added to the medium after 38 min and measurements were made from 40-100 min. Zero perfusion time on the figure corresponds to the sample taken at 40 min. Each point is the mean of 5-8 determinations and vertical bars show ± S.E.M. Unpublished work of Snell and Walker.

in the latter case tracer amounts of alanine were used whereas the perfusions were carried out with loading concentrations of alanine. This implies that there is a restriction on the capactiy to convert high concentrations of alanine to pyruvate in the pathway of gluconeogenesis.

PRODUCTS OF ALANINE METABOLISM IN THE PERFUSED LIVERS OF DEVELOPING NEONATAL RATS

Carbon Products

Since there is considerable evidence that alanine is one of the major substrates for gluconeogenesis *in vivo* (see reviews by Exton, 1972; Felig, 1973; Snell and Walker, 1973b), the fate of alanine was investigated further in the perfused livers of developing neonatal rats. Figure 2 shows the time-course of changes in a number of metabolites during the metabolism of 10 mM-alanine by perfused livers from 10 day-old and adult rats. In the adult the disappearance of alanine and the formation of glucose was linear with time and the formation of lactate was linear up to 45 min after the addition of substrate. In the neonatal rat although alanine disappearance was linear over the whole time-period, lactate disappeared in an exponential fashion and glucose formation tailed off considerably with time. This is consistent with lactate making a large contribution to the observed glucose formation.

Table II gives a more detailed analysis of both carbon and nitrogen products of alanine metabolism. Alanine uptake from the perfusion medium is greater in

Table II. Alanine metabolism in the perfused liver of developing neonatal rats.

Metabolites in medium	Age (days)			
	10	20	30	Adult
Alanine	–110	–112	–73	–75
Glucose	25	28	15	16
Lactate	–20	19	7	6
Pyruvate	<1	3	2	2
Urea	11	30	25	26
NH_3	13	14	4	3
Glucose and lactate + pyruvate (as C_3)	30	77	39	39
Urea + NH_3 (as NH_3)	35	73	54	55
Excess of alanine removed over C_3 formed	80	35	34	36
Excess of alanine removed over NH_3 formed	75	39	19	20

Livers from 16-20 h-starved rats were perfused with an initial concentration of 10 mM-alanine. Metabolite changes in the medium were measured by standard enzymic methods. The values are changes in μmoles (means of 5-8 determinations) in the perfusion medium during 60 min perfusion expressed per g of liver. Unpublished work of Snell, Lord and Walker.

Table III. Activities of enzymes of gluconeogenesis during neonatal development.

Enzyme	Neonatal age (days)			
	10	20	30	Adult
Alanine aminotransferase (EC 2.6.1.2.)[a]	1.9	3.8	7.5	20.5
Pyruvate carboxylase (EC 6.4.1.1.)[b]	12.9	11.9	-	12.5
Phosphoenolpyruvate carboxykinase (EC 4.1.1.32.)[c]	4.1	3.0	1.8	1.8
Fructose diphosphatase (EC 3.1.3.11.)[c]	2.5	2.1	1.9	1.3
Glucose 6-phosphatase (EC 3.1.3.9.)[c]	22.8	18.3	11.8	10.2

Activities are expressed as μmol/min per g of liver and are taken from [a]Snell and Walker, 1972; and [b]Ballard and Hanson, 1967; and [c]Vernon and Walker, 1968.

the suckling and weanling rats (10 and 20 days *post partum*) than in young weaned and adult rats. The total carbon products recovered account for only 27% of the alanine removed in the 10 day-old rat, 69% in the 20 day-old, and about 52% in the older, weaned animals. So despite the increased uptake of alanine, the 10 day-old rat appears to be restricted in the capacity to convert this into the carbon products measured and this restriction is apparently not present in the 20 day-old. Glucose appears to be the major carbon product of alanine metabolism at all the ages. As previously anticipated a significant proportion of the glucose carbon in the 10 day-old rat is derived from lactate (20 out of the 50 3-carbon units formed), but at all other ages alanine appears to be the source.

As pointed out earlier this restriction on the conversion of alanine to glucose in the 10 day-old rat can best be explained by a deficiency of enzymic capacity to convert alanine into pyruvate. The enzyme catalysing this reaction is alanine aminotransferase (EC 2.6.1.2) and the pattern of development of the enzyme shows a low activity throughout the suckling period with an increase to the adult value beginning at about the time of weaning (Snell and Walker, 1972). Table III shows the activities of alanine aminotransferase and the key enzymes on the pathway of gluconeogenesis from pyruvate as assayed *in vitro*. In the adult rat total alanine aminotransferase activity is far in excess of the other enzymes along the pathway, whereas in the 10 day-old rat aminotransferase activity is the lowest of the recorded activities. Thus, despite the capacity of the enzymes of a gluconeogenesis from pyruvate being enhanced in the 10 day-old rat (reading Table III horizontally) the low activity of alanine aminotransferase at this age restricts the conversion of loading amounts of alanine to glucose. At 20 days *post partum* alanine aminotransferase activity is double that in the 10 day-old animal and is no longer the rate-limiting activity so that the enhanced capacity of the other gluconeogenic enzymes can be fully exploited. At 30 days *post partum* and in the adult further increases in alanine aminotransferase activity occur and the activities of the other gluconeogenic enzymes have all decreased so reducing the capacity of the gluconeogenic pathway to handle saturating amounts of substrate. These enzymic changes explain very well the changing pattern of alanine metabolism observed in Table II. However, as pointed out earlier the activities of enzymes as assayed *in vitro* under optimal conditions may bear little quantitative relation to the activities operating *in vivo*, so the interpretation offered must be viewed with some caution.

Furthermore the actual pathway operating *in vivo* may contain many more complexities than is obvious from a simple consideration of the sequence of reactions involved in the pathway. Nowhere is this more true than in the pathway of gluconeogenesis and Fig.3 shows the possible subcellular interrelationships which might operate in the transformation of alanine to glucose. It must be stressed that this scheme is only one of a number of possible alternatives which can be devised to account for the different subcellular locations of the enzymes of gluconeogenesis, and also those of ureogenesis, and the reported impermeability of the mitochondrial membrane to certain anions and nucleotides. It seeks to combine the known facts that pyruvate carboxylation in the rat occurs almost entirely in the mitochondria and that both carbon and reducing power are required in the cytosol for the gluconeogenic sequence from oxaloacetate to glucose. In addition the metabolism of alanine also gives rise to nitrogen products of which NH_3 and urea are the most significant. The liberation of NH_3 by glutamate dehydrogenase and its incorporation into citrulline via carbamyl phosphate formation and reaction with ornithine take place within the mitochondria, whereas the incorporation of a second molecule of NH_3 in the form of aspartate and the remaining enzymic steps leading to the liberation of urea take place in the cytosol. Another feature included in the scheme is the occurrence of isoenzymic forms of alanine aminotransferase which in the suckling rat are equally distributed between the cytosol and the mitochondria (Snell and Walker, 1972).

The localisation of pyruvate carboxylase within the mitochondria means that pyruvate formed by transamination in the cytosol has to be transported into the mitochondria in order to be carboxylated. The rate of carboxylation of pyruvate by intact isolated mitochondria from developing rats (Snell, 1974b) is, in fact, only 10-20% of the activity of the solubilised pyruvate carboxylase assayed *in*

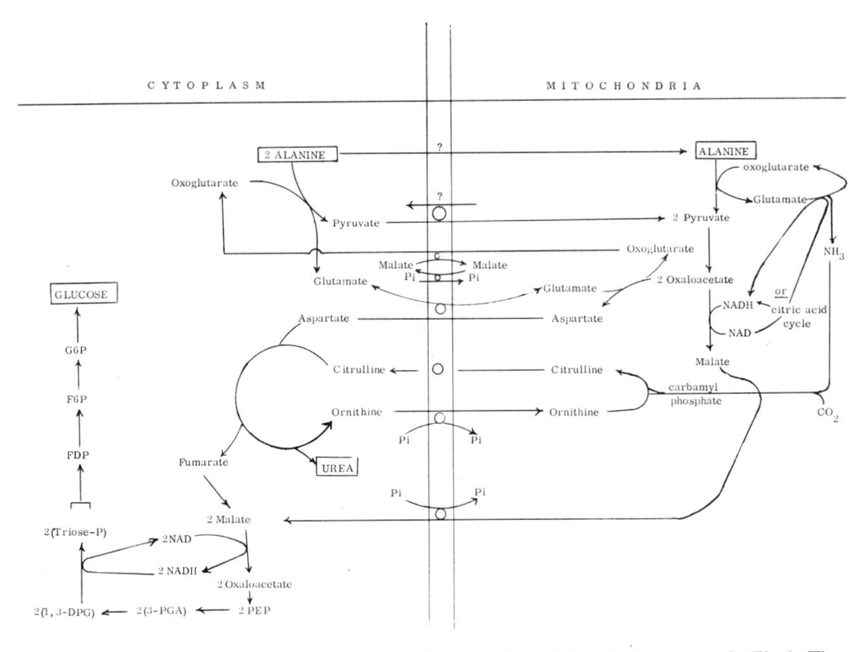

Fig.3 Proposed scheme for gluconeogenesis from alanine. Abbreviations are as in Fig.1. The scheme shows the net conversion of 2 molecules of alanine to one molecule each of glucose and urea.

Table IV. Effect of cortisol on alanine metabolism in the perfused liver of 10 day-old rats.

Metabolites in medium	+ NaCl	+ Cortisol
Alanine	–105	–120
Glucose	23	32
Lactate and pyruvate	–20	24
Alanine aminotransferase	2.05	12.2

10 day-old rats were injected intraperitoneally with 0.25 mg cortisol acetate or saline (control) and sacrificed 24 h later for liver perfusion with 10 mM-alanine as substrate or for assay of alanine aminotransferase activity in liver homogenates as described by Snell and Walker (1972). Other details are as in Table II. The values are means of 4 determinations in each case. Unpublished work of Snell.

vitro (reported in Table III). This suggests that the recently described pyruvate translocator (Papa *et al.*, 1971; Halestrap and Denton, 1974) may be the rate-limiting step in the overall reaction *in vivo*. Other such examples are also evident from an examination of the scheme in Fig.3.

On this basis the interpretation of the patterns of alanine metabolism during development (Table II) in terms of enzymic activities measured in *in vitro* assays (Table III) may not be justified. However, further evidence that gluconeogenesis

Table V. Urea and ammonia formation from alanine in perfused livers of developing rats.

Age (days)	Rate of Formation (μmol/min/g of liver)	
	Urea	Ammonia
10	0.19 (6)	0.22 (5)
20	0.49 (8)	0.23 (7)
30	0.42 (8)	0.06 (6)
Adult	0.44 (4)	0.05 (4)

Livers from 16-20 h-starved rats were perfused with an initial concentration of 10 mM-alanine as substrate. The number of determinations is given in parentheses. Unpublished work of Lord, Snell and Walker.

from alanine in the 10 day-old rat is indeed restricted by the activity of alanine aminotransferase is shown in Table IV. In the suckling rat it is possible to accelerate the development of alanine aminotransferase, which normally increases in activity beginning at about 15 days *post partum*, with a single intraperitoneal injection of cortisol (Snell and Walker, 1972). Ten day-old rats treated in this way showed 6-fold increase in alanine aminotransferase activity and a marked increase in glucose formation from alanine from 26 3-carbon units in the saline-treated controls to 40 3-carbon units in the cortisol-treated animals.

Nitrogen Products

Table II also gives details of the formation of nitrogen products of alanine metabolism in the perfused livers of neonatal rats. The actual rates of urea and NH_3 formation in perfused livers are given in Table V. The rate of urea formation is low in the 10 day-old rat but increases to the adult value by 20 days *post partum.* In contrast the rate of NH_3 formation is high in both the 10 and 20 day-old rat but decreases to the adult value by 30 days *post partum.* The relatively low rate of urea formation in the perfused liver of the 10 day-old rat is in good agreement with measurements of urea turnover *in vivo* (R. Lord, K. Snell and D.G. Walker, unpublished work). The decay of specific radioactivity with time in plasma urea was measured following a single intraperitoneal injection of a tracer amount of [^{14}C] urea (5 μCi/100 g body wt.) into rats of different ages. The calculated rates of turnover of the body urea pool in 5, 10 and 30 day-old rats were 1.53, 1.69 and 4.29 μmol/min/100 g body wt., respectively.

Returning again to the data in Table II, the total nitrogen products recovered account for 32%, 65% and 74% of the alanine removed in the 10 day-old, 20 day-old and older rats, respectively. In the 10 day-old rat the nitrogen products appear equally as urea and NH_3, in the 20 day-old rat urea formation is increased but NH_3 formation is still significant, and in the older rats urea formation is about the same as at 20 days but NH_3 formation has decreased to low levels.

An explanation of this data is both complex and speculative in nature and involves a consideration of the data in Table II, the scheme of alanine metabolism in Fig.3 and the relevant enzyme activities which are given in Table VI. It is apparent from Table VI that in the 10 day-old rat the low alanine aminotransferase activity is close to being one of the rate-limiting steps in the overall formation of urea from alanine. Nevertheless the lowest activity is observed for argininosuccinate synthetase and this is the rate-limiting enzyme in the urea cycle itself.

Table VI. Activities of enzymes of ureogenesis during neonatal development.

Enzyme	Neonatal age (days) 10	20	30	Adult
Alanine aminotransferase (EC 2.6.1.2.)[a]	1.9	3.8	7.5	20.5
Aspartate aminotransferase (EC 2.6.1.1.)[b]	213	151	–	132
Glutamate dehydrogenase (EC 1.4.1.3.)[c]	5.9	6.1	5.2	5.9
Carbamoylphosphate synthetase (EC 2.7.2.2.)[d]	9.7	12.5	–	13.0
Ornithine carbamoyltransferase (EC 2.1.3.3.)[d]	245	273	–	290
Argininosuccinate synthetase (EC 6.3.4.5.)[e]	0.37	0.57	–	0.95
Argininosuccinate lyase (EC 4.3.2.1.)[e]	1.7	2.6	–	2.6
Arginase (EC 3.5.3.1.)[f]	200	520	1220	950

Activities are expressed as μmol/min per g of liver and are taken from [a]Snell and Walker, 1972; [b]Herzfeld and Greengard, 1971; [c]in direction of deamination, Hommes and Richters, 1969; [d]Raiha and Suihkonen, 1968; [e]Miller and Chu, 1970; [f]Greengard *et al.*, 1970.

Thus, the low activity of argininosuccinate synthetase in the 10 day-old rat will restrict the operation of the urea cycle and the formation of urea. It will also lower the level of arginine (an intermediate in the urea cycle between the release of fumarate and the release of urea) and a consequence of this will be a decreased synthesis of acetylglutamate (for which arginine is an activator). Acetylglutamate is an obligatory co-factor for carbamyl phosphate synthetase and so a lack of acetylglutamate will prevent the condensation of CO_2 and NH_3 and lead to an accumulation of NH_3 (see Krebs *et al.*, 1973 for further details and references). Another factor limiting the operation of the urea cycle in the 10 day-old rat may be a lack of aspartate for condensation with citrulline. Although the aspartate aminotransferase activity is high in the 10 day-old rat (in both the mitochondrial and cytosol compartments, Herzfeld and Greengard, 1972), because of the enhanced activity of the pathway for gluconeogenesis from oxaloacetate and the low rate of pyruvate formation (from alanine by alanine aminotransferase) for oxaloacetate synthesis, the availability of oxaloacetate for transamination with asparate may be restricted. The accelerated pathway for gluconeogenesis from pyruvate may have another consequence in pulling the alanine aminotransferase reaction over towards glutamate formation and thus encouraging NH_3 formation. Measurements of the reactants and products of alanine aminotransferase in freeze-clamped liver of 10 day-old rats show that the reaction is indeed displaced from equilibrium *in vivo* (K. Snell, unpublished work). The presence of alanine aminotransferase activity in equal amounts in both the cytosol and the mitochondria in the 10 day-old rat (Snell and Walker, 1972) may disturb the normal co-ordination of ammonia and asparate formation from urea synthesis which operates in the adult when alanine transamination is predominantly cytosolic.

In the 20 day-old rat there is an increase in both the provision of alanine amino nitrogen via increased alanine aminotransferase activity and of the capacity for urea synthesis via increased argininosuccinate synthetase activity (Table VI). The result of this is an increase in urea formation from alanine. However, flow through the gluconeogenic pathway from pyruvate is still accelerated and the isoenzymic activities of alanine aminotransferase are still fairly equally distributed between the mitochondria and the cytosol (Snell and Walker, 1972) so that similar considerations may apply to account for the accumulation of ammonia as for the 10 day-old rat.

In the older animals argininosuccinate synthetase is increased somewhat further, alanine aminotransferase activity is now high and oxaloacetate is no longer being directed towards glucose synthesis. Most of the cellular alanine aminotransferase activity is located in the cytosol; co-ordination of NH_3 and aspartate formation for urea synthesis is thus facilitated and the accumulation of NH_3 no longer occurs.

RELEVANCE OF THE PERFUSION DATA TO ALANINE METABOLISM *IN VIVO*

The main conclusions reached in the study with perfused livers may be summarised and interpreted as follows. The capacity of the liver to take up alanine is enhanced in the 10 day-old and 20 day-old rats. In the 10 day-old rat this enhanced uptake is not associated with the metabolism of the alanine towards glucose or urea formation. It is tempting to speculate that this increased uptake in the absence of increased catabolism is due to a preservation of the intact alanine molecule for use in protein synthesis in the rapidly growing neonatal rat. In support of this is the observation that the short-fall of carbon products and that of nitrogen products over alanine utilised were quantitatively very similar (Table II); this is consistent with the utilization of both carbon and nitrogen of the alanine molecule either directly, or indirectly after conversion to other amino acids, for incorporation into protein. The restriction by alanine aminotransferase of glucose synthesis from alanine may only operate under conditions of excessive alanine loading, since *in vivo* estimates of gluconeogenesis at normal plasma alanine levels show an enhancement in the suckling rat (see earlier). Thus the activity of alanine aminotransferase would place an upper limit on the amount of alanine which could be metabolised for gluconeogenesis and any extra alanine above that limit would be preferentially directed towards protein synthesis. In the case of urea synthesis the enzymic limitation is imposed not only by alanine aminotransferase but also by the urea cycle enzymes argininosuccinate synthetase and argininosuccinate lyase. This means that alanine amino nitrogen will be conserved and this is consistent with the observed pattern of urea formation *in vivo* (see earlier), the observed value of 75% for nitrogen retention in the carcass of the suckling rat (Miller, 1970), and the observation that in the suckling rat arginine is preferentially directed towards protein synthesis rather than towards urea formation (Drotman and Campbell, 1972). The enhanced ammoniagenesis observed in the 10 and 20 day-old rats compared to later ages may be physiologically appropriate in terms of nitrogen conservation, since the alanine amino nitrogen may then be retained in a form that is potentially reutilisable for biosynthetic processes rather than being eliminated from the body in the form of urea.

In the 20 day-old rat the rise in alanine aminotransferase raises the limitation imposed on gluconeogenesis in the 10 day-old rat so that alanine is preferentially directed towards glucose synthesis. At this age the rat is in the process of weaning on to a solid food diet and so the provision of amino acids from dietary protein will also be increasing (see Snell and Walker, 1972). In this sense more amino acids can therefore be spared for carbohydrate synthesis, but because weaning is also associated with increases in the dietary carbohydrate supply the requirement for glucose synthesis to maintain blood glocuse levels will be diminished and the

Table VII. **Effects of various agents on plasma alanine concentration and tissue alanine aminotransferase activity in neonatal rats.**

Treatment	Plasma alanine (μmol/ml)	Alanine aminotransferase activity (μmol/min/g of tissue)	
		Liver	Skeletal muscle
Saline (control)	0.25	1.93	1.46
Quinolinate	0.37	1.91	1.35
3-Mercaptopicolinic acid	0.36	1.89	1.48
L-Cycloserine	0.26	0.16	0.05

Ten day-old rats were starved overnight and injected intraperitoneally with quinolinate, 3-mercaptopicolinic acid and *L*-cycloserine at doses of 250, 100 and 25 mg/kg, respectively. Control animals received an appropriate volume of 0.9% saline. Animals were sacrificed after 30 min, blood was collected and the plasma after deproteinisation was used for the determination of alanine by an enzymic method (Williamson, 1970). Liver and skeletal muscle were rapidly dissected and homogenised, and alanine aminotransferase activity was assayed as described by Snell and Walker (1972). Each value is the mean of 4-6 determinations each determination was on tissue from a single animal.

gluconeogenic pathway will be used instead for the storage of carbohydrate in the form of glycogen. Measurements of liver glycogen synthesis and liver glycogen concentration show that weaning is associated with an increasing deposition of glycogen until the adult value is reached at about 30 days *post partum* (see Snell and Walker, 1973b).

At 30 days *post partum* the normal adult situation is reached. Dietary protein supply is sufficiently adequate to provide excess of amino acids for protein synthesis and body growth, the carbohydrate intake is sufficient to maintain normoglycaemia and there is sufficient excess of both carbohydrate precursors and carbohydrate to maintain body glycogen stores.

THE ROLE OF THE ALANINE CYCLE IN THE NEONATAL RAT

In recent years evidence has accumulated supporting the hypothesis that alanine is released from extrahepatic tissues *in vivo* and that this alanine originates by *de novo* synthesis in muscle involving the transamination of glucose-derived pyruvate. The alanine is then transported by the circulation to the liver where it is reconverted to glucose by gluconeogenesis. This sequence of events is analogous to the Cori cycle involving lactate and has been termed the alanine cycle (Mallette *et al.*, 1969; Felig *et al.*, 1970).

The supply of amino groups required for pyruvate amination in muscle can only be provided directly by glutamate (Rowsell, 1956; Rowsell and Corbett, 1958) via alanine aminotransferase. Other aminotransferases which catalyse 2-oxoglutarate amination, however, are present in muscle (Rowsell and Corbett, 1958; Krebs, 1972), and so these enzymes acting in concert with alanine aminotransferase can bring about amino transfer from many amino acids to pyruvate.

In order to assess the possible role of alanine transfer from muscle *in vivo* in the neonatal rat, various inhibitors were injected and plasma concentrations of alanine and tissue activities of alanine aminotransferase were measured (Table VII). Animals were starved overnight so that the circulating levels of alanine could only

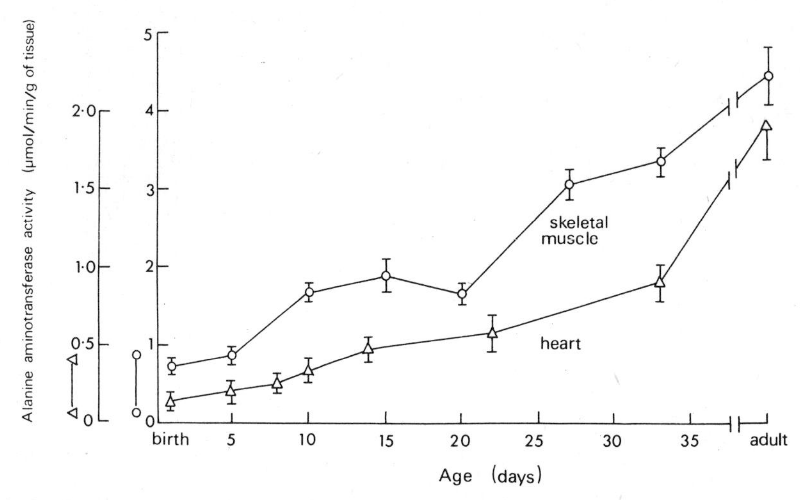

Fig.4 Alanine aminotransferase activity in heart and skeletal muscle during the development of neonatal rats. Each point is the mean of 5-8 determinations (vertical bars show ± S.D.) and each determination was on the pooled tissue from 2-3 animals (< 10 days *post partum*) or from individual animals (> 10 days *post partum*). Alanine aminotransferase activity was assayed in homogenates as described by Snell and Walker (1972).

be derived from endogenous sources. Inhibition of gluconeogenesis in the liver by the administration of quinolinate or 3-mercaptopicolinate was due to a block in the conversion of oxaloacetate to phosphoenolpyruvate as shown by a rise in liver aspartate and malate levels and a fall in phosphoenolpyruvate levels (K. Snell, unpublished work). Conversion of pyruvate to glucose was therefore blocked in animals injected with these inhibitors. The rise in plasma alanine that occurred at 30 min after injection (Table VII) indicates two things. Firstly, that at normal plasma levels of alanine the fasted neonatal rat is actively metabolising the amino acid along the gluconeogenic pathway. Secondly, the rise in plasma alanine indicates that endogenous production of alanine does occur in these rats. Liver and muscle activities of alanine aminotransferase were not affected by these agents. *L*-Cycloserine on the other hand caused a marked inhibition of muscle and liver aminotransferase activities but had no effect on plasma alanine levels (Table VII). Thus removal of alanine along the pathway of gluconeogenesis in the liver was blocked but so too was the endogenous supply of alanine from muscle tissue. It appears from these results that the transfer of alanine from muscle tissue to the liver for conversion to glucose is a process that can occur in the neonatal rat.

On the other hand the extent to which this occurs in the suckling rat as compared to the adult is not clear. An estimate of the capacity of muscle to aminate pyruvate at different ages was obtained by assaying for alanine aminotransferase activity. The development curves obtained in both heart muscle and skeletal muscle are shown in Fig.4. For both tissues aminotransferase activity was less than half of the adult value throughout the suckling period; for skeletal muscle the rise in adult levels began at around the time of weaning. One of the most important sources of amino groups for alanine synthesis in muscle is probably the branched-chain amino acids, which are preferentially catabolised in muscle rather

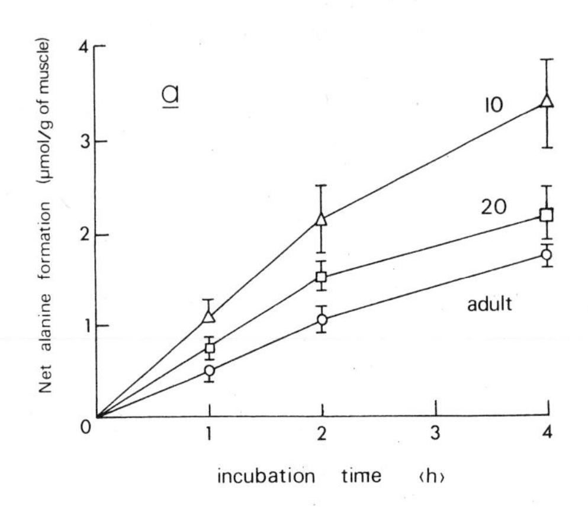

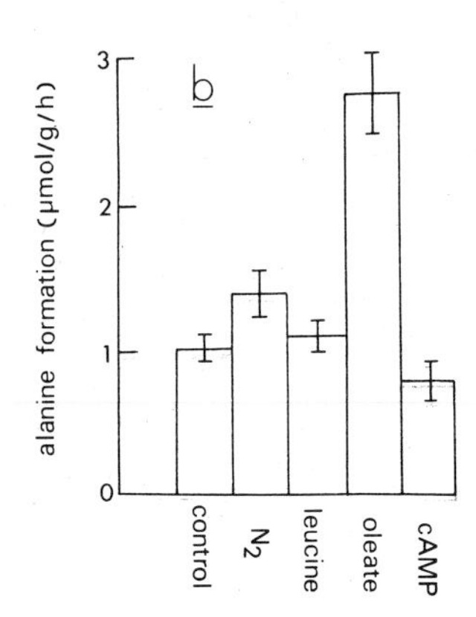

Fig.5 Alanine formation in homogenates of skeletal muscle from developing neonatal rats. (a) Time-course of net formation of alanine (alanine present at zero-time has been subtracted) in homogenates from 10 (△), 20 day old (□) and adult (○) rats. (b) Effect of agents added *in vitro* on the rate of alanine formation in homogenates from 10 day-old rats. The oleate was complexed with bovine serum albumin before addition; incubations with equivalent concentrations of albumin alone gave control rates of alanine formation. Vertical bars show ± S.D. of the mean of 3-4 determinations. Other details are as in Table IX.

Table VIII. Leucine aminotransferase activity in skeletal muscle of adult and neonatal rats.

Age	Leucine aminotransferase activity (μmol/min/g of tissue)
10 days *post partum*	0.16 ± 0.01 (4)
Adult (12 weeks)	0.22 ± 0.02 (4)

Ten percent homogenates of hind limb skeletal muscle were prepared in 0.1 M-potassium phosphate buffer, pH 7.8, and assayed for leucine aminotransferase activity as described by Taylor and Jenkins (1966). The results of determinations on tissue from a single animal (adult) or pooled from 3 animals (10 days) are given as mean values ± S.E.M.

than in liver (Miller, 1962). The transfer of the amino nitrogen to pyruvate occurs by the sequential action of the branched-chain amino acid aminotransferase (leucine aminotransferase, EC 2.6.1.6.) and alanine aminotransferase. Measurements of leucine aminotransferase activity in skeletal muscle showed a lower value at 10 days *post partum* than in the adult rat (Table VIII). These results suggest that the transfer of amino groups from leucine to form alanine may be limited in the neonate compared to the adult.

Endogenous alanine formation was assessed *in vitro* by incubating homogenates of muscle from fed rats of different ages in air at 37° and measuring the production of alanine with time (see Ozand *et al.*, 1973). Alanine formation was linear for the first 2 h of incubation but began to diminish somewhat thereafter (Fig.5a). Initial rates of alanine formation at the different ages are given in Table IX. The increased alanine formation observed in the neonatal rats does not appear to correlate with the rates of amino transfer measured under optimal conditions at different ages (Fig.4 and Table VIII). This implies that endogenous formation of pyruvate may

Table IX. Alanine formation in muscle homogenates during development.

Age (days)	Rate of alanine formation (μmol/h/g of tissue)
10	1.06 ± 0.14 (6)
20	0.74 ± 0.06 (5)
Adult	0.56 ± 0.04 (6)

Hind limb muscle was rapidly dissected and homogenised in Krebs Bicarbonate Ringer pH 7.4 (10%). After filtering the homogenate through gauze it was diluted with an equal volume of Ringer and incubated under air at 37°C in a shaking water bath. Samples were deproteinised with 2M-$HClO_4$ at 0, 2 and 4 h and alanine determined in the extracts by an enzymic method (Williamson, 1970). Values for the rate of alanine formation at various ages are given above as means ± S.E.M. with the number of determinations in parentheses; each determination was on tissue sampled from a single animal (20 days *post partum* or adult) or pooled tissue from 3 animals (10 days *post partum*).

be the rate-limiting step in the overall synthesis of alanine rather than amino transfer in these conditions. In agreement with this interpretation is the observation that the addition of 10 mM-leucine to incubations of muscle homogenates from 10 day-old rats had no appreciable effect on alanine formation (Fig.5b). However, anaerobic incubations under N_2 resulted in higher rates of alanine formation compared to controls. This is consistent with an increased glycolytic supply of pyruvate derived from anaerobically-induced glycogenolysis. The addition of 2 mM-oleate to muscle homogenate incubations produced a marked stimulation of alanine formation. This action may be the result of an inhibitory effect of fatty acids on pyruvate oxidation (Garland *et al.*, 1964) leading to redirection of pyruvate towards alanine synthesis and a stimulatory effect of fatty acids on the oxidation of branched-chain amino acids (Buse *et al.*, 1972) thereby increasing the supply of amino groups for transamination with pyruvate. The small inhibitory effect of 10 μM-dibutyryl cyclic AMP on alanine formation could be related to the stimulation of pyruvate oxidation by epinephrine and glucagon (presumably acting via cyclic AMP) observed in muscle preparations from fed rats (Buse *et al.*, 1973).

Although the measurements of endogenous alanine formation suggest that the neonatal rat has a higher rate than the adult, taking into account the increasing muscle mass per unit body weight observed in developing rats (Cheek *et al.*, 1965) the rates of alanine formation are much the same when expressed per unit body weight at each age (viz., 26.5, 19.0 and 22.4 μmol/h/100 g body wt. for 10, 20 day-old and adult rats). Thus the muscle mass of the suckling rat seems to maintain a basal capacity for alanine formation comparable with that of the fed adult despite a diminished enzymic potential for amino transfer. The actual rate of alanine formation by the muscle *in vivo* will, of course, depend on the rates at which amino and carbon substrates are made available within the tissue. The low enzymic capacity for amino transfer in the neonatal rat may serve to restrict excessive loss of amino nitrogen from the muscle under situations of massive protein catabolism. This may be one of the factors which limits the contribution of body protein to the loss in body weight of 24 h-starved neonatal rats before 20 days *post partum* (Hahn and Koldovsky, 1966). The quantitative aspect of the role of the alanine cycle in the suckling rat *in vivo* is still not clear. However, measurements of whole body recycling of glucose using a combination of [6-^{14}C]-

and [6-^{3}H]-glucose in tracer amounts *in vivo* suggest that this is less than 10% of the total turnover of the glucose pool in the fed suckling rat (Walker and Snell, 1973). The estimation of recycling by this method will have included contributions from both the Cori cycle (involving lactate) and the alanine cycle since both precursors have pyruvate as a common intermediate in their conversion to glucose.

PATHOLOGICAL DISTURBANCES OF GLUCONEOGENESIS

Since gluconeogenesis is such a complex process and involves such close interrelationships with other aspects of cellular metabolism it is extremely difficult to predict the clinical symptoms and pathological consequences of disturbances in its normal functioning. These difficulties are compounded by the paucity of knowledge of the fine details of the metabolic pathways involved and of the mechanisms which regulate and control their operation under normal circumstances. This is particularly true when we consider man and, more especially, the developing infant. The relevance of the considerations I have presented for the developing neonatal rat to the human situation is far from clear, although the basic biochemical features are probably common to both species. One clinical feature which is probably present in most situations where there are dysfunctions of gluconeogenesis is hypoglycaemia. This will be especially so in the suckling infant if, as is the case in the rat (Snell and Walker, 1973b), the supply of dietary carbohydrate in the milk is insufficient to meet the body requirements for glucose. Indeed hypoglycaemia in the paediatric age group is a commonly observed symptom in a variety of disorders. Many of these disorders are associated with defects in the endocrine system and hypoglycaemia is an indirect result of the hormonal disturbances. This aspect of neonatal abnormalities which lead to hypoglycaemia has been well-reviewed recently (Adam, 1971; Exton, 1972; Pagliara *et al.*, 1973).

Relatively few specific enzyme deficiencies have been positively identified as causing defects in gluconeogenesis which result in hypoglycaemia. Glucose 6-phosphatase deficiency (type I glycogen storage disease) and fructose-1,6-diphosphatase deficiency are probably the best described examples of enzymic defects in gluconeogenesis (see Pagliara *et al.*, 1973). Hommes has described a form of Leigh's disease associated with a marked reduction of hepatic pyruvate carboxylase activity (Hommes *et al.*, 1968; Hommes, 1973). Hypoglycaemia has not been a common finding in this condition and this may be due to the fact that a number of gluconeogenic precursors can by-pass this block by feeding into the gluconeogenic pathway at the level of oxaloacetate (Snell and Walker, 1973b) or 2-phosphoglycerate (Snell, 1974a). Haymond *et al.* (1973) have described a case of thiamine-unresponsive maple syrup urine disease associated with severe fasting hypoglycaemia. Their data suggested a block in gluconeogenesis from alanine but without any defect in the pathway from pyruvate to glucose. The infusion of alanine into their patient caused a rise in blood glutamate, lactate and glutamine levels. This suggests that alanine aminotransferase activity may not be impaired; but that there may be a deficiency in glutamate dehydrogenase which prevents the regeneration of 2-oxoglutarate from the glutamate formed by transamination (see Fig.3). The 2-oxoglutarate for continued transamination might then have to be drawn from the citric acid cycle causing a drain on the intermediates of the cycle and leading to oxoloacetate deficiency and impairment of gluconeogenesis. Regeneration of 2-oxoglutarate could also occur by transamination of glutamate with oxaloacetate (by aspartate aminotransferase) which would also lead to a deficiency of oxaloacetate. The role, if any, of the elevated levels of branched-chain amino acids in the hypoglycaemia is not known. Finally, it has been suggested that some cases of ketotic hypoglycaemia may be due to enzyme deficiencies of the alanine cycle, since they seem to be associated with an impaired release of glucogenic amino acids into the circulation but with no impairment of glucose formation when these precursors were infused (Pagliara *et al.*, 1973).

ACKNOWLEDGEMENTS

Part of the original work reported was carried out in collaboration with Professor D.G. Walker in the Department of Biochemistry, University of Birmingham and was supported by a grant from the Wellcome Trust. I am grateful to Professor Walker for this encouragement and to Anne Phillips and Robin Lord for their skilled assistance.

REFERENCES

Adam, P.A.J. (1971). *Advan. Metab. Disord.* **5**, 183-275.

Ballard, F.J. and Hanson, R.W. (1967). *Biochem. J.* **104**, 866-871.

Ballard, F.J. and Oliver, I.T. (1963). *Biochim. biophys. Acta* **71**, 578-588.

Buse, G.M., Biggers, J.F., Frederica, K.H. and Buse, J.F. (1972). *J. biol. Chem.* **247**, 8085-8096.

Buse, G.M., Biggers, J.F., Drier, C. and Buse, J.F. (1973). *J. biol. Chem.* **248**, 697-706.

Cheek, D.B., Powell, G.K. and Scott, R.E. (1965). *Bull. John Hopkins Hosp.* **116**, 378-387.

Clark, M.G., Bloxham, D.P., Holland, P.C. and Lardy, H.A. (1973). *Biochem. J.* **134**, 589-597.

Clark, M.G., Bloxham, D.P., Holland, P.C. and Lardy, H.A. (1974). *J. biol. Chem.* **249**, 279-290.

Drotman, R.B. and Campbell, J.W. (1972). *Am. J. Physiol.* **222**, 1204-1212.

Exton, J.H. (1972). *Metab. clin. Exp.* **21**, 945-990.

Felig, P. (1973). *Metab. clin. Exp.* **22**, 179-207.

Felig, P., Pozefsky, T., Martiss, E. and Cahill, G.F. (1970). *Science* **167**, 1003-1004.

Friedmann, B.F., Goodman, E.H., Saunders, H.L., Kostos, V. and Weinhouse, S. (1971). *Metabolism* **20**, 2-12.

Garland, P.B., Newsholme, E.A. and Randle, P.J. (1964). *Biochem. J.* **93**, 665-678.

Greengard, O., Sahib, M.K. and Knox, W.E. (1970). *Arch. biochem. Biophys.* **137**, 477-482.

Hahn, P. and Koldovský, O. (1966). "Utilization of Nutrients during Postnatal Development", pp.1-177. Pergamon Press, Oxford.

Halestrap, A. and Denton, R.M. (1974). *Biochem. J.* **138** , 313-316.

Haymond, M., Karl, I.E., Feigin, R.D., DeVivo, D. and Pagliara, A.S. (1973). *Pediat. Res.* **7**, 500-508.

Hems, R., Ross, B.D., Berry, M.N. and Krebs, H.A. (1966). *Biochem. J.* **101**, 284-292.

Herzfeld, A. and Greengard, O. (1971). *Biochim. biophys. Acta* **237**, 88-98.

Hommes, F.A. (1973). *In:* "Inborn Errors of Metabolism" (F.A. Hommes and C.J. Van den Berg, eds) pp.127-130. Academic Press, London.

Hommes, F.A. and Richters, A.R. (1969). *Biol. Neonat.* **14**, 359-364.

Hommes, F.A., Polman, H.A. and Reerinck, J.D. (1968). *Archs. Dis. Childh.* **43**, 423-426.

Krebs, H.A. (1972). *Advan. Enzyme Regul.* **10**, 397-420.

Krebs, H.A., Hems, R. and Lund, P. (1973). *Advan. Enzyme Regul.* **11**, 361-377.

Mallette, L.E., Exton, J.H. and Park, C.R. (1969). *J. biol. Chem.* **244**, 5713-5723.

Miller, L.L. (1962). *In:* "The Amino Acid Pools" (J.T. Holden, ed) pp.708-721. Elsevier, Amsterdam.

Miller, S.A. (1970). *Fedn Proc.* **29**, 1497-1502.

Miller, A.L. and Chu, P. (1970). *Enzym. Biol. Clin.* **11**, 497-503.

Newsholme, E.A. and Start, C. (1973). "Regulation in Metabolism", pp.71-73. John Wiley & Sons, London.

Newsholme, E.A., Carbtree, B., Higgins, S.J., Thornton, S.D. and Start, C. (1972). *Biochem. J.* **128**, 89-97.

Ozand, P.T., Tildon, J.T., Wapnir, R.A. and Cornblath, M. (1973). *Biochem. Biophys. Res. Commun.* **53**, 251-257.

Pagliara, A.S., Karl, I.E., Haymons, M. and Kipnis, D.M. (1973). *J. Pediat.* **82**, 365-379, 558-577.
Papa, S., Francavilla, A., Paradies, G. and Meduri, B. (1971). *FEBS Lett.* **12**, 285-288.
Raiha, N.C.R. and Suihkonen, J. (1968). *Biochem. J.* **107**, 793-797.
Rowsell, E.V. (1956). *Biochem. J.* **64**, 246-252.
Rowsell, E.V. and Corbett, K. (1958). *Biochem. J.* **70**, 7p.
Snell, K. (1974a). *Biochem. J.* **142**, 433-436.
Snell, K. (1974b). *Int. J. Biochem.* 5 *(in press).*
Snell, K. and Walker, D.G. (1972). *Biochem. J.* **128**, 403-413.
Snell, K. and Walker, D.G. (1973a). *Biochem. J.* **132**, 739-752.
Snell, K. and Walker, D.G. (1973b). *Enzyme* **15**, 40-81.
Taylor, R.T. and Jenkins, W.T. (1966). *J. biol. Chem.* **241**, 4391-4395.
Vernon, R.G. and Walker, D.G. (1968). *Biochem. J.* **106**, 321-329.
Vernon, R.G. and Walker, D.G. (1972). *Biochem. J.* **127**, 521-529.
Vernon, R.G., Eaton, S.W. and Walker, D.G. (1968). *Biochem. J.* **110**, 725-731.
Walker, D.G. and Snell, K. (1973). *In:* "Inborn Errors of Metabolism" (F.A. Hommes and C.J. Van den Berg, eds) pp.97-117. Academic Press, London.
Williamson, D.H. (1970). *In:* "Methoden der Enzymatischen Analyse" (H.U. Bergmeyer, ed) p.1094. Verlag-Chemie, Weinheim.
Yeung, D. and Oliver, I.T. (1967). *Biochem, J.* **103**, 744-748.

DISCUSSION

Greengard: You mentioned that in the 10 day-old rat cortisol treatment results both in increased gluconeogenesis and in a 20-fold increase in the level of alanine aminotransferase. This convinces me that indeed this enzyme limits the utilization of alanine. However, others may not be so easily convinced; they could say that cortisol may achieve this in some other manner. Therefore, would it not be worth doing a control experiment in which you would give cortisol to a 40 day-old animal, in which the same dose (per unit body weight) would not change the level of alanine aminotransferase, and it should not increase the utilization of alanine for glucose synthesis either.

Snell: Yes, that is certainly a worthwhile experiment to do. Perhaps I should emphasize that I did measure FDP-ase and G6P-ase levels in the animals which were pretreated with cortisol. These activities were not significantly different from those of control animals. So the increased glucose formation is not due to an increase in those particulaı rate-limiting enzymes of the gluconeogenic pathway.

Land: One must be very cautious in suggesting that optimal enzyme activities are an indication of what is rate limiting in a metabolic pathway. What you really want to measure is flux rates. You showed that the activity of arginine succinate synthetase is very low. You suggested that this could limit urea synthesis and alanine utilization. What I would like to suggest is that urea synthesis could be limited by the flux through alanine aminotransferase. Ammonia may sequester α-keto-glutarate by the glutamate dehydrogenase, which is present in very large amounts in 10 day-old animals. This may have two effects. You would alter the equilibrium of alanine aminotransferase in the wrong direction, firstly by having a high level of glutamate but also by lowering the α-keto-glutarate level, the partner for alanine utilization.

Snell: I showed in fact that ammonia formation is raised in the neonatal rat, and you are suggesting now that this could cause a rise in glutamate?

Land: Yes. Do you have high levels of glutamate in your tissues, which stops the utilization of alanine?

Snell: Yes, that is in fact exactly what occurs. Measuresments of hepatic concentration of alanine and α-oxoglutarate and glutamate and pyruvate show that the aminotransferase is displaced from equilibrium. There is indeed accumulation of glutamate, which answers your point.

Tager: What happens to all the alanine that disappears?

Snell: The excess of alanine removed over carbon recovered is very similar, if not identical, to the excess of alanine removed over the nitrogen products, suggesting that nitrogen and carbon are going hand in hand. And that suggests to me that protein synthesis might be the sort of synthetic pathway into which alanine is directed, because the alanine molecule *in toto* is being used, but I have no evidence for it directly.

REGULATION OF PYRUVATE METABOLISM IN FETAL RAT LIVER

R. Berger and F.A. Hommes

University of Groningen, School of Medicine, Department of Pediatrics
Laboratory of Developmental Biochemistry, 10 Bloemsingel
Groningen, The Netherlands

INTRODUCTION

In adult tissue energy is produced mainly by three metabolic processes: the breakdown of carbohydrates and fatty acids which are ultimately oxidized to carbon dioxide and water in the tricarboxylic acid cycle.

Fetal tissue is characterized by an enzyme pattern that in many respects differs from that of adult tissue. The contribution of each of these pathways to fetal energy metabolism may therefore not be the same as in the adult. This is reflected e.g. by the very low activity of fatty acid oxydation in fetal liver (Bailey and Lockwood, 1973). Carbohydrate in the form of glucose, which is transported from the maternal to the fetal circulation across the placenta serves therefore as the main fuel for the fetus.

There are some indications that, towards the end of gestation, the fetus is hypoxic; during the process of birth the fetus has to cope with several periods of ischaemia (Ballard, 1971; Ballard *et al.*, 1971). However, the energy supply must continue even in the absence of oxygen.

As pyruvate is an end product of glycolysis and at the same time a substrate for the tricarboxylic acid cycle, regulation of its metabolism may be important in the switch between aerobic and anaerobic glycolysis. In this study pyruvate metabolism in fetal rat liver has been investigated using suspensions of isolated hepatocytes and mitochondria. Experimental procedures have been described previously (Berger and Hommes, 1973, 1974).

PYRUVATE METABOLISM IN FETAL HEPATOCYTES

Fetal hepatocytes produce lactate and pyruvate, derived from either added glucose or endogenous glycogen, when incubated in Krebs-Ringer phosphate solution (Berger and Hommes, 1974). In Fig.1 it is shown that almost linear rates of lactate and pyruvate accumulation are obtained in the aerobic as well as in the anaerobic state. It can be seen that in the aerobic state significant amounts of lactate accumulate; thus in these cells the mitochondria cannot oxidize all the pyruvate, derived from glycolysis.

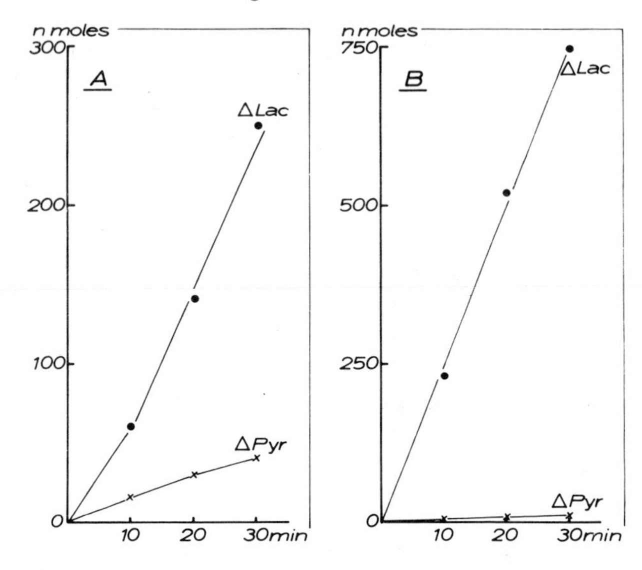

Fig.1 Time course of lactate and pyruvate accumulation during glycolysis in isolated fetal rat liver cells. A: aerobic state. B: antimycin-inhibited state induced by adding antimycin (10 μg/ml) to the medium. Prot. 3.7 mg. Cells were incubated in Krebs-Ringer phosphate solution, final volume 2 ml.

Table I. Fluxes through pyruvate dehydrogenase (PDH) and lactate dehydrogenase (LDH) during glycolysis in isolated fetal hepatocytes.

Exp.	Lactate		Sp. ac.	CO_2	CO_2	Flux	
	nmoles	dpm	dpm/nmoles	dpm	nmoles	LDH	PDH
1	410	10.200	25	1522	61	2.8	0.41
2	215	16.160	75	2600	35	2.2	0.36

Hepatocytes were prepared from 20-21 days old fetuses according to Berger and Hommes (1974). The cells were incubated (4.9 mg prot. in Exp.1 and 3.2 mg prot. in Exp.2) in Krebs-Ringer phosphate solution containing 0.1 mM (3.4-^{14}C)-glucose. The incubations were carried out in stoppered Warburg vessels. Reaction time 30 min, temp. 37° C. The flux is given in nmoles/min × mg prot.).

To determine which part of pyruvate is oxidized in the mitochondria and which part is reduced to lactate, fetal hepatocytes were incubated in the presence of (3,4-^{14}C) -glucose. As a result of glycolysis, pyruvate is labelled at the C-1 position and this carbon atom is split off in the pyruvate dehydrogenase reaction and appears as $^{14}CO_2$. Furthermore (1-^{14}C)-lactate is formed. By measuring the specific activity of lactate, the fluxes through lactate dehydrogenase and pyruvate dehydrogenase can be calculated. The results of such an experiment are given in Table I. The flux through pyruvate dehydrogenase is about 0.4 nmoles/min × mg prot., while lactate accumulates with a mean rate of 2.5 nmoles/min × mg prot.

Thus only 10-15% of pyruvate, produced by glycolysis, is oxidized by the mitochondria; the major part is reduced to lactate.

The question arises which factors limit the oxidation of pyruvate in these cells. One possibility is that the pyruvate dehydrogenase complex is inhibited by one of the mechanisms which regulates the flux through this enzyme. Inhibition can occur by phosphorylation of the pyruvate dehydrogenase component of the enzyme complex (Linn *et al.*, 1969) or by accumulation of the products acetyl-CoA and NADH (Tsai *et al.*, 1973). A low rate of pyruvate oxidation in hepatocytes can furthermore be due to low enzyme content. Therefore regulation of pyruvate metabolism was investigated in more detail by studying pyruvate oxidation in isolated fetal rat liver mitochondira.

OXIDATION OF PYRUVATE (PLUS MALATE) IN FETAL RAT LIVER MITOCHONDRIA

Pyruvate oxidation in fetal liver mitochondria was studied using different respiratory states. These states are characterized as follows (LaNoue *et al.*, 1972): mitochondrial respiration was stimulated by the addition of ADP + glucose + hexokinase (state 3), FCCP or FCCP plus oligomycin. Under these conditions the energy state of the mitochondria is rather low while the nicotinamide adenine dinucleotides are in the oxidized form. If no additions are made to the mitochondria, the respiration is inhibited (state 4 according to LaNoue *et al.* (1972), state 2 according to Chance and Williams (1955)). In this state the intramitochondrial ATP/ADP ratio is high, the nicotinamide adenine dinucleotides are reduced. In the presence of oligomycin respiration is also inhibited; if glucose plus hexokinase is added, the mitochondria can be depleted from ATP resulting in a low intramitochondrial ATP/ADP ratio (LaNoue *et al.*, 1972). In this way the oxidation of pyruvate plus malate can be studied under different conditions of energy and redox state.

In Table II the results of an experiment are given where fetal rat liver mitochondria were incubated in the presence of (3-^{14}C-)pyruvate (plus malate), in different respiratory states. When (3-^{14}C-) pyruvate is used, it takes several turns of the Krebs-cycle before the label appears as $^{14}CO_2$. Moreover, by the addition of cold malate, the label is diluted into the large malate pool, so little is lost from the suspension. In such a way all Krebs-cycle intermediates become equally labelled and the incorporation of ^{14}C into the malate pool is a measure of the flux through the Krebs-cycle (LaNoue *et al.*, 1970). It can be seen that in the respiration inhibited states (Table II, lines 2 and 3) depletion of ATP results in a lower incorporation of ^{14}C into malate but incorporation of ^{14}C into the complex peak (consisting of acetate, aspartate, glutamate and β-OH-butyrate, which do not resolve under these conditions) is strongly stimulated. Furthermore in the latter case less ^{14}C is recovered in pyruvate.

Thus the effect of ATP-depletion of the mitochondria results in a diminished flux through the cycle and in a stimulation of pyruvate utilisation by alternative pathways.

The accumulation of ketone bodies during the oxidation of pyruvate plus malate in different respiratory states was studied in the experiment shown in Table III. As can be seen an appreciable synthesis of ketone bodies occurred in state 3 while in the presence of FCCP or FCCP plus oligomycin the accumulation of β-OH-butyrate and acetoacetate is diminished. Ketone bodies accumulate also

Table II. Incorporation of ^{14}C into amino acids and tricarboxylic acid cycle intermediates during oxidation of 3-^{14}C-pyruvate plus malate in fetal rat liver mitochondria.

Conditions	-OH-but. + ac. + asp. + glu.	succinate (dpm)	malate	pyruvate
State 3	394.000	81.900	198.000	623.000
State 4	265.000	30.000	152.000	1,169.000
Oligom. + ADP	547.000	33.600	54.000	926.000

Incubation medium (1 ml) consisted of KCl 15 mM, $MgCl_2$ 5 mM, EDTA 2 mM, Tris-HCl 50 mM, K-phosphate 25 mM, glucose 33 mM, sucrose 50 mM, pyruvate 1 mM, malate 1 mM; final pH 7.4. Time 10 min. Temperature 25°C. Sp. act. pyruvate: 2248 dpm/nmol. Mit. protein 7.9 mg. The reaction components were separated by column chromatography on a Dowex 1X10 column according to LaNoue *et al.* (1970).

Table III. Accumulation of β-OH-butyrate and acetoacetate during oxidation of pyruvate plus malate in fetal rat liver mitochondria.

Conditions	β-OH-butyrate (nmoles)	Acetoacetate
State 3	9.9	22.6
State 4	6.5	10.2
FCCP (0.2 μM)	3.3	9.7
FCCP + oligom. (5γ)	3.4	4.5
Oligom. + ADP	21.8	11.9

Reaction conditions are given in Table II. Mit. protein 2.9 mg.

in the respiration-inhibited states; there is a striking increase in ketogenesis when the intramitochondrial ATP/ADP ratio is lowered by the addition of an ATP-trapping system (Table III, lines 2 vs. 5). From these experiments it can be concluded that in the respiration-inhibited states a low ATP/ADP ratio inside the mitochondria leads to an increase in pyruvate utilization. As no effect of ATP or ADP on the enzymes of ketone body synthesis is known, an effect on the pyruvate dehydrogenase complex becomes likely.

Regulation of the pyruvate dehydrogenase complex by phosphorylation and dephosphorylation is depicted schematically in Fig.2. The enzyme complex is active in the non-phosphorylated form. Inactivation occurs by phosphorylation of one of the serine residues in the pyruvate dehydrogenase component of the enzyme complex (Reed *et al.*, 1972), in the presence of ATP. This reaction is catalyzed by a specific kinase which is inhibited by ADP and pyruvate (Linn *et al.*, 1969), thiamine-diphosphate (Roche and Reed, 1972) and by inorganic pyrophosphate (Wieland *et al.*, 1972). Activation of the pyruvate dehydrogenase complex is accomplished by hydrolysis of the phosphoserine ester, catalyzed by a specific phosphatase. The pyruvate dehydrogenase phosphatase requires both Mg^{2+} and Ca^{2+} for optimal activity (Siess and Wieland, 1972). As the pyruvate dehydrogenase complex is localized in the matrix of the mitochondria (Smoly *et al.*, 1970) the activity of the enzyme may be affected by the intramitochondrial

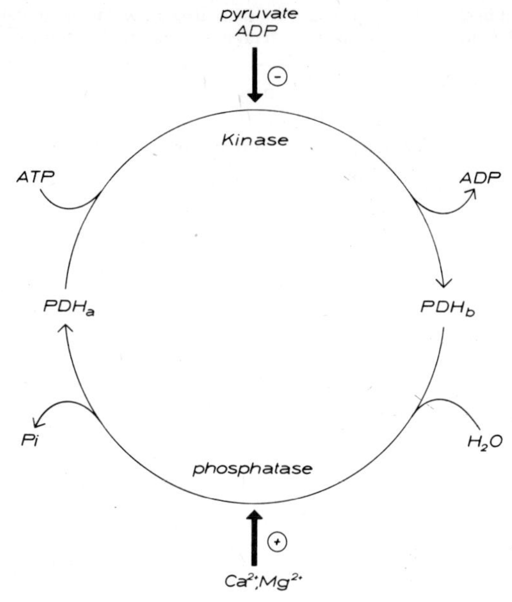

Fig. 2 A schematic representation of the regulation of pyruvate dehydrogenase by phosphorylation and dephosphorylation.

Table IV. Intramitochondrial ATP and ADP levels during oxidation of pyruvate plus malate and acetylcarnitine plus malate in fetal rat liver mitochondria.

Conditions	pyruvate + malate			acetylcarnitine + malate		
	ATP	ADP	ATP/ADP	ATP	ADP	ATP/ADP
State 3	1.7	0.9	1.9	1.4	1.1	1.3
State 4	0.9	0.6	1.5	0.7	0.6	1.1
FCCP (0.2 μM)	0.7	0.7	1.0	0.3	0.7	0.4
FCCP + oligom. (5γ)	0.7	0.8	0.9	0.5	0.5	1.0
oligom. + ADP	0.6	1.6	0.4	0.4	1.3	0.3

Reaction conditions are given in Table II. Mit. protein 3-7 mg. At the end of the incubation mitochondria were centrifuged through a silicon oil layer into perchloric acid, according to Harris and Van Dam (1968). Values are given in nmoles/mg protein.

concentration of pyruvate, free magnesium and free calcium and by the ATP/ADP ratio of the matrix.

In the experiment shown in Table IV the intramitochondrial ATP/ADP during oxidation of pyruvate plus malate was measured. For comparison the intramitochondrial ATP and ADP contents were also measured in the presence of acetylcarnitine plus malate as substrates.

With pyruvate plus malate as substrates elevated ATP/ADP ratios are obtained during state 3 as well as during state 4 oxidation. In the presence of an ATP-trapping system (+ADP) the intramitochondrial ATP content is low, whereas the

Table V. Active and inactive forms of pyruvate dehydrogenase during oxidation of pyruvate plus malate and acetylcarnitine plus malate in fetal rat liver mitochondria.

Conditions	pyruvate + malate		acetylcarnitine + malate	
	PDH_a			
	nmoles/min x mg	% total act.	nmoles/min x mg	% total act.
State 3	2.0	65	1.7	36
State 4	1.5	49	0.5	10
FCCP (0.2 μM)	2.3	75	2.0	42
FCCP + oligom. (5γ)	1.2	37	0.3	6
Oligom. + ADP	3.1	100	4.2	87

Reaction conditions are given in Table II. Mit. protein 5.8 mg and 4.5 mg respectively. Pyruvate dehydrogenase was assayed according to Portenhauser and Wieland (1972).

level of ADP is high resulting in a low ATP/ADP ratio. In the presence of FCCP and FCCP plus oligomycin intermediate ATP/ADP ratios are obtained. If acetylcarnitine (plus malate) is used as a substrate, the intramitochondrial ATP/ADP ratios are about the same as with pyruvate (plus malate) as substrate.

Results of measurements of the amount of pyruvate dehydrogenase present in the active form (expressed as percentage of total activity) during oxidation of pyruvate plus malate are given in Table V. In state 3 65% of pyruvate dehydrogenase is in the active form; this value is slightly increased by the addition of FCCP; in the presence of FCCP plus oligomycin pyruvate dehydrogenase is largely in the inactive form.

During state 4 oxidation of pyruvate about half of the enzyme complex is active; upon depleting the mitochondria of ATP, pyruvate dehydrogenase complex becomes fully activated.

If pyruvate as a substrate is replaced by acetylcarnitine, the changes in the amount of active pyruvate dehydrogenase among the various respiration states are similar but in all cases relatively more pyruvate dehydrogenase complex is found in the inactive form. This is readily explained by the inhibitory effect of pyruvate on pyruvate dehydrogenase kinase (F. Hucho *et al.*, 1972). From Tables IV and V it can be concluded that in the respiratory inhibited states (lines 2 and 5 of both tables), a relationship exists between the intramitochondrial ATP/ADP ratio on the one hand and the amount of pyruvate dehydrogenase present in the active form on the other hand. Thus in the presence of a high ATP/ADP ratio, pyruvate dehydrogenase is inhibited; lowering the ATP/ADP ratio (high ADP level) results in an inhibition of the kinase by ADP and therefore in an activation of the pyruvate dehydrogenase complex. (It is assumed that the phosphatase has in both cases the same activity.)

However, in the respiration-stimulated states such a relationship does not seem to hold. For instance, during state 3 oxidation of pyruvate plus malate as well as that of acetylcarnitine plus malate the ATP/ADP ratios are rather high, but nevertheless a substantial amount of pyruvate dehydrogenase complex is found to be in the active form. Furthermore, with pyruvate plus malate as a substrate similar ATP/ADP ratios are obtained in the presence of FCCP or FCCP plus oligomycin (Table IV); in spite of this the amount of active pyruvate dehydrogenase

Table VI. Intramitochondrial levels of NAD and NADH during oxidation of pyruvate plus malate in fetal rat liver mitochondria.

Conditions	NAD	NADH	Σ	NAD/NADH
State 3	2.48	0.22	2.70	11.2
State 4	1.34	0.47	1.80	2.9
FCCP (0.2 μM)	2.12	0.22	2.34	9.7
FCCP + oligom. (5γ)	2.10	0.14	2.24	15.0
Oligom. + ADP	1.04	1.60	2.64	0.6

Reaction conditions are given in Table II. Values are given in nmoles/mg protein.

Table VII. Intramitochondrial CoA-SH and acetyl-CoA levels during oxidation of pyruvate plus malate in fetal rat liver mitochondria.

Conditions	CoA-SH	Acetyl-CoA	Σ	CoA-SH/acetyl-CoA
State 3	0.73	0.50	1.23	1.46
State 4	1.03	0.42	1.45	2.45
FCCP (0.2 μM)	1.15	0.38	1.53	3.02
FCCP + oligom. (5γ)	1.11	0.31	1.42	3.58
Oligom. + ADP	0.31	1.15	1.46	0.27

Reaction conditions are given in Table II. Mit. protein 4.1 mg. Values are given in nmoles.

complex differs by a factor two (Table V). With acetylcarnitine plus malate as a substrate the difference between FCCP and FCCP plus oligomycin is even greater (Table V), but this may in part be explained by the higher ATP/ADP ratio in the latter case (Table IV).

These results indicate that in these cases the interconversion of pyruvate dehydrogenase is not solely determined by the intramitochondrial ATP/ADP ratio but that other factors may play a role as well. In this respect (energy-dependent) changes in the concentration of free magnesium and calcium in the matrix, independent of those caused by alterations in the amount of complexing agents (ATP, citrate, phosphate) may be responsible for the observed discrepancies between the intramitochondrial ATP/ADP ratio and the amount of active pyruvate dehydrogenase complex. During active respiration, adult liver mitochondria are able to accumulate Ca^{2+} (Lehninger *et al.*, 1967) and Mg^{2+} (Kun *et al.*, 1974). The matrix is discharged from these cations in the presence of FCCP (see also Scarpa, 1974).

Walajtys provided strong evidence that at least the calcium content of the matrix is sufficient to activate pyruvate dehydrogenase phosphatase under these condition (Walajtys *et al.*, 1974). However in view of the high K_m of the phosphatase for Mg^{2+} (Hucho, 1974), regulation of the interconversion by energy-dependent changes in the concentration of free Mg^{2+} may play a role in the presence of FCCP or FCCP plus oligomycin, if it is assumed that cation transport across the fetal mitochondrial inner membrane behaves (qualitatively) similarly to the adult case.

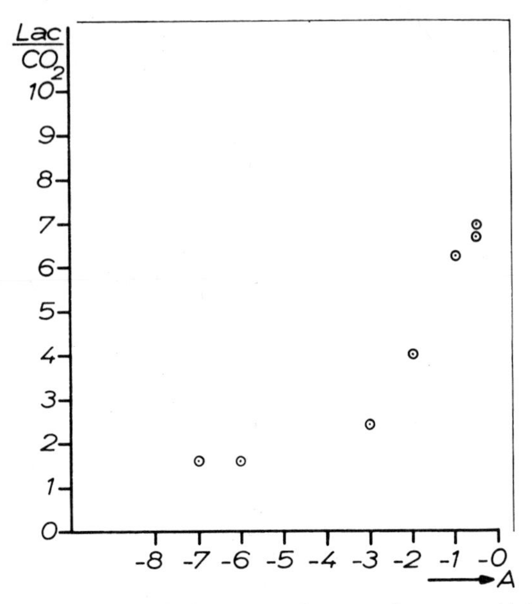

Fig.3 The distribution of glycolytically generated pyruvate between oxidation in the mitochondria and reduction to lactate as a function of age.

In vitro pyruvate dehydrogenase complex is inhibited by NADH competitively with NAD, and by acetyl-CoA, competitively with CoA (Tsai *et al.*, 1973). Thus the enzyme activity may be regulated, *in vivo*, by the ratios of free CoA/acetyl-CoA and by free NAD/NADH of the mitochondrial matrix. In Table VI the results are given of measurements of total NAD and NADH in the various respiration states. In state 3 as well as in the presence of FCCP or FCCP plus oligomycin the mitochondrial nicotinamide adenine dinucleotides are highly oxidized resulting in a high NAD/NADH ratio. The reverse is observed in the respiration-inhibited states.

The results of measurements of total CoA and acetyl-CoA in fetal rat liver mitochondria are given in Table VII. Relatively high CoA-SH/acetyl-CoA ratios are observed in state 4, in the presence of FCCP and FCCP plus oligomycin; during state 3 oxidation this ratio is somewhat lowered. Upon ATP depletion a large portion of CoA-SH is in the esterified form, resulting in a high level of acetyl-CoA and a low CoA-SH/acetyl-CoA ratio.

From Tables VI and VII it can be seen that in the respiration-stimulated states the level of NADH and acetyl-CoA is low. Under these conditions the activity of pyruvate dehydrogenase is not limited by product inhibition. In state 4 significant amounts of ketone bodies are produced (Table III), the NAD/NADH ratio is low (Table IV) and pyruvate dehydrogenase complex is for 50% in the active form. In going to the ATP-depleted state the flux through the Krebs cycle is lowered (Table II) probably caused by the low oxaloacetate concentrations in this highly reduced state. Furthermore the CoA-SH/acetyl-CoA is decreased compared to that in state 4 (Table VII); this cross-over observed between CoA and acetyl-CoA indicates that the level of acetyl-CoA is mainly determined by the rate of formation. The observed two-fold stimulation of ketogenesis in the oligomycin-inhibited state is therefore completely due to the two-fold increase in pyruvate dehydrogenase activity (Table V).

From these experiments it can be concluded that in fetal liver mitochondria the activity of the pyruvate dehydrogenase complex is mainly determined by the phosphorylation-dephosphorylation mechanism; regulation by the intramitochondrial NAD/NADH and CoA-SH/acetyl-CoA ratio takes place only to a minor extent.

DEVELOPMENTAL ASPECTS OF PYRUVATE METABOLISM IN FETAL RAT LIVER

It was already observed by Van Rossum (1963) that liver slices, prepared from 16-18 days old fetuses, show a large Pasteur effect (defined as the ratio of the rate of lactate production under anaerobic and aerobic conditions). This ratio decreases from about 12 (16-18 days old fetuses) to about 6 in liver slices from newborn animals. The relative contribution of the oxidative pathways of pyruvate metabolism in fetal energy metabolism may therefore change during development.

Changes in the distribution of pyruvate, derived from glycolysis, between oxidation in the mitochondria and reduction to lactate, has been studied in hepatocytes prepared from fetuses of different ages. The results are shown in Fig.3. The ratio of the rate of lactate production and carbon dioxide evolution has been plotted as a function of age. As has been stated above, towards the end of gestation most of the pyruvate is converted to lactate while only 10-15% is oxidized in the mitochondria. In hepatocytes prepared from younger fetuses however, oxidative metabolism of pyruvate becomes more important. It was found that the decrease in the lactate/CO_2 ratio was mainly caused by a lower rate of lactate accumulation; the flux through pyruvate dehydrogenase remained fairly constant during development. The question arises what limits pyruvate oxidation in fetal hepatocytes. Measurements of the total activity of pyruvate dehydrogenase in hepatocytes isolated from the livers of term-fetuses, showed that this activity could just account for the observed rate of carbon dioxide production (not shown).

Furthermore it was found that 80% of the pyruvate dehydrogenase complex was in the active form. It seems that pyruvate oxidation in fetal hepatocytes is limited by the low content of the pyruvate dehydrogenase complex, the enzyme working at almost maximal capacity, rather than by an extensive inhibition of the enzyme complex.

Thus in term fetuses, energy is produced mainly by "anaerobic" glycolysis. In this way the production of ATP is much less dependent on the availability of oxygen.

ACKNOWLEDGEMENTS

These investigations were supported (in part) by the Netherlands Foundation for Chemical Research (S.O.N.) with financial aid from the Netherlands Foundation for the Advancement of Pure Research (Z.W.O.).

The expert technical assistance of Mrs. I. Rietsema-Broekema is gratefully acknowledged. Our thanks are due to Professor J.M. Tager for helpful discussion.

REFERENCES

Bailey, E. and Lockwood, E.A. (1973). *Enzyme* **15**, 239-253.
Ballard, F.J. (1971). *Biochem. J.* **124**, 265-274.
Ballard, F.J. and Philippidis, H. (1971). *In:* "Regulation of Gluconeogenesis" (H.D. Söling and B. Willms, eds) pp.66-81. Thieme Verlag.
Berger, R. and Hommes, F.A. (1973). *Biochim. biophys. Acta* **314**, 1-7.
Berger, R. and Hommes, F.A. (1974). *Biochim. biophys. Acta* **333**, 535-545.
Chance, B. and Williams, G.R. (1955). *J. biol. Chem.* **217**, 383-438.
Harris, E.J. and Van Dan, K. (1968). *Biochem. J.* **106**, 759-766.
Hucho, F. (1974). *Eur. J. Biochem.* **46**, 499-505.
Kun, E., Dummel, R.J. and Battaglia, W.L. (1974). *Fedn Proc.* **33**, **Abs. 187**.
LaNoue, K.F., Nicklas, W.J. and Williams, J.R. (1970). *J. biol. Chem.* **245**, 102-111.
LaNoue, K.F., Bryla, J. and Williamson, J.R. (1972). *J. biol. Chem.* **247**, 667-679.
Lehninger, A.L., Carafoli, E. and Rossi, C.S. (1967). *Adv. Enz. Reg.* **29**, 259-320.
Linn, T.C., Pettit, F.H. and Reed, L.J. (1969). *Proc. natn. Acad. Sci. U.S.A.* **62**, 234-241.
Reed, L.J., Linn, T.C., Hucho, F., Namihira, G., Barrera, C.R., Roche, T.E., Pelley, J.W. and Randall, D.D. (1972). *In:* "Metabolic Interconversion of Enzyme" (O.H. Wieland, C. Helmreich and H. Holtzer, eds) pp.281-291. Springer Verlag, Heidelberg.
Roche, T.E. and Reed, L.J. (1972). *Biochem. Biophys. Res. Comm.* **48**, 840-846.
Scarpa, A. (1974). *Biochem.* **13**, 2789-2794.
Siess, E. and Wieland, O.H. (1972). *Eur. J. Biochem.* **26**, 96-105.
Smoly, J.M., Kuylenstierna, B. and Ernster, L. (1970). *Proc. natn. Acad. Sci. U.S.A.* **66**, 125-131.
Tsai, C.S., Burgett, M.W. and Reed, L.J. (1973). *J. biol. Chem.* **248**, 8348-8352.
Van Rossum, G.D.V. (1963). *Biochim. biophys. Acta* **74**, 15-32.
Walajtys, E.J., Gottesman, D.P. and Williamson, J.R. (1974). *J. biol. Chem.* **249**, 1857-1865.
Wieland, O.H., Siess, E., Funcke, J.H. v., Patzelt, C., Schirmann, A., Löfflet, G. and Weiss, L. (1972). *In:* "Metabolic Interconversion of Enzymes" (O.H. Wieland, C. Helmreich and H. Holtzer, eds) pp.292-309. Springer Verlag, Heidelberg.

DISCUSSION

Clark: Did you measure the pyruvate concentration in these mitochondria?

Berger: I measured the total amount of pyruvate present in the mitochondria by centrifuging them through a silicone oil layer. It varies between the two states, between 1 and 2 nmoles per milligram of protein, that is assuming 1 microlitre per milligram of protein, millimolar.

Clark: Does it vary when you are using two different substrates?

Berger: I did not measure pyruvate when I used acetylcarnitine as substrate.

Clark: The reason I ask is that Mowbray in London has shown recently that carnitine derivatives are very good counter transporters for internal pyruvate. And if in fact that was occurring here, then you might expect the pyruvate dehydrogenase to be switched off.

Tager: Have you made any measurements of calcium and magnesium in the mitochondria under the two conditions where you have a difference in the activation of pyruvate dehydrogenase, in spite of the fact that the ATP to ADP ratio is similar?

Berger: No, I did not, because it is very difficult to evaluate the concentrations of free magnesium and calcium from total amounts. Isolated mitochondria contain a substantial amount of calcium but it is probably bound as a phosphate complex. Furthermore, magnesium is complexed to citrate and ATP. Adult mitochondria are known to contain a high amount of magnesium. Some of that must be present in the free uncomplexed form. It has not been measured in the fetal mitochondria, but I expect an effect of magnesium in view of the recent determination of the

K_m magnesium of the phosphatase. It was found to be 2 millimolar (Hucho *et al.*, *Arch. biochem. Biophys.* **151**, 328 (1972)), but recently it has been measured by Hucho (*Eur. J. Biochem.* **46**, 499 (1974)) and he found it to be 20 millimolar. Small changes in the concentration of free magnesium can therefore indeed affect the phosphatase.

Cremer: I was particularly interested in your last statement on pyruvate oxidation which is limited by the lower amount of pyruvate dehydrogenase. Have you looked at the post-natal change of this enzyme?

Berger: We have not followed the activity of the enzyme after birth. Knowles and Ballard (*Biol. Neonat.* **24**, 4 (1974)) have studied the development of pyruvate dehydrogenase in liver and found a low activity just after birth, it increases 5 days after birth and then it starts to increase. Most of the PDH is in the inactive form, about 10 to 20% is in the active form. I have no data on the hepatocytes. But he also found a very low activity of PDH just after birth.

SYSTEMS RELATIONSHIPS AND THE CONTROL OF METABOLIC PATHWAYS IN DEVELOPING BRAIN

N.Z. Baquer, P. McLean and A.L. Greenbaum

Courtauld Institute of Biochemistry
The Middlesex Hospital Medical School, London, England
and
Department of Biochemistry, University College London
London, England

INTRODUCTION

The studies of Pette and his associates (Pette *et al.*, 1962a,b; Pette, 1966; Bass *et al.*, 1969; Staudte and Pette, 1972) have emphasized the value of a consideration of the relationships of the constant and specific proportion enzymes in different tissues as a means of defining the enzymic adjustments necessary to achieve specialized metabolic function. This approach involves the assumption that changes of enzyme content of a tissue are related to the appearance of new metabolic potentialities and, from this point of view, appears to be particularly well-suited to the study of developing tissues where it may be considered that the enzyme profile of the adult is related to the function of the tissue and that the foetal and early post-natal stages are, metabolically, less differentiated and their enzymic make-up represents stages in the progress of development towards full functional competence.

The application of the principle of constant and specific enzyme proportionality to the study of metabolic adaptation in brain presents special difficulties due to the heterogeneity of the tissue and to the existence of multiple metabolic compartments (see Van den Berg, 1973; Balázs *et al.*, 1973). Nevertheless, the application of the method to brain may be worthwhile when the enzymes are being considered on a developmental basis in that it should be possible to relate the deviation of a particular enzyme, or enzyme group, from the normal ontological pattern either to the appearance of new cell types or to the development of new functional abilities. Greengard (1971) has already drawn attention to the fact that some brain enzymes show a definite pattern of emergence and that their appearance occurs in definite groups or "clusters".

The present study extends an already extensive, but somewhat selective, literature of enzyme measurements in brain (see, for example, Van den Berg,

1974) by making a systematic study of the enzymes of the two major pathways of carbohydrate metabolism, the tricarboxylic acid cycle, the lipogenic pathway and a number of related enzymes, in groups of rats at definite time intervals during the developmental period of the first 20 days *post-partum*. The data so obtained has been examined first on the basis described above to establish the relationships between the pathways, as well as within the individual pathways, as the brain undergoes functional differentiation. This analysis has served to reveal the sequential pattern of emergence of pathways (and, in some cases, their decay) and, combined with a study of the metabolism of specifically labelled substrates, has allowed an examination of the special role of the pathways in relation to function.

The pentose phosphate pathway with its multiple roles, in nucleic acid synthesis, lipogenesis, glycoprotein formation, hydroxylation reactions and the detoxication of hydrogen peroxide, must play a changing role during development. It is probably significant that brain has a particularly high proportion of bound forms of pentose phosphate pathway enzymes among tissues so far studied (Baquer and McLean, 1972) and it is apparent that compartmentation is involved in the complex control mechanisms, a view strengthened by the observations of Appel and Parrot (1970) on neurotransmitter activation of the pentose phosphate pathway in synaptosomes. The marked neurological disturbances produced by 6-aminonicotinamide, a potent inhibitor (via lethal synthesis of a NADP analogue) of 6-phosphogluconate dehydrogenase (Herken *et al.*, 1969), also attests to the vital role of the pathway in adult brain. For this reason, particular emphasis is given in this chapter to the control and function of the pentose phosphate pathway in brain.

The present chapter is divided between two aspects - firstly, a study of the developmental changes in the relationships between the enzymes of the two alternative pathways of glucose utilization of the tricarboxylic acid cycle and of lipogenesis and, secondly, an investigation into factors controlling the activity of the pentose phosphate pathway in synaptosomes.

SYSTEMS RELATIONSHIPS

1. *Enzyme Studies*

It has been shown in a number of laboratories (Crane and Sols, 1953; Johnson and Whittaker, 1963; Bagdasarian and Hulanicka, 1965; Baquer and McLean, 1972; MacDonnell and Greengard, 1974) that a substantial fraction of the enzymes of the glycolytic and pentose phosphate pathways is associated with particulate fractions in brain. Although this association may be artefactual and dependent on the medium used for preparation (Vallejo *et al.*, 1970; Clarke and Masters, 1972) it, nevertheless, militates against the use of fractionated extracts in the study of systems relationships and it was for this reason that the initial studies reported here were all performed upon whole homogenates with the enzymes fully released by freezing and thawing or by Triton-X-100. This approach does not permit an analysis of the contribution of different cell types or compartments but does serve to establish the broad trends among the major systems operating at different time during brain development.

The results in this section (Figs 1 to 3) are presented in the form shown as activities found in whole tissue homogenates of brain cerebrum, expressed as percentages of the activities found in similar homogenates of brain taken from rats on the first day *post-partum*. The different baseline parameter used here, the early-undifferen-

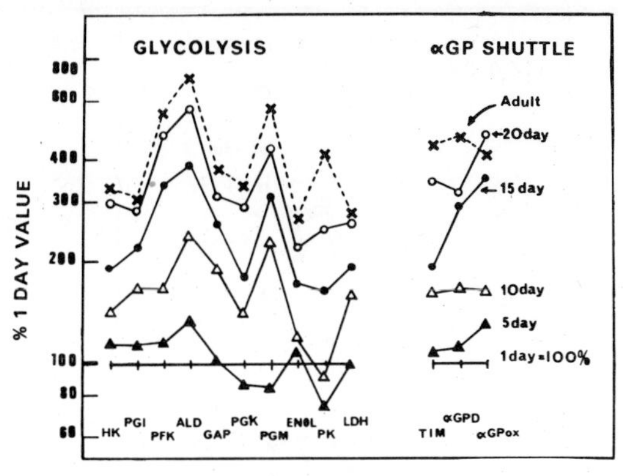

Fig.1 Developmental changes in enzymes of the glycolytic pathway and α-glycerophosphate shuttle in rat brain. The values at 5 days, ▲; 10 days, △; 15 days, ●; 20 days, ○ and adult rat brain, X--X; are given as a percentage of the activities at 1 day and are shown on a logarithmic scale. The activities of the enzymes were determined using whole homogenates of cerebral hemispheres with the enzymes fully activated or released by freezing and thawing or treatment with Triton-X-100, 1% final concentration for 30 min. For details of methods, see Baquer *et al.* (1973). Each point represents the mean of not less than six values. The activities at 1 day, as μmoles/g brain/min at 25° are:- *Glycolytic route:* HK, hexokinase, 4.43; PGI, phosphoglucose isomerase, 34.2; PFK, phosphofructokinase, 6.52; ALD, aldolase, 5.34; GAP, glyceraldehyde 3-phosphate dehydrogenase, 29.8; PGK, phosphoglycerate kinase, 35.7; PGM, phosphoglyceromutase, 19.3; ENOL, enolase, 15.9; PK, pyruvate kinase, 44.3; LDH, lactate dehydrogenase, 44.6. *α-Glycerophosphate shuttle:* TIM, triose phosphate isomerase, 92.2; αGPD, α-glycerophosphate dehydrogenase, 0.26; αGPOX, α-glycerophosphate oxidase, 0.33.

tiated state, rather than the fully-developed, adult, profile as used by Flexner *et al.* (1953) and by MacDonnell and Greengard (1974) was chosen as it seemed to emphasize the breakpoints and the different speeds of development of the enzymes more clearly.

A. Developmental changes in enzymes of the glycolytic pathway.

The time-sequence of the development of the enzymes of the glycolytic route is shown in Fig.1. It is apparent from this figure that, while each enzyme follows a different time course in its approach to the value ultimately attained in the adult (shown by the dispersion of points in a vertical axis), a characteristic pattern emerges as early as 5 days *post-partum*, i.e. at a time when only a fraction of the adult activity has been attained. A second fact which emerges from Fig.1 is that three enzymes show a disproportionate increase in activity.

Among the reactions leading to the formation of glyceraldehyde-3-phosphate, phosphofructokinase and aldolase are outstanding and in this context it should be noted that the low activity of aldolase and its tightly-bound form in brain may make this rise especially significant (Lowry and Passonneau, 1964; Masters and Holmes, 1972). In the group of enzymes between glyceraldehyde-3-phosphate dehydrogenase and lactate dehydrogenase, the greatest increase is of phosphoglyceromutase, an enzyme with a low first order rate constant, defined by Lowry and Passonneau (1964) as the ratio of V_{max}:K_m. The glycerophosphate shuttle, which is of importance in the movement of cytosolically-generated NADH to the mitochondrial compartment, shows parallel changes to the glycolytic route. Pyruvate kinase appears somewhat anomalous and out of phase with the other enzymes in

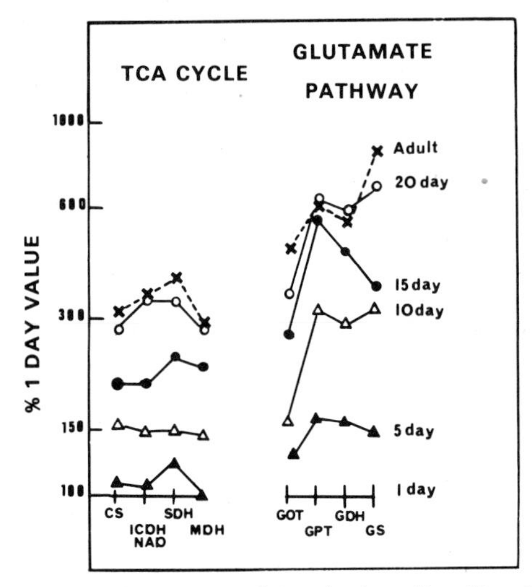

Fig. 2 Developmental changes in enzymes of the tricarboxylic acid cycle and of enzymes concerned with glutamate metabolism in rat brain. The symbols used for the different developmental stages and the mode of presentation are as given in Fig.1. The activities at 1 day as μmoles/g brain/min at 25° are: *Tricarboxylic acid cycle (TCA):* CS, citrate synthase, 4.97; ICDH-NAD, isocitrate dehydrogenase-NAD linked, 0.74; SDH, succinate dehydrogenase, 0.79; MDH, malate dehydrogenase, 201. *Glutamate pathway:* GOT, glutamate-oxaloacetate transaminase, 20.9; GPT, glutamate-pyruvate transaminase, 1.44; GDH, glutamate dehydrogenase, 2.91; GS, glutamine synthetase, 0.12.

that it shows no significant change until 15 days after birth and then rises very sharply indeed to the adult level. This peculiar behaviour of pyruvate kinase has also been observed by MacDonnell and Greengard (1974).

B. Developmental changes in enzymes of the tricarboxylic acid cycle.

The changes in the activity of the enzymes of the tricarboxylic acid cycle are shown in Fig.2. It is clear that, here again, the enzymes show a coordinated pattern of increase similar to that found for the glycolytic enzymes and, indeed, it can be shown (as Staudte and Pette (1972) have already demonstrated for a number of other tissues) that a close correlation exists between the changes of activity of hexokinase and citrate synthase, the initiating enzymes of the two pathways. What is, perhaps, more striking in Fig.2 is the lack of correlation between the enzymes of the tricarboxylic acid cycle and the enzymes of related amino acid metabolism, particularly those related to glutamate. This may arise from either of two aspects; first, the numerous roles filled by glutamate in brain and, second, the specific compartmentation of some of these functions (see, Berl, 1973). Glutamate may act as a putative neurotransmitter, elevating both cAMP and cGMP (Ferendelli *et al.*, 1974), as a substrate (Haslam and Krebs, 1963) or as a precursor for glutamine (see Van den Berg, 1973) or of acetyl groups (D'Adamo and D'Adamo, 1968) and it may fulfil these different functions in different compartments of the cell or even in different cells. Thus, a more detailed study of the enzyme distributions than the one attempted here would be required to establish the relationship of enzyme activity to physiological function for the group of tricarboxylic acid cycle enzymes with those metabolizing glutamate.

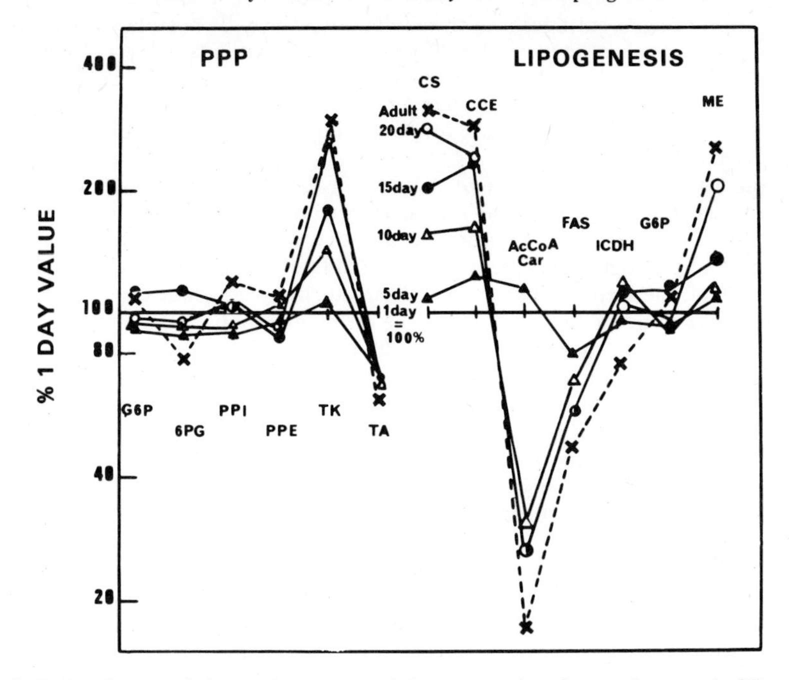

Fig.3 Developmental changes in enzymes of the pentose phosphate pathway and of lipogenesis in rat brain. The symbols used for the different developmental stages and the mode of presentation are as given in Fig.1. The activities at 1 day as μmoles/g brain/min at 25° are: *Pentose phosphate pathway (PPP):* G6P, glucose 6-phosphate dehydrogenase, 1.73; 6PG, 6-phosphogluconate dehydrogenase, 0.77; PPI, pentose phosphate isomerase, 0.93; PPE, pentose phosphate epimerase, 2.13; TK, transketolase, 0.18; TA, transaldolase, 1.56. *Lipogenesis:* CS, citrate synthase, 4.97; CCE, citrate cleavage enzyme, 0.16; AcCoA car, acetyl CoA carboxylase, 0.023; FAS, fatty acid synthetase, 0.087; ICDH, isocitrate dehydrogenase-NADP linked, 5.82; G6P, glucose 6-phosphate dehydrogenase, 1.73; ME, malic enzyme, 0.97.

C. *Developmental changes in the enzymes of the pentose phosphate pathway and of lipogenesis.*

Figure 3, which shows the developmental pattern for the enzymes of the pentose phosphate pathway and of lipogenesis, stands in sharp contrast to the patterns revealed for the glycolytic route and tricarboxylic acid cycle. From this figure it may be seen that while five of the enzymes involved show no appreciable change over the whole period studied, the sixth, transketolase, shows a strong developmental pattern similar to that found in the pathways discussed above. This aberrant behaviour of a single enzyme emphasizes that the pentose phosphate pathway should be viewed, as Horecker (1962) has insisted, as two independent routes of pentose phosphate formation from hexosemonophosphate; one (which includes the two dehydrogenases) is oxidative and the other (via transaldolase and transketolase) is non-oxidative. Pentose phosphate isomerase and pentose phosphate epimerase are common to both segments. Thus, the figure shows that the enzymes of the oxidative segment remain essentially constant throughout development but, in contrast, transketolase, the rate-limiting enzyme of the non-oxidative sequence, increases sharply. Since it has been established (see McIlwain, 1971) that nucleic acid turnover continues into the adult state, it may be postulated that the non-oxidative route for the formation of pentose phosphate may play a role in this process. A parallel situation exists in ascites tumour cells which have a remarkably similar enzyme profile to brain (Baquer *et al.*, 1973) and in which the non-oxidative route of pentose phosphate formation is of considerable importance (Gumaa and McLean, 1969).

The quantitative histochemistry of enzymes of the pentose phosphate pathway in the central nervous system of the rat has been studied by Kuhlman and Lowry (1956) and by Kauffman (1972). They showed that while the non-oxidative enzymes and 6-phosphogluconate dehydrogenase did not vary among the different regions studied, the activity of glucose 6-phosphate dehydrogenase changed in correspondence with the lipid content of the different regions. They were also able to show a possible association of transaldolase with nerve cell bodies based on the high activity in the cellular granular layer of the cerebellum (Kauffman, 1972).

The high proportion of glucose 6-phosphate and 6-phosphogluconate dehydrogenases and functional activity of the pentose phosphate pathway found in association with the large particle fraction and synaptosomes (Baquer and McLean, 1972; Appel and Parrot, 1970, and see below) suggests diverse roles for the pathway in brain. Thus, direct comparison of the total activity of the oxidative segment of the pathway with only one related function, i.e. lipogenesis, is hazardous and, in fact, no direct correlation could be found for parallel changes in the enzymic activity of the pentose phosphate pathway and lipogenesis, although isotopic studies with specifically labelled glucoses showed a remarkable coordination (see Fig.6).

Figure 3 also shows changes of the enzymes of the lipogenic pathway together with those of potential donors of reductive potential for this process. The two key enzymes of lipogenesis, acetyl CoA carboxylase and fatty acid synthase, both fall sharply from the 5th day *post partum*, but the three major NADPH-generating systems do not follow this pattern, indeed they each show an individual profile. Glucose 6-phosphate and 6-phosphogluconate dehydrogenase remain virtually constant in activity from 1 to 20 days *post partum*, NADP-linked isocitrate dehydrogenase decreases with time while malic enzyme rises over the same period. These differing patterns suggest that each of these systems may be linked to different acceptor systems and may be subject to different control processes.

Another important feature of the enzyme changes in the lipogenic pathway that emerges from Fig.3 is the lack of coordination in the pattern of change of citrate cleavage enzyme and the lipogenic enzymes. The increase of pyruvate dehydrogenase (Wilbur and Patel, 1974; Cremer and Teal, 1974), of citrate synthase and of citrate cleavage enzyme during development would lead to a potential for increased acetyl CoA formation, a potential which is not used for lipid formation as myelination decreases. It is more probable that the increased capacity to produce acetyl groups is related to the formation of acetylcholine and serotonin and to the synthesis of a variety of acetylated compounds (e.g. N-acetylaspartate and glutamate) which are found in brain in unusually high concentration and all of which increase markedly after the 10th day *post partum*. The developmental changes shown in Fig.3 are in accord with the findings of Volpe and Kishimoto (1972) on fatty acid synthase, with Loverde and Lehrer (1973) on NAD- and NADP-linked isocitrate dehydrogenases, and with the results of Wilbur and Patel (1974) on citrate synthase activity. The present results showing an in-

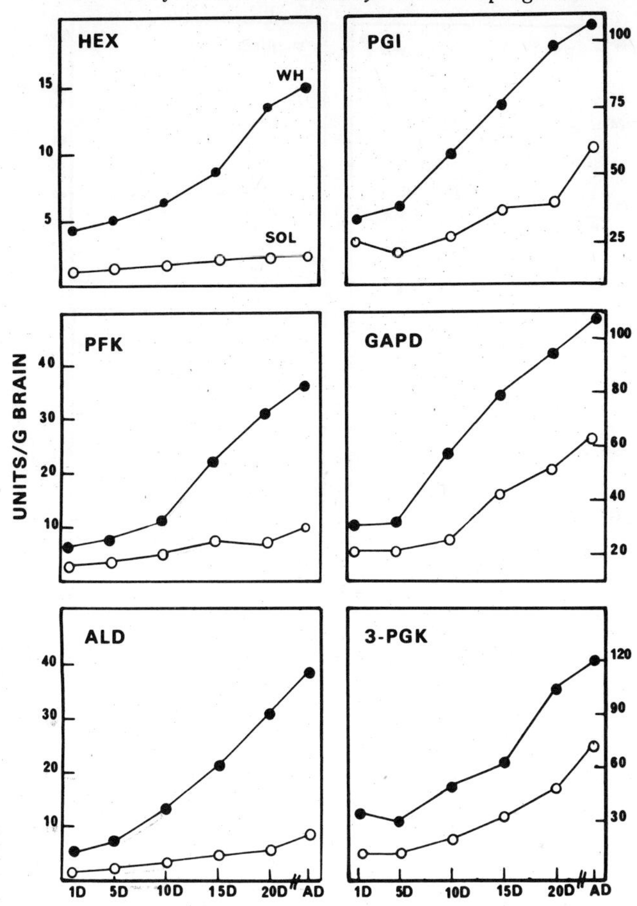

Fig.4A Developmental changes in glycolytic enzymes in rat brain; the contribution of soluble and particulate fractions. Reactions in the conversion of glucose to 3-phosphoglycerate. Closed circles represent whole homogenate values, open circles the activity of high speed supernatant fractions (100,000 g/45 min). Linear scales and absolute units are used. Each point represents the mean of not less than 6 values. The stage of development is indicated on the abscissa. A unit is 1 μmole substrate converted/min at 25°. The enzymes are grouped according to the relative contribution of cytosolic and particulate fractions to the increase observed in the whole homogenate. *Group I:* Particulate fraction increases more rapidly than soluble fraction. HEX, hexokinase; PFK, phosphofructokinase; ALD, aldolase. *Group II:* Particulate and soluble fractions make approximately equal contribution to the total overall increase. PGI, phosphoglucose isomerase; GAPD, glyceraldehyde 3-phosphate dehydrogenase; 3PGK, 3-phosphoglycerate kinase.

crease of citrate cleavage enzyme with age is in contrast to those of D'Adamo and D'Adamo (1968) and of Buckley and Williamson (1973); these differences may be explained by the use of supernatant fractions from brain extracts by these authors and of whole homogenates in the present experiments. Separate fractionation studies have revealed that more than 50% of the citrate cleavage enzyme activity

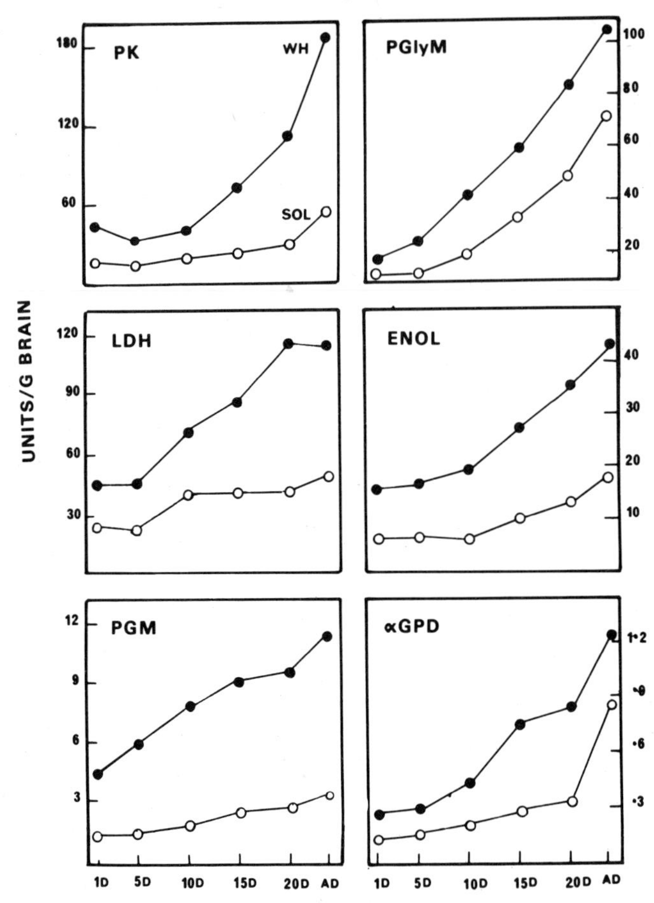

Fig.4B Developmental changes in glycolytic enzymes in rat brain; the contribution of soluble and particulate fractions. Reactions in the conversion of 3-phosphoglycerate to lactate. The symbols and mode of presentation are as in Fig.4A. *Group I:* Particulate fraction increases more rapidly than soluble fraction. PK, pyruvate kinase; LDH, lactate dehydrogenase; PGM, phosphoglucomutase. *Group II:* Particulate and soluble fractions make approximately equal contribution to the total overall increase. PGLYM, phosphoglyceromutase; ENOL, enolase; αGPD; α-glycerophosphate dehydrogenase.

was associated with particulate fractions of the brain (Baquer *et al.*, unpublished experiments).

D. The significance of the compartmentation of hexokinase.

The association of hexokinase with the particulate fraction of brain has been known since the early work of Crane and Sols (1953) and there have been numerous studies on the subcellular localization of this enzyme in brain (Bachelard, 1967; Craven and Basford, 1969; Craven *et al.*, 1969; Vallejo *et al.*, 1970; Mayer and Hübscher, 1971; Wilson, 1972), on the changes in different compartments during development (Kellogg *et al.*, 1974; MacDonnell and Greengard, 1974) and on the kinetic properties of the bound and free forms of brain hexokinase (Copley and Fromm, 1967; Wilson, 1968; Newsholme *et al.*, 1968; Bachelard and Goldfarb, 1969; Thompson and Bachelard, 1970).

The present results are in agreement with the findings that, during development, a large, and increasing, proportion of the hexokinase is found in association with the particulate fraction of the brain (Fig.4A). Discussion of the significance of this bound form has largely centred on the kinetic properties of the bound and free forms and the concensus of opinion would seem to suggest that the particulate enzyme has a lower K_m for ATP, a higher K_i for glucose 6-phosphate and that, at physiological levels of ATP, glucose 6-phosphate and inorganic phosphate, the association of hexokinase with the membrane, as opposed to its existence in the free form, would confer biological advantage (Copley and Fromm, 1967; Wilson, 1968; Newsholem *et al.*, 1968), an advantage that would be greatly increased as development progressed and a greater proportion of the cellular hexokinase came to be in the bound form.

Another facet of this problem, hitherto less adequately considered, is the possible advantages which might be conferred on the mitochondria, in terms of control of its ATP/ADP level, by the binding of hexokinase to its surface. It has been shown by Vallejo *et al.* (1970) that the bound brain hexokinase is located outside from the actractyloside barrier but it is, nevertheless, apparent from the studies of Vallejo *et al.* (1970) and of Rose and Warms (1967) that the bound hexokinase is able to use the intramitochondrially-generated ATP in the absence of this inhibitor.

The recent studies of Wieland and Portenhauser (1974), showing a direct relationship between the ATP/ADP ratio of mitochondria and the inactive (phospho)/active (dephospho) forms of pyruvate dehydrogenase, suggest the possibility that bound hexokinase, by virtue of its location, may preferentially use mitochondrial ATP, thus lowering the intramitochondrial ATP/ADP quotient and favouring the formation of the dephosphorylated, active, form of pyruvate dehydrogenase. Regulatory advantages would stem from such a system for, not only would this translocation of ATP serve to link glycolytic flux to mitochondrial pyruvate dehydrogenase flux, but it would also tend to conserve cytosolic ATP, which may be geared to the NA^+/K^+ ATP'ase activity and ion transport. The bound form of hexokinase may also be of significance in relation to the association of the pentose phosphate pathway enzymes with particulate fractions of the brain (Fig.5) although whether hexokinase and the enzymes of the pathway are located in the same regions of the brain or are bound at similar sites in subcellular membrane fractions is not yet clear.

E. *Compartmentation of enzymes of the glycolytic route and of the pentose phosphate pathway.*

The developmental changes in the enzymes of the glycolytic pathway in two different cell fractions are shown in Figs 4A and B. It is apparent that changes in the soluble fraction do not account for the overall increase in activity. Examination of the compartmental values for the glycolytic enzymes reveals two broad groups: (i) those in which the changes in the soluble fraction make only a small contribution to the overall increase from birth to the adult stage; these include hexokinase, phosphofructokinase, aldolase, pyruvate kinase, lactate dehydrogenase and phosphoglucomutase; and (ii) those in which the soluble and particulate fractions make approximately equal contributions at all stages; these include phosphoglucoseisomerase, glyceraldehyde 3-phosphate dehydrogenase, 3-phosphoglycerate kinase, phosphoglyceromutase, enolase and α-glycerophosphate dehydrogenase.

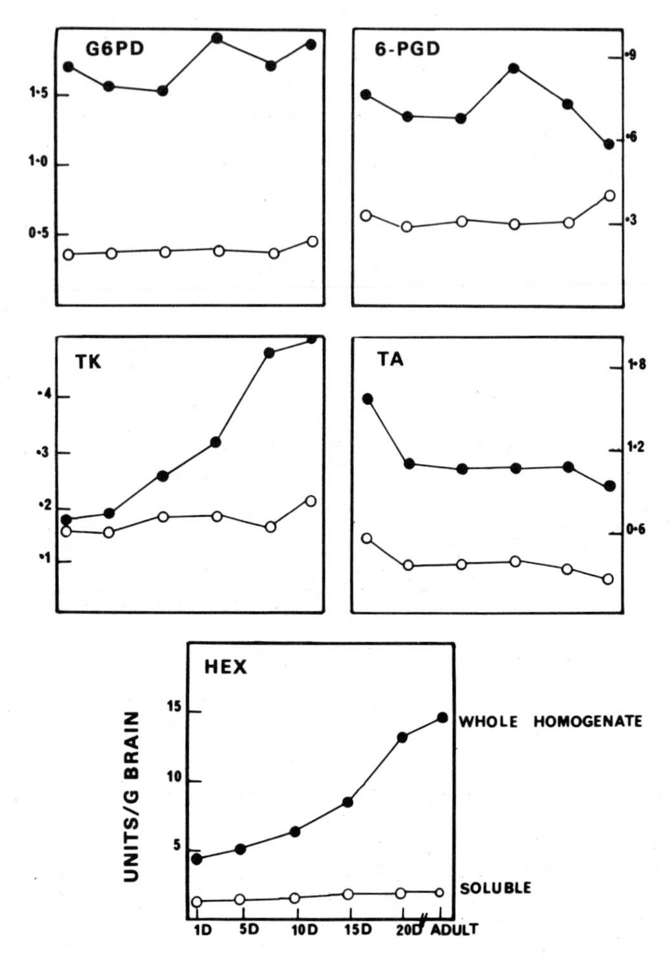

Fig.5 Developmental changes in pentose phosphate pathway enzymes in rat brain; the contribution of soluble and particulate fractions. The symbols and mode of presentation are as in Fig.4. G6PD, glucose 6-phosphate dehydrogenase; 6PGD, 6-phosphogluconate dehydrogenase; TK, transketolase; TA, transaldolase; HEX, hexokinase.

Each enzyme of the pentose phosphate pathways shows a unique developmental compartmentation profile (see Fig.5). Glucose 6-phosphate dehydrogenase has an almost constant activity in both soluble and particulate fractions and the former comprises less than 30% of the whole homogenate activity. 6-phosphogluconate dehydrogenase, while showing a similar constancy in development, has a rather higher contribution of soluble fraction activity which rises to 50% by the adult stage. Transketolase changes in a manner similar to the glycolytic enzymes of group (i), i.e. the total activity rises throughout development, an increase which is accounted for largely by changes in the particulate fraction. The profile of change of transaldolase is in complete contrast with transketolase and it is clear that the transketolase/transaldolase quotient will change substantially during development. This raises interesting questions with regard to the regulation of the non-oxidative reactions of the pentose phosphate pathway during development.

An evaluation of the full significance of the distribution of enzymes among subcellular fractions must await further knowledge of changes of kinetic properties on binding, factors regulating the degree of binding and a more detailed study of the distribution between different cell types and subcellular and membrane fractions. Some progress in this direction has already been achieved and it is possible to cite, for example, the studies of Clarke and Masters (1972) who have shown that the extent of binding of aldolase is markedly affected by metabolites such as lactate, fructose diphosphate and inorganic phosphate. The quantitative histochemical techniques developed by Lowry and his school are also powerful tools in the further elucidation of these problems (Kauffman, 1972; Kato and Lowry, 1973; Lehrer and Maker, 1973).

2. Isotope Incorporation Studies

There have been extensive studies on the utilization of ^{14}C-labelled glucose by brain slices (see, Batázs, 1970; O'Neill, 1974). Despite some criticism of the use of brain slices because of alteration in the profile of metabolism with respect to lactate formation (Hotta, 1962; McIlwain and Bachelard, 1971; Hostetler *et al.*, 1970), this preparation has proved extremely useful in investigating pathways of metabolism and retains many of the specialized functions of the tissue (Katz and Chase, 1970). The data in Fig.6 show developmental changes in the oxidation of glucose, pyruvate and acetate and are given here for comparison with the enzyme profiles.

It is apparent that the utilization of glucose by both the glycolytic route and by the tricarboxylic acid cycle increases during development as shown by the increase in the formation of $^{14}CO_2$ from [3,4-^{14}C] glucose, an index of the activity of the sum of the glycolytic route and the pyruvate dehydrogenase complex, and by the increase in the oxidation of carbon-6 of glucose, which is liberated as $^{14}CO_2$ only after completion of the entire sequence of glycolytic and tricarboxylic acid cycle reactions. It is assumed here that the glucoronate pathway plays a negligible role in the catabolism of glucose by brain (Balázs, 1970; M.Sochor, unpublished observations). These findings are in accord with the enzyme profiles shown in Figs 1 to 3 which show a marked increase of the enzymes of both glycolysis and the tricarboxylic acid cycle during development. More direct evidence on the rise of Krebs cycle activity during development can be derived from the data showing an increased rate of $^{14}CO_2$ formation from [1-^{14}C] - and [2-^{14}C] -pyruvate, a substrate which directly enters the tricarboxylic acid cycle, by-passing all the regulatory features involved in the glycolytic span. It should be noted, in passing, that there is a clear 2:1 relationship between the rate of $^{14}CO_2$ formation from [1-^{14}C] pyruvate relative to that from [2-^{14}C] pyruvate, suggesting that, of two acetyl CoA units produced by the pyruvate dehydrogenase complex, only one is oxidized and the other sequestered in an accumulated product such as glutamate or an N-acetyl derivative (Cremer, 1964; Gaitonde *et al.*, 1965; Sano, 1970; Van den Berg, 1973; Tuček and Cheng, 1974). Again these isotope data appear to be in broad agreement with the enzymic profiles which show increased activity of pyruvate dehydrogenase (Wilbur and Patel, 1974; Cremer and Teal, 1974).

The technique of determining control points in a metabolic pathway by presenting the pathway with substrates which enter before or after enzymes of sus-

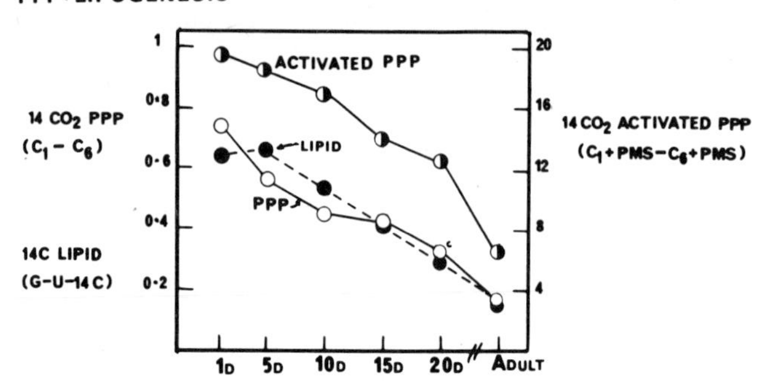

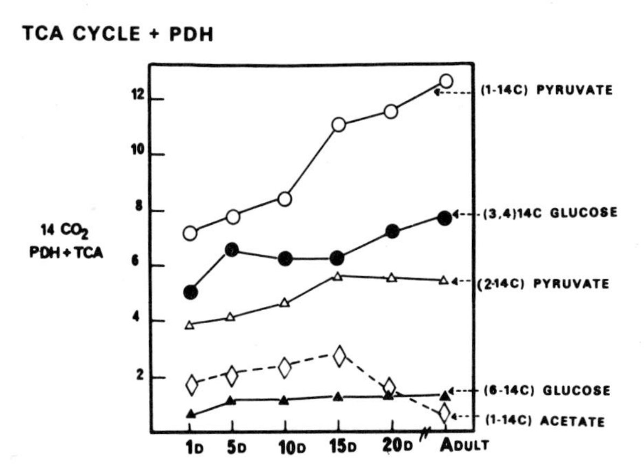

Fig. 6 The utilization of ^{14}C glucose, pyruvate and acetate by rat brain (slices) at different stages of development and the effect of phenazine methosulphate on the activation of the pentose phosphate pathway. 250 μg of brain slices were incubated in a medium containing 4.5 ml Krebs-Ringer bicarbonate with 0.5 μC ^{14}C-labelled substrate and 20 mM cold substrate. Incubations with glucose labelled on C1 or C6 were carried out in the presence or absence of 0.1 mM phenazine methosulphate (PMS) to obtain the values for the activated pentose phosphate pathway (PPP). Gas phase O_2 /CO_2, 95/5; time of incubation 1 h. $^{14}CO_2$ and 14C-lipid were estimated as previously described (Lagunas, McLean and Greenbaum, 1970). Each point represents the mean of not less than 6 determinations. The stage of development is given along the abscissa. *Pentose phosphate pathway and lipogenesis.* The results are given as μg atoms glucose carbon used/g brain/h at 37°. The PPP is assessed from the difference in $^{14}CO_2$ yields from [1-^{14}C] glucose and [6-^{14}C] glucose (abbreviated to C_1 - C_6), the activated PPP is the difference between $^{14}CO_2$ yields from 1 to 6 labelled glucose incubated in the presence of PMS. Lipid synthesis was assessed from the incorporation of [u-^{14}C] glucose in a total lipid fraction. *Tricarboxylic acid cycle (TCA) and pyruvate dehydrogenase (PDH).* The results are given as μg atoms glucose carbon converted to $^{14}CO_2$ /g brain/h at 37°. The [1-^{14}C] pyruvate, [2-^{14}C] pyruvate and [1-^{14}C] acetate values were made equivalent to the radiochemical yeilds from labelled glucose by halving the results, i.e. normalized to the same specific activity/3 carbon unit.

pected importance, has been widely used since the early studies of Chaikoff (1953). This principle can be applied in the present instance by a comparison of the oxidation rates of carbon-6 of glucose with [2-^{14}C] pyruvate and [1-^{14}C] acetate at different times *post-natal.* The release of the end product, $^{14}CO_2$, in each case, involves the enzymes of the tricarboxylic acid cycle. In the case of glucose, superimposed on the control exerted by enzymes of the cycle, there is some control at hexokinase and phosphofructokinase as well as a possible limitation at pyruvate dehydrogenase by manipulation of the relative amounts of the enzyme in the active and inactive forms. The oxidation of pyruvate, used at a concentration of 20 mM, is not subject to any regulatory influences by enzymes of the glycolytic pathway and should, by virtue of the high substrate level employed, be metabolised by pyruvate dehydrogenase working in its fully-activated state (Portenhauser and Wieland, 1972). Acetate enters the Krebs cycle below the pyruvate dehydrogenase complex via the acetate thiokinase reaction and will thus by-pass regulation at pyruvate dehydrogenase and be subject only to control by the enzymes of the cycle itself. With these considerations in mind, it is interesting to note that the ratio $^{14}CO_2$ from [2-^{14}C] pyruvate/$^{14}CO_2$ from [6-^{14}C]-glucose remains essentially constant throughout the development period studied which must, presumably, reflect a close integration in the developmental profile of the glycolytic and tricarboxylic acid cycle enzymes. On the other hand, the ratio $^{14}CO_2$ from [2-^{14}C] pyruvate/$^{14}CO_2$ from [1-^{14}C] acetate remains constant only in the earlier stages of development. After 15 days the ratio increases and in the adult is 4 times the value found at 1 day *post-natal.* It is possible that the constancy in early development reflects a greater ability by the brain of suckling rat to use fat and ketone bodies as fuel and the switchover in the adult to a metabolite profile more heavily dependent on glucose - again this is reflected in the enzyme profile of ketone body metabolism (Itoh and Quastel, 1970; Buckley and Williamson, 1973).

The pentose phosphate pathway presents a picture of greater complexity in that the isotopic flux data is not totally in accord with the enzyme data. During development, the oxidation of carbon-1 of glucose remains substantially constant while that of carbon-6 increases 2-fold. If the difference between the $^{14}CO_2$ yields 1- and 6-labelled glucose is taken, as a first approximation, to be an index of the activity of the pentose phosphate pathway, then, as seen in Fig.6, the activity of the pathway apparently drops some 5-fold during the period from birth to adult. This is not in agreement with the enzymic profile which shows that the enzymes of the oxidative segment of the pathway remain substantially constant over this period while the activity of the rate-limiting enzyme of the non-oxidative sequence, transketolase, increases. These differences presumably reflect the use of different preparations for the two measurements. It would appear that in brain slices, where integrated control mechanisms are operative, rate-limitation is imposed on the pathway by external factors, the most probable of which being the $NADP^+$/NADPH quotient (Negelein and Haas, 1935; Gumaa *et al.*, 1971; Eggleston and Krebs, 1974). Examination of the data shows that the parallelism between the $^{14}CO_2$ yield from the pentose phosphate pathway and the rate of incorporation of carbon-1 of glucose into lipid (Fig.6) is exceptionally striking, as is the correlation between glucose incorporation into lipid and the decline in the activity of acetyl CoA carboxylase and fatty acid synthetase (Fig.3). Thus, it would seem that the decline in lipogenesis, one of the major systems utilizing NADPH, is associated with the fall in the measured activity of the pentose phosphate pathway in brain slices. This is in keeping both with the known inhibition of glucose-6-phosphate and 6-phosphogluconate dehydrogenases by NADPH and with the limitation of the activity of this route by the availability of $NADP^+$.

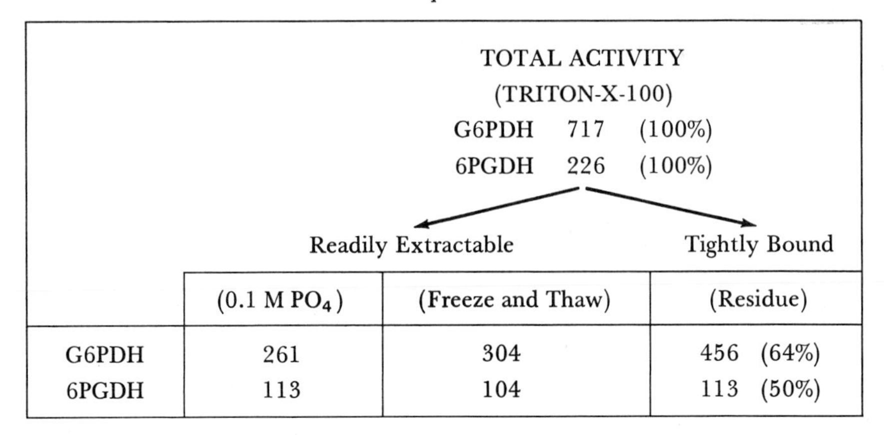

TOTAL ACTIVITY
(TRITON-X-100)
G6PDH 717 (100%)
6PGDH 226 (100%)

	Readily Extractable		Tightly Bound
	(0.1 M PO_4)	(Freeze and Thaw)	(Residue)
G6PDH	261	304	456 (64%)
6PGDH	113	104	113 (50%)

Fig. 7 Association of pentose phosphate pathway enzymes with the large particle fraction (LPF) of brain - 10 days (mUnits/g brain). The large particle fraction (LPF) was prepared by the method of Appel and Parrot (1970). The LPF diluted 10-fold with 0.1 M potassium phosphate buffer pH 7.2 containing 0.1 mM dithiothreitol to remove readily extractable enzymes as described by Klingenberg (1967). Total activity was determined after treatment with Triton-X-100, 1% final concentration for 1 h.

It may also be seen from Fig.6 that the maximal potential activity of the pentose phosphate pathway, as shown by the effect of the artificial electron acceptor phenazine methosulphate, is strikingly in excess of that actually expressed at all stages of development. The stimulation by phenazine methosulphate, approximately 30-fold, is amongst the highest in the tissues so far studied and it is apparent that brain has a high reserve capacity for reductive potential in terms of the activity of the pentose phosphate pathway.

It would be of interest to speculate on whether there is an intermittent high requirement for NADPH related to the specialized functions of the brain. It would therefore seem more important to view the significance of pentose phosphate pathway activity in brain in the light of its high potential activity and marked compartmentation rather than in relation to the lower contribution to the total oxidative metabolism.

THE ROLE OF THE PARTICLE-BOUND PENTOSE PHOSPHATE PATHWAY

A substantial fraction of the total cell pentose phosphate pathway enzyme is associated with the large particle fraction in brain (Baquer and McLean, 1972). Figure 7 shows the distribution of the two dehydrogenases of the pathway between the readily-extractable and tightly-bound forms of these enzymes in the particulate fraction. Approximately 50% of the activity of each of the dehydrogenases is recovered in the tightly-bound form, with the remainder being extracted with 0.1 M phosphate buffer by the method of Klingenber (1967) or by freezing and thawing. This distribution could indicate two 'compartments' for the enzymes of the pentose phosphate pathway within this fraction. Wilson (1972) has already shown that hexokinase in brain also shows a similar readily-extractable and a tightly-bound, latent, form.

Table I shows the profile of utilization of specifically labelled glucoses by syn-

Table I. Effect of serotonin on the utilization of specifically labelled glucose by synaptosomes.

$^{14}CO_2$ Yield (μg atoms glucose carbon)	Synaptosomes	Synaptosomes + Serotonin	Brain Slices
Medium, with half Ca^{++}			Krebs-Ringer
[1-^{14}C] glucose	0.71	2.99	0.94
[3,4-^{14}C] glucose	4.36	4.57	4.22
[6-^{14}C] glucose	0.32	0.31	0.64
[U-^{14}C] glucose	1.98	2.35	2.00
C1/C6	2.2	9.6	1.5
Medium, no Ca^{++}			
[1-^{14}C] glucose	0.63	3.30	
[6-^{14}C] glucose	0.36	0.30	
C1/C6	1.8	11.0	

The Krebs-Ringer bicarbonate medium used was modified to contain Na^+:K^+ ratio of 5:1 and was further adjusted to contain either half the normal calcium or no calcium at all. Glucose was present at 10 mM final concentration, the final volume was 2.0 ml. To each plastic incubation tube was added 0.25 μC ^{14}C-labelled glucose and large particle fraction equivalent to 6 mg protein. The gas phase was O_2:CO_2 95:5, the time of incubation 15 min at 37°. The values for synaptosomes are given as μg atoms glucose carbon yield/g protein/15' at 37°. Brain slices were incubated in normal Krebs-Ringer bicarbonate medium with 20 mM glucose as described in Fig.6. For comparison the $^{14}CO_2$ yields have been normalised to equivalent yields of $^{14}CO_2$ from [U-^{14}C] glucose (this approximates to the output from 1 g slices for 40 min). Stage of development, 10 days *post partum*.

aptosomes. Here it may be seen that there is evidence for the operation of the pentose phosphate pathway (C_1/C_6 quotient of 2.2); for the glycolytic pathway and pyruvate dehydrogenase (high [3,4-^{14}C] glucose conversion to $^{14}CO_2$) and for tricarboxylic acid cycle activity (as shown by the conversion of [6-^{14}C] glucose to $^{14}CO_2$). The profile in synaptosomes is closely similar to that obtained with brain slices (for comparison, the brain slice figures have been normalized to a similar yield of $^{14}CO_2$ from uniformly labelled glucose). When serotonin is added, there is an increase only in the [1-^{14}C] glucose oxidation, about 4-fold, as seen in the C_1/C_6 quotient, results similar to those of Appel and Parrot (1970). There is no evidence from this data that any other oxidative pathway is modified by serotonin, although this would not necessarily show if there were an increase in a route such as the conversion of glucose to glucosamine.

Further evidence on the regulation and compartmentation of the pentose phosphate pathway in synaptosomes is shown in Fig.8. When a washed synaptosomal fraction is incubated with [1-^{14}C] glucose alone, then addition of a range of artificial electron acceptors and substrates which can utilize NADPH (oxidized glutathione, for example) can stimulate the pentose phosphate pathway, indicating

14CO 2 FORMATION FROM (1–14C)GLUCOSE

BY SYNAPTOSOMES 20 DAY BRAIN

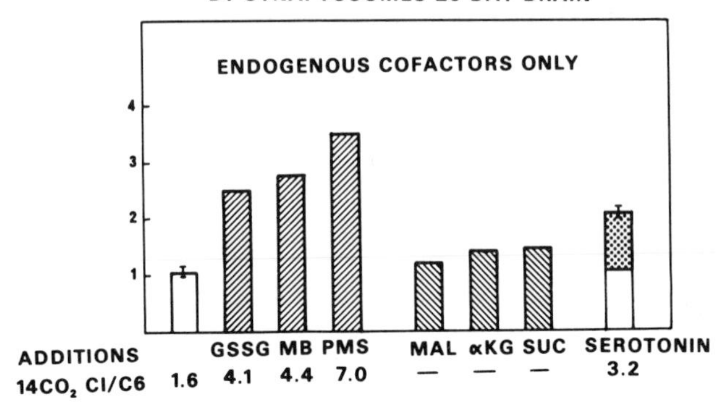

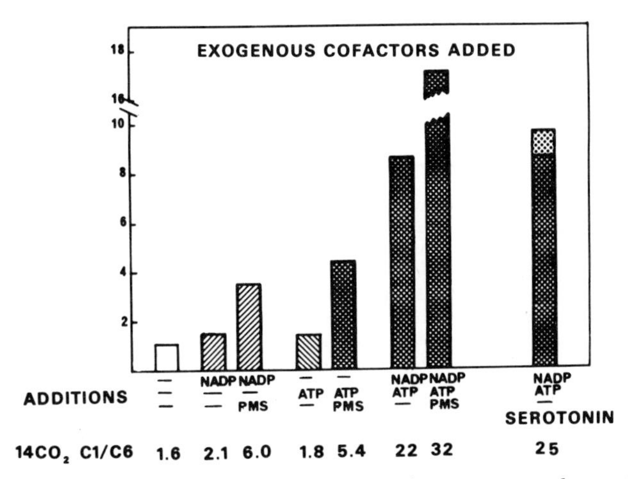

Fig.8 Regulation of the pentose phosphate pathway in synaptosomes by serotonin and by modification of the redox state and ATP level. The incubation conditions were as described in Table I. Modified Krebs-Ringer bicarbonate medium with Na^+:K^+ ratio of 5:1 and half the normal Ca^+ was used, the LPF was prepared from 20 day rat brain. The final concentrations of the additives were as follows: oxidized glutathione (GSSG) 5 mM; methylene blue (MB), 0.05 mM; phenazine methosulphate (PMS), 0.1 mM; malate (MAL), 25 mM; α-ketoglutarate (αKG), 25 mM; succinate (SUC), 25 mM; serotonin, 1 mM; $NADP^+$, 1.0 mM; ATP, 6.0 mM. The quotient of $^{14}CO_2$ yields from [1-^{14}C] glucose/[6-^{14}C] glucose (C1/C6) is given at the foot of each column.

that the $NADP^+$/NADPH quotient is vital in the control of the activity of this pathway. This is shown in the C_1/C_6 quotients given under each column and is, again, in agreement with the results of Appel and Parrot (1970). When Krebs cycle intermediates are added to promote the formation of intramitochondrial ATP, then less effect is observed, indicating that ATP availability is not a major limitation. The increase above basal activity, attributable to the addition of serotonin, is shown in the last column of this figure. These data lead to two conclusions; first, that 'internal' and 'external' pentose phosphate pathway routes exist

Table II. Effect of cAMP and agents stimulating cAMP formation in brain on the pentose phosphate pathway in synaptosomes.

Additions	$^{14}CO_2$ from [1-^{14}C] glucose (μg atom/g protein/15 min)	Increase in adenyl cyclase Hungen & Roberts (1973) (μmoles/g protein/15 min)
None, basal activity	1.08 ± 0.09	—
Serotonin	2.03 ± 0.13 **	+ 0.1
Noradrenaline	2.34 ± 0.15 **	+ 2.6
Histamine	0.98	+ 0.9
Acetylcholine + eserine	0.93 ± 0.16	+ 0.2
Adenosine	0.95	
Adenine	0.88	
cAMP + theophylline	0.83 ± 0.08	

The incubation medium was as described in Table I with half the normal Ca^{++} content. The final concentration of the neurotransmitters and related substances used: serotonin, 1 mM; noradrenaline, 1 mM; histamine, 0.3 mM; acetylcholine + eserine, 0.1 mM each; adenosine, 0.125 mM; adenine, 0.125 mM; dibutyryl cAMP, 0.2 mM; theophylline, 5 mM. The concentration used in the quoted experiments of Hungen and Roberts (1973) was 0.5 mM for each compound; their values have been recalculated to 15 min basis for comparison with our $^{14}CO_2$ data. The $^{14}CO_2$ data is given as the mean value ± SEM, the asterisks indicate Fishers P value of < 0.01.

and, second, that the 'internal' pathway appears to have a major limitation in the supply of $NADP^+$. In this context, 'internal' and 'external' are intended to have a specific connotation. The 'internal pathway' is that occurring inside the limiting membrane of the organelle and therefore complete with its complement of enzymes and cofactors, i.e. requiring no supplementation. The 'external pathway' represents the existence of the enzymes of the pathway outside the limiting membrane of the organelle which are exposed to the incubation medium and are, therefore, deprived of cofactors. This 'external pathway' cannot act upon glucose unless supplemented with ATP and $NADP^+$.

When the combined internal and external systems are examined by the addition of exogenous $NADP^+$ and ATP, together with [1-^{14}C] glucose, then there is a dramatic increase in the observed activity of the pentose phosphate pathway. Even with this system, there is still a limitation imposed by the rate of reoxidation of NADPH, as shown by the further stimulation found on addition of phenazine methosulphate, giving a 20-fold increase in total pathway activity. When serotonin is added to synaptosomes in the presence of glucose, ATP and $NADP^+$, then there is a small, but significant, increase of the pathway, the magnitude of which is almost exactly equivalent to that caused by serotonin when only glucose was present. This may be interpreted as showing that serotonin promotes the activity of the pentose phosphate pathway only in the internal system not freely available to added ATP and $NADP^+$ and that there is an effective compartmentation and localization of the effect of this transmitter substance.

It seemed reasonable to suppose that if serotonin were, indeed, acting to increase the activity of the pentose phosphate pathway by virtue of its neurotransmitter properties, then it would probably do so via the agency of cAMP,

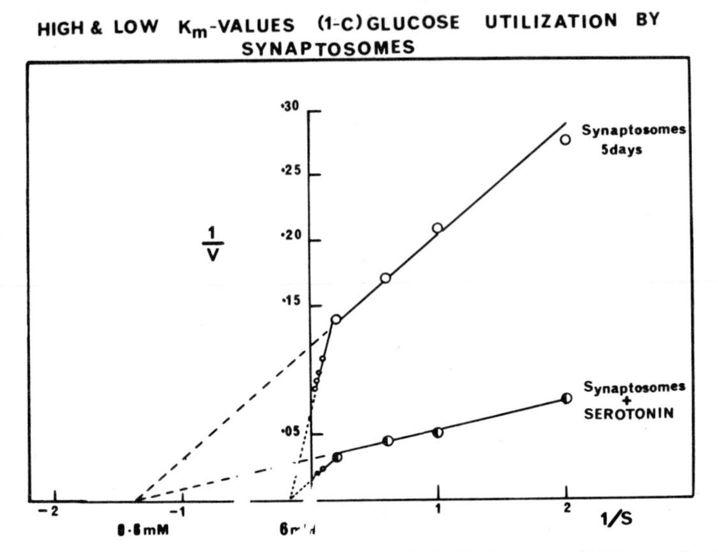

Fig. 9 The effect of serotonin on the conversion of [1-^{14}C] glucose to $^{14}CO_2$ as a function of glucose concentration. The incubation conditions were as described in Table I. Krebs-Ringer bicarbonate medium modified to contain Na^+:K^+ ratio of 5:1 and with half the normal Ca^{++} was used, the LPF from 5 day rat brain was used. The concentration of serotonin was 1 mM. Other conditions were as described in Table I.

which has been widely implicated in this type of process (see Greengard and Kebabian, 1974; Hungen *et al.*, 1974). Table II shows the effect of a number of neurotransmitters and related substances on the activation of the pentose phosphate pathway and the results of Hungen and Roberts (1973) on the action of some of these compounds on the adenylate cylcase activity of the large particle fraction of brain. As may be seen, there is no correlation between activation of adenylate cyclase and the increase of pentose phosphate pathway activity, the discrepancy between the action of serotonin and noradrenaline on the two systems being particularly striking. This lack of correlation is further emphasized by the failure of adenosine and adenine, compounds both known to raise the cAMP content of brain slices (Shimizu *et al.*, 1973), to increase the activity of the pentose phosphate pathway in synaptosomes.

Moreover, it might be anticipated that cAMP would mimic the effect of serotonin on the pentose phosphate pathway. As shown in Table II, dibutyryl cAMP, in the presence of theophylline, had no effect on the rate of oxidation of carbon-1 of glucose. Calcium ions are involved in the process of neurotransmission and are critical in the adenylate cyclase-cAMP response (Rahamimoff and Alnaes, 1973). Incubation of synaptosomes in a medium free from added Ca^{2+} did not alter the stimulation of the pentose phosphate pathway by serotonin.

It was also possible that neurotransmitters might cause a change in membrane function such as to alter transport of glucose with subsequent effects on the pathways of glucose metabolism. Diamond and Fishman (1973) have recently described low- and high-K_m systems for the transport of deoxyglucose into synaptosomes but found that neither noradrenaline, acetylcholine or cAMP affected the uptake of the deoxysugar. Measurements of the conversion of [1-^{14}C] glucose to $^{14}CO_2$ as a function of concentration in the presence and absence of serotonin are shown in Fig.9. While the presence of the low- and high-K_m systems are readily seen in this figure and serotonin increased the rate of oxidation via the pentose phosphate pathway at all concentrations of glucose studied, there is, nevertheless, no change in either of the overall apparent K_m values.

A further function linking pentose phosphate pathway activity with neurotransmitter action might be located in the process of lipid formation for the maintenance of membrane structure. ^{14}C-glucose is incorporated into lipid by synaptosomal fractions to a limited extent (0.2 μmoles/g protein/h) but this rate was not altered by the addition of serotonin to the incubation medium.

Thus, although there was a clear-cut stimulation of the pentose phosphate pathway by serotonin and noradrenaline, no positive correlation had been established, under the present conditions, between activation of the pathway, activation of adenylate cyclase and the process of neurotransmission. Further, because of the linking of the pentose phosphate pathway to the $NADP^+$/NADPH redox state, it is necessary to identify the hydrogen acceptor system involved in any process in which the pentose phosphate pathway is stimulated.

The concentration of serotonin and noradrenaline used in a large number of published experiments with synaptosomes, studying a range of parameters, has often been in the range 0.05 to 1.0 mM (Hungen and Roberts, 1973; Appel and Parrot, 1970; Diamond and Fishman, 1973) and, in this study, similar concentrations (0.1 to 1.0 mM) have been used. Such concentrations may be regarded as substrate levels and the possibility must be considered that it is the further metabolism of serotonin itself which might provide the NADPH-utilizing system.

The breakdown of serotonin and noradrenaline in brain by monoamine oxidase yeilds an aldehyde, ammonia and H_2O_2 (Kapeller-Adler, 1970). One route for the detoxication of hydrogen peroxide is via glutathione peroxidase, glutathione reductase and the two dehydrogenases of the pentose phosphate pathway. This route is illustrated in Fig.10. Activity of each enzyme of this sequence can be detected in the large particle fraction of brain (de Marchena *et al.*, 1974; Achee *et al.*, 1974; Baquer *et al.*, unpublished results). The amounts present are sufficient to account for a rate of oxidation of serotonin which is consistent with the increased rate of utilization of NADPH and consequent stimulation of the oxidation of glucose by the pentose phospahte pathway. Values for the activities of the relevant enzymes in the large particle fraction of brain are given in Fig.10. It should be emphasized that this route involving glutathione peroxidase is particularly favourable for the activation of the pentose phosphate pathway since not only does it utilize NADPH, thus simultaneously decreasing the inhibition on glucose 6-phosphate and 6-phosphogluconate dehydrogenases by removal of NADPH and provision of $NADP^+$ for the oxidative process, but it also yields oxidized glutathione, which is a powerful agent in releasing 6-phosphogluconate dehydrogenase from inhibition by NADPH (Eggleston and Krebs, 1974), thus having a dual push-pull action on the pentose phosphate pathway.

The further metabolism of the aldehyde produced from the monoamine may also involve reductive processes using NADPH (Eccleston *et al.*, 1966, 1969). The catabolism of serotonin to 5-hydroxytrytophol by reduction of the intermediate aldehyde has been demonstrated to occur in rat brain preparations as well as the oxidation to hydroxyindole acetic acid. The finding that serotonin and noradren-

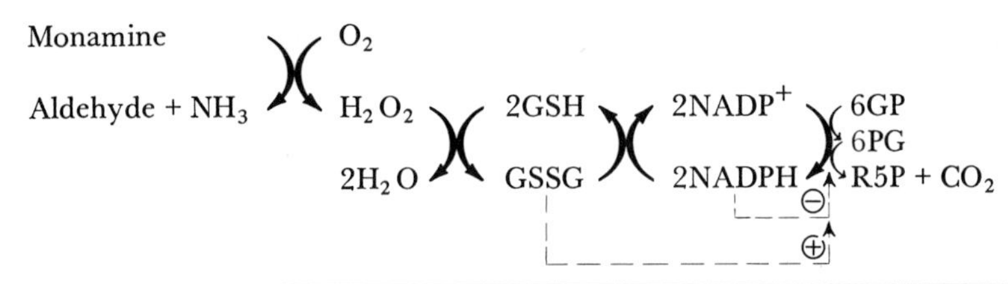

	Enzyme activity milliunits/mg protein	Brain fraction Species and age
Glucose 6-phosphate dehydrogenase	7.3	LPF, Rat, 20 days
6-phosphogluconate dehydrogenase	3.2	LPF, Rat, 20 days
NADP-GSSG reductase	8.6	LPF, Rat, 20 days
Glutathione peroxidase	circa 10	Total brain, various species (de Marchena *et al.*, 1974)
Monoamine oxidase	3.3	Mitochondria, Rabbit, adult, (Achee *et al.*, 1974)

Fig.10 Proposed pathway linking serotonin degradation by monoamine oxidase to the activation of the pentose phosphate pathway via H_2O_2 formation.

aline stimulated the pentose phosphate pathway while histamine did not, suggests that this may be related to the presence of high levels of monoamine oxidase and low levels of diamine oxidase in rat brain (Burkard *et al.*, 1963; Snyder and Hendley, 1968).

It is suggested therefore that one vital function of the pentose phosphate pathway in synaptosomes is to provide for the rapid detoxication of hydrogen peroxide. The necessity for the protection of brain tissue from peroxidative damage has been established (de Marchena *et al.*, 1974; Zeman and Dyken, 1969). The products of peroxidative damage to lipids and proteins accumulate in brain with age and at an accelerated rate in some diseases, leading to massive neuronal death (Zeman and Dyken, 1969). The present findings are in accord with previous proposals that the pentose phosphate pathway is of functional significance in the maintenance of glutathione in the reduced state which, in turn, serves to protect membrane and cellular sulphydryl groups and the integrity of the synaptic plasma membrane (Hotta and Seventko, 1968; Appel and Parrot, 1970).

The design of experiments in which exogenous serotonin or noradrenaline are added to synaptosomal preparations or brain slices to study effects on carbohydrate metabolism is complicated by the on-going metabolism of the neurohormone by such processes as oxidative deamination and O-methylation. Considerations on these lines may account for the high concentrations of the hormones generally employed, although their use may lead to problems of interpretation. Devices which might permit reducing the amount of hormone to more pharmacological levels would include the use of monoamine oxidase inhibitors while the use of the L and D forms of adrenaline and noradrenaline would allow the separation of specific from non-specific effects (Hungen and Roberts, 1973). However, even these are not without problems. The complexities inherent in the use of inhibitors on the uptake, release, binding and route of metabolism of neurohormones in brain slices *in vitro* has been fully discussed by Katz and Chase (1970).

This brief survey of some developmental changes in the pentose phosphate pathway and the investigation of some aspects of the functional activity of the pathway in synaptosomes suggests the versatile and changing role of the pathway during development. In early developmental stages it provides ribose for RNA and DNA synthesis and the reductive potential for lipogenesis and myelination. It is suggested from the enzyme profile that, in the later stages of development, the pentose phosphate pathway may provide ribose for the continuing turnover of RNA by the non-oxidative route via transketolase. The pentose phosphate pathway may provide the reductive potential for the synthesis of neurotransmitter substances, a process which is greatly accelerated by 10th to 12th day after birth. It is also interesting to speculate whether the particulate form of the pentose phosphate pathway is of significance in driving the glutamate dehydrogenase reaction in favour of reductive amination. Finally, it would seem clear that the reductive potential of the pathway is used in the degradation of neurotransmitters, serotonin and catacholamines in particular, in relation to the protection against the hydrogen peroxide formed in monoamine oxidase reactions. The huge reserve of potential activity of this pathway in brain (30 to 40-fold greater than actual flux) is a measure of the importance of this function.

ACKNOWLEDGEMENTS

We wish to thank our colleagues, Miss Jula White and Mr. John S. Hothersall, for permission to include some of their unpublished data. We are also grateful for the generous support of the Wellcome Trust.

REFERENCES

Achee, F.M., Togulga, G. and Galay, S. (1974). *J. Neurochem.* **22**, 651-661.
Appel, S.H. and Parrot, B.L. (1970). *J. Neurochem.* **17**, 1619-1626.
Bachelard, H.S. (1967). *Biochem. J.* **104**, 286-292.
Bachelard, H.S. and Goldfarb, P.S.G. (1969). *Biochem. J.* **112**, 579-586.
Bagdasarian, G. and Hulanicka, D. (1965). *Biochem. biophys. Acta* **99**, 367-369.
Balázs, R. (1970). *In:* "Handbook of Neurochemistry" (A. Lajtha, ed) Vol.III, pp.1-36. Plenum Press, New York and London.
Balazs, R., Patel, A.J. and Richter, D. (1973). *In:* "Metabolic Compartmentation in Brain" (R. Balázs and J.E. Cremer, eds) pp.167-184. MacMillan Press, London.
Baquer, N.Z. and McLean, P. (1972). *Biochem. biophys. Res. Commun.* **46**, 167-174.
Baquer, N.Z., McLean, P. and Greenbaum, A.L. (1973). *Biochem. biophys. Res. Commun.* **53**, 1282-1288.
Bass, A., Brdiczka, D., Eyer, P., Hofer, S. and Pette, D. (1969). *Eur. J. Biochem.* **10**, 198-206.
Van den Berg, C.J. (1973). *In:* "Metabolic Compartmentation in Brain" (R. Balazs and J.E. Cremer, eds) pp.137-166. MacMillan Press, London.
Van den Berg, C.J. (1974). *In:* "Biochemistry of the Developing Brain" (W. Himwich, ed) Vol. II, pp.149-197. Marcel Dekker Inc., New York.
Berl, S. (1973). *In:* "Biochemistry of the Developing Brain" (W. Himwich, ed) Vol.I, pp.219-252. Marcel Dekker Inc., New York.
Buckley, B.M. and Williamson, D.H. (1973). *Biochem. J.* **132**, 653-656.

Burkhard, W.P., Gey, K.F. and Pletscher, A. (1963). *J. Neurochem.* **10**, 183-186.
Chaikoff, I.L. (1953). *Harvey Lect. (1951-52)* **47**, 99-124.
Clarke, F.M. and Masters, C.J. (1972). *Arch. biochem. Biophys.* **153**, 258-265.
Copley, M. and Fromm, H.J. (1967). *Biochemistry* **6**, 3503-3509.
Crane, R.K. and Sols, A. (1953). *J. biol. Chem.* **203**, 273-292.
Craven, P.A. and Basford, R.E. (1969). *Biochemistry* **8**, 3520-3525.
Craven, P.A., Goldblatt, P.J. and Basford, R.E. (1969). *Biochemistry* **8**, 3525-3532.
Cremer, J.E. (1964). *J. Neurochem.* **11**, 165-185.
Cremer, J.E. and Teal, H.M. (1974). *FEBS Lett.* **39**, 17-20.
D'Adamo, A.F. and D'Adamo, A.P. (1968). *J. Neurochem.* **15**, 315-323.
Diamond, I. and Fishman, R.A. (1973). *J. Neurochem.* **20**, 1533-1542.
Eccleston, D., Moir, A.T.B., Reading, H.W. and Ritchie, I.M. (1966). *Brit. J. Pharmacol. and Chemother.* **28**, 367-377.
Eccleston, D., Reading, H.W. and Ritchie, I.M. (1969). *J. Neurochem.* **16**, 274-276.
Eggleston, L.V. and Krebs, H.A. (1974). *Biochem. J.* **138**, 425-435.
Ferrendelli, J.A., Chang, M.M. and Kinscherf, D.A. (1974). *J. Neurochem.* **22**, 535-540.
Flexner, L.B., Belknap, E.L. and Flexner, J.B. (1953). *J. Cell. Comp. Physiol.* **42**, 151-166.
Gaitonde, M.K., Dahl, D.R. and Elliott, K.A.C. (1965). *Biochem. J.* **94**, 345-352.
Greengard, O. (1971). *Essays in Biochemistry* **7**, 159-205.
Greengard, O. and Kebabian, J.W. (1974). *Fedn Proc.* **33**, 1059-1067.
Gumaa, K.A. and McLean, P. (1969). *Biochem. biophys. Res. Commun.* **35**, 86-93.
Gumaa, K.A., McLean, P. and Greenbaum, A.L. (1971). *Essays in Biochemistry* **7**, 39-86.
Haslam, R.J. and Krebs, H.A. (1963). *Biochem. J.* **88**, 566-578.
Herken, H., Lange, K. and Kolbe, H. (1969). *Biochem. biophys. Res. Commun.* **36**, 93-100.
Horecker, B.L. (1962). "Ciba Lectures in Microbial Biochemistry; Pentose Metabolism in Bacteria", p.30. John Wiley & Sons Inc., New York and London.
Hostetler, K.Y., Landau, B.R., White, R.J., Albin, M.S. and Yashar, D. (1970). *J. Neurochem.* **17**, 33-39.
Hotta, S.S. (1962). *J. Neurochem.* **9**, 43-51.
Hotta, S.S. and Seventko, J.M. (Jnr.). (1968). *Arch. biochem. Biophys.* **123**, 104-108.
von Hungen, K. and Roberts, S. (1973). *Eur. J. Biochem.* **36**, 391-401.
von Hungen, K., Roberts, S. and Hill, D.F. (1974). *J. Neurochem.* **22**, 811-819.
Itoh, T. and Quastel, J.H. (1970). *Biochem. J.* **116**, 641-655.
Johnson, M.K. and Whittaker, V.P. (1963). *Biochem. J.* **88**, 404-409.
Kapeller-Adler, R. (1970). "Amino Oxidases and Methods for Their Study". Wiley-Interscience, New York.
Kato, T. and Lowry, O.H. (1973). *J. Neurochem.* **20**, 151-163.
Katz, R.I. and Chase, T.N. (1970). *Adv. in Pharmacol. and Chemother.* **8**, 1-30.
Kauffman, F.C. (1972). *J. Neurochem.* **19**, 1-9.
Kellogg, E.N., Knull, H.R. and Wilson, J.E. (1974). *J. Neurochem.* **22**, 461-463.
Klingenberg, M. (1967). *In:* "Methods in Enzymology" (R.W. Estabrook and M.E. Pullman, eds) Vol. **10**, pp.3-18.
Kuhlman, R.E. and Lowry, O.H. (1956). *J. Neurochem.* **1**, 173-180.
Lagunas, R., McLean, P. and Greenbaum, A.L. (1970). *Eur. J. Biochem.* **15**, 179-190.
Lehrer, G.M. and Maker, H.S. (1973). *In:* "Metabolic Compartmentation in the Brain" (R. Balázs and J.E. Cremer, eds) pp.235-244. MacMillan, London.
Loverde, A.W. and Lehrer, G.M. (1973). *J. Neurochem.* **20**, 441-448.
Lowry, O.H. and Passonneau, J.V. (1964). *J. biol. Chem.* **239**, 31-42.
MacDonnell, P.C. and Greengard, O. (1974). *Arch. biochem. Biophys.* **163**, 644-655.
McIlwain, H. (1971). *Essays in Biochemistry* **7**, 127-158.
McIlwain, H. and Bachelard, H.S. (1971). "Biochemistry and the Central Nervous System" 4th ed., p.51. Churchill Livingston, Edinburgh and London.
de Marchena, O., Guarnieri, M. and McKhann, G. (1974). *J. Neuorchem.* **22**, 773-776.
Masters, C.J. and Holmes, R.S. (1972). *Biol. Revs.* **47**, 309-361.
Mayer, R.J. and Hubscher, G. 91971). *Biochem. J.* **124**, 491-500.
Negelein, E. and Haas, E. (1935). *Biochem. Z.* **282**, 206-220.
Newsholme, E.A., Rolleston, F.S. and Taylor, K. (1968). *Biochem. J.* **106**, 193-201.
O'Neill, J. (1974). *In:* "Biochemistry of the Developing Brain" (W. Himwich, ed) Vol.2, pp. 69-121. Marcel Dekker Inc., New York.

Pette, D. (1966). *In:* "Regulation of Metabolic Processes in Mitochondira" (J.M. Tager, S. Papa, E. Quagliarielle and E.C. Slater, eds) pp.28-49. Elsevier Publishing Co., Amsterdam.
Pette, D., Klingenberg, M. and Bücher, Th. (1962). *Biochem. biophys. Res. Commun.* **7**, 425-428.
Pette, D., Luh, W. and Bücher, Th. (1962). *Biochem. biophys. Res. Commun.* **7**, 419-424.
Portenhauser, R. and Wieland, O. (1972). *Eur. J. Biochem.* **31**, 308-314.
Rahamimoff, R. and Alnaes, E. (1973). *Proc. natnl. Acad. Sci. U.S.A.* **70**, 3613-3616.
Rose, I.A. and Warms, J.V.B. (1967). *J. biol. Chem.* **242**, 1635-1645.
Sano, I. (1970). *Inter. Rev. of Neurobiol.* **12**, 235-263.
Shimizu, H., Takenoshita, M., Huang, M. and Daly, J.W. (1973). *J. Neurochem.* **20**, 91-95.
Snyder, S.H. and Hendley, E.D. (1968). *J. Pharmacol.* **163**, 386-392.
Staudte, H.W. and Pette, D. (1972). *Comp. Biochem. Physiol.* **41B**, 533-540.
Thompson, M.F. and Bachelard, H.S. (1970). *Biochem. J.* **118**, 25-34.
Tuček, S. and Cheng, S.C. (1974). *J. Neurochem.* **22**, 893-914.
Vallejo, C.G., Marco, R. and Sebastian, J. (1970). *Eur. J. Biochem.* **14**, 478-485.
Volpe, J.J. and Kishimoto, Y. (1972). *J. Neurochem.* **19**, 737-753.
Wilbur, D.O. and Patel, M.S. 91974). *J. Neurochem.* **22**, 709-715.
Wilson, J.E. (1968). *J. biol. Chem.* **243**, 3640-3647.
Wilson, J.E. (1972). *Arch. biochem. Biophys.* **150**, 96-104.
Wieland, O.H. and Portenhauser, R. (1974). *Eur. J. Biochem.* **45**, 577-588.
Zeman, W. and Dyken, P. (1969). *Pediatrics* **44**, 570-583.

DISCUSSION

Tyson Tildon: I would like to ask Dr. Baquer if there is some significance to the biphasic curve found for the oxidation of [1-^{14}C]-glucose in the synaptosomes. It looked like substrate activation. Could that be an explanation?

Baquer: That is a possibility. It has been reported [Diamond and Fishman (1973) ref.25] that there are 2 K_m's for the transport of 2-deoxyglucose. The present data supports the existence of two K_m's. The use of [1-^{14}C]-glucose and measurement of $^{14}CO_2$ means that it is not possible to state which step is concentration dependent, transport, phosphorylation or the oxidative steps of the pentose phosphate pathway. It may be significant that the high apparent K_m value of 6 mM is close to the blood sugar value.

Tyson Tildon: Was this inhibited by ATP?

Baquer: We have not tried the inhibition by ATP.

Tager: I was very interested in the suggestion you made, that in tissues which have a high oxidative potential, hexokinase is bound in the mitochondria, which serves to keep the ADP to ATP ratio in the mitochondria high and therefore activate pyruvate dehydrogenase. Is anything known about where this hexokinase is situated in the mitochondria?

Baquer: Wilson [Wilson, 1972 (Ref.82)] has investigated brain mitochondrial and synaptosomal hexokinase. He has shown that part of the hexokinase can be extracted with osmotic treatment. Triton liberates the latent form. He has found that between 30% and 50% is latent. The evidence in the literature suggests that the tightly bound form of hexokinase is outside the atractyloside barrier [Vallejo *et al.*, 1970 (Ref.78)].

Tager: Would this distinguish between the mitochondrial and the synaptosomal enzyme?

Baquer: Yes, the synaptosomes and mitochondria were separated on a gradient.

Tager: Maybe John Land has information on this.

Land: We have a preparation of mitochondria which is essentially free from synaptosomal contamination. We measured hexokinase on these preparations. A very large percentage of the total brain hexokinase is found in these preparations, which is not extractable with high concentration of KCl. It is tightly bound and it appears to be on the outer membrane of the mitochondria.

Van den Berg: You showed that the rate of $^{14}CO_2$ production from [1-^{14}C]-acetic acid increased and then decreased during development. This is also observed *in vivo* when the incorporation of acetic acid into amino acids is measured. It is surprising that acetyl-CoA synthase does not follow this pattern. It starts to increase after birth and continues to increase until it reaches the adult value. Do you have any idea why the CO_2 production from acetic acid goes down with age?

Baquer: Probably the brain switches over to more glucose utilization. However, this is a very interesting problem. We have found in preliminary experiments that the particulate form of acetyl-CoA synthetase increases 6-fold from birth to 20 days while the soluble form increases only about 2-fold. On this basis one would certainly expect an increase in the rate of mitochondrial oxidation of [1-^{14}C]-acetate. This discrepancy raises the interesting question of control of alternative pathways utilizing acetyl CoA.

Van den Berg: But acetate, I think, does not seem to be metabolized by all the mitochondria, like β-hydroxybutyrate. Therefore the switch from oxidation of β-hydroxybutyrate to glucose will probably not be visible with acetic acid.

Hommes: The suggestion you made about the removal of amines by monoamine oxidase and the coupling of this process to the pentose phosphate pathway cycle is an extremely interesting one. Has the presence or absence of superoxide dismutase in mitochondria of the brain been demonstrated? Because this enzyme might use up quite a bit of H_2O_2.

Baquer: Not that I know of.

Cremer: You mentioned that you measured the ^{14}C incorporation from glucose, into the total lipid fraction. Can you tell us something about the relative rates of lipogenesis and the pentose pathway? Is there any direct relationship between the two?

Baquer: They matched very well.

REGULATORY FACTORS IN GLUCOSE AND KETONE BODY UTILIZATION BY THE DEVELOPING BRAIN

J.E. Cremer, H.M. Teal and D.F. Heath

M.R.C. Toxicology Unit, Medical Research Council Laboratories Woodmansterne Road, Carshalton, Surrey, UK

INTRODUCTION

The quantitatively important contribution made by ketone bodies as oxidizable substrates for the brain is now well established. Persson *et al.* (1972) found that in young children fasted for relatively short periods, 6 to 12 h, the concentrations of ketone bodies, acetoacetate plus 3-hydroxybutyrate, in blood rose to values of 0.4 to 1.4 mM and arteriovenous difference measurements across the brain showed the uptake of ketone bodies to be proportional to the blood concentration and greater than values reported for adults (Gottstein *et al.*, 1971). These findings have been confirmed by Kraus *et al.* (1974) who extended their investigations to include A-V difference measurements of oxygen, glucose, lactate, pyruvate, glycerol and free fatty acids. Changes in the latter two metabolites could not be detected during passage through the brain in human infants. Of the glucose removed from the blood at least 26% was returned as lactate plus pyruvate. The remainder of the glucose, together with the ketone bodies removed, would account for the total oxygen consumed if combustion of these substrates were complete.

A study on regulatory factors of substrate utilization by the brain, to be of relevance to the human infants, should take into account the metabolism of both glucose and ketone bodies. The suitability of the young suckling rat as an experimental animal model has been described by Williamson and Buckley (1973) in their contribution to an earlier symposium in this series. The demonstrable high rate of ketone body utilization by the young rat brain appears to correlate with high activities in the brain of those enzymes which catalyse the initial steps in ketone body oxidation. Both the activities of the enzymes and the capacity of the brain to utilize ketone bodies decline in adulthood. The same authors raised the question of whether ketone bodies provide carbon for biosynthetic processes in the developing brain, especially lipogenesis and cholesterol synthesis, as well as acting as sources of energy.

In the present chapter results of further studies on the suckling rat brain are reported. These include quantitative aspects of glycolysis, pyruvate and ketone body oxidation. Estimates of rates *in vivo* are compared with activities and properties of the relevant enzymes and measured *in vitro*, with particular reference to pyruvate dehydrogenase. This enzyme is at a pivotal position in carbohydrate and ketone body oxidation. Data on relative rates of incorporation of labelled carbon from glucose and ketone bodies into brain lipids are also presented and discussed.

ESTIMATED RATES OF GLUCOSE AND KETONE BODY UTILIZATION

Estimates, from an analysis of isotopic data, of rates of utilization by the brain of glucose and ketone bodies in the 18 day old rat have been reported recently by Cremer and Heath (1974). In this study the radioactive precursors used were ^{14}C labelled glucose, lactate and D-3-hydroxybutyrate given by intraperitoneal injection. Time-curves for the specific radioactivities of these substrates in the blood were obtained. The incorporation of label into various brain metabolites, particularly tricarboxylate cycle acids and the closely associated amino acids, was measured and quantitatively related to the blood radioactivity data. Details of the experimental procedures and mathematical analyses are given by Cremer and Heath (1974) and only the main findings will be presented here and discussed in relation to possible regulatory factors.

In the 18 day old suckling rats used, the concentration of blood glucose was similar in all individuals with an overall mean ± S.D. value of 5.88 ± 0.53 mM. Ketone body concentrations (acetoacetate plus 3-hydroxybutyrate), however, varied considerably between animals of different litters and ranged from 0.44 to 2.15 mM. Blood lactate (in mixed blood samples) was also variable giving values between 1 and 3 mM. The variation in the concentration of blood ketone bodies that occurred under apparently normal physiological conditions allowed confirmation by isotopic tracer methods of the previously observed proportionality between blood concentration and rate of net uptake by the brain (Hawkins *et al.*, 1971).

Data from arteriovenous difference measurements in both children and young rats have shown that when ketone bodies are oxidized by the brain there is a net output of lactate from brain to blood. Two implications arise from this in terms of regulatory processes. One is that the rate of pyruvate oxidation via acetyl-CoA is slower than any preceding step in the glycolytic pathway. The other is that the 'blood brain barrier' must be readily permeable to pyruvate and lactate. Both these points will be considered in more detail.

An outline of the main events that occur during metabolism of glucose and ketone bodies by the brain is given in Scheme 1. Before discussing the various enzymatic steps the question of lactate permeability will be considered. Reports in the literature on the permeability of brain capillaries to pyruvate and lactate are conflicting. Crone and Sørensen (1970) concluded from experiments on adult dogs that these two substrates did not pass readily. However, Oldendorf (1972; 1973) using a similar technique, based on the rapid uptake of isotopic tracers injected into the carotid artery, found evidence in adult rats of a fairly rapid uptake of both pyruvate and lactate. We have been unable to find data for similar studies on young rats, but using a somewhat different experimental approach the evidence is that transport is rapid (Cremer and Heath, 1974).

An estimate of the minimum rate of exchange of lactate between the blood and the brain was based on the combined data obtained following the intraperitoneal injection of either [^{14}C] glucose or [^{14}C] lactate. With [^{14}C] glucose it

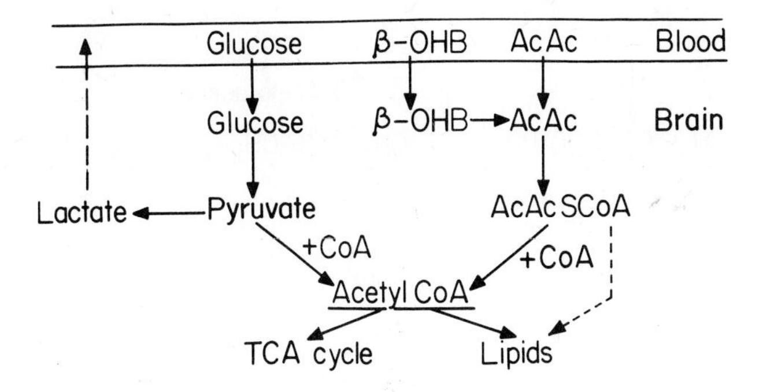

Scheme 1

could be shown that a fraction of the brain glucose reached rapid isotopic equilibrium with blood glucose and was the precursor of a labelled pool of brain pyruvate plus lactate. Although these two metabolites were in rapid isotropic equilibrium with each other, their specific radioactivity did not approach the theoretical value relative to that of the precursor brain glucose, but reached a lower constant ratio. Similarly, following the injection of [^{14}C] lactate there was an initial rapid increase in the specific radioactivity of brain lactate, but the value remained low relative to that of blood lactate giving a constant ratio of only 0.36. Checks on the labelling of blood lactate following an injection of [^{14}C] glucose and *vice versa*, showed that the extent of cross-labelling was low during the first 10 minutes of the experiments. Therefore, a likely explanation of the low specific radioactivities of brain lactate was that in experiments with [^{14}C] glucose, unlabelled blood lactate freely exchanged with the labelled lactate formed in the brain from glucose thereby diluting away the label from the brain to the blood. The converse would occur when injected [^{14}C] lactate entered the brain from blood and was constantly diluted with unlabelled lactate formed in the brain from glucose. From an analysis of these and other data, the following conclusions could be made: 1) [^{14}C] glucose and [^{14}C] lactate in the blood labelled a common pool of lactate in the brain. 2) This lactate pool exchanged label rapidly with the pyruvate pool from which acetyl-CoA was formed. 3) The rate of input of blood lactate into the brain lactate pool labelled was at least 2μmol/min g and exchange of lactate molecules across the brain capillaries must be in excess of this value.

This rapid rate of exchange of lactate causes a substantial fraction of the label in lactate that originated as [^{14}C] glucose in the brain to be lost to the blood. If this loss is not taken into consideration in studies using [^{14}C] glucose in young animals, rates of metabolic steps beyond the formation of acetyl-CoA will be grossly underestimated. The specific radioactivity of the pyruvate feeding into the tricarboxylate cycle in the brain can be estimated from the combined data obtained with [^{14}C] glucose and [^{14}C] lactate (Cremer and Heath, 1974).

As has been shown for the adult rat (Cremer, 1971) and for 10 day old and adult mice (Van den Berg, 1973) the pattern of incorporation of label into brain amino acids is very similar following injections of either [^{14}C] glucose, [^{14}C]-3-hydroxybutyrate or [^{14}C] acetoacetate. These, together with the more recent data on young rats (Cremer and Heath, 1973), can be taken as evidence that both classes of substrates feed into a common pool of acetyl-CoA. This is oxidized via

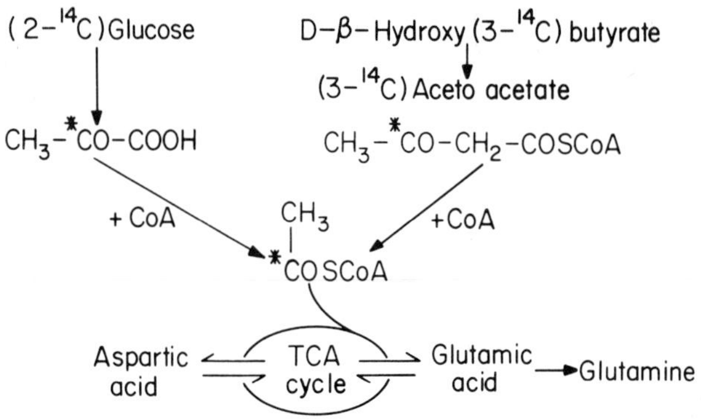

(2-^{14}C)Glucose

CH_3-*CO-COOH

D-β-Hydroxy(3-^{14}C) butyrate

(3-^{14}C) Aceto acetate

CH_3-*CO-CH_2-COSCoA

+ CoA

+CoA

CH_3-*COSCoA

Aspartic acid ⇌ TCA cycle ⇌ Glutamic acid → Glutamine

Scheme 2

Table I. Estimate of rates from a compartmental analysis of isotopic data in the 18 day old rat.

	μmol/min g of brain
R_g pyruvate from glucose	0.96
R_p acetyl-CoA from pyruvate	0.74 (0.94 - 0.52)
R_k acetyl-CoA from ketone bodies	0.30 (0.11 - 0.53)
R citrate cycle rotation	1.05

Blood ketone body concentrations ranged from 0.44 to 2.15 mM with a mean value of 1.24 mM. Values for R_p and R_k are those estimated at the mean and, in parentheses, the lowest and highest blood concentrations of ketone bodies. Data are from Cremer and Heath (1974).

a tricarboxylate cycle from which label passes to a large fraction of brain glutamate (Scheme 2).

In a steady state the rate of formation of acetyl-CoA will equal its rate of oxidation and in normal, conscious animals kept under identical experimental conditions this rate can be expected to show only a small variation between individuals Therefore, although the rates of acetyl-CoA formation from either pyruvate or ketone bodies may vary in a reciprocal relationship between different animals, the sum of these two rates will remain constant. Estimates of rates in the 18 day old rat brain are given in Table I.

The estimated value for pyruvate formation from glucose, i.e. glycolysis, is subject to greater error than that for the other values. The rate, R_g, could not be less than 0.82 μmol/min g and the most probable value is given in Table I. When this rate is compared with the rates at which pyruvate was converted to acetyl-CoA, it is obvious that in most animals an excess of pyruvate was produced. This is entirely consistent with the A-V difference measurements of Hawkins *et al.* (1971) if the excess pyruvate is rapidly reduced to lactate which then passes to the blood.

This situation is very different from that of the fed adult where the rate of glycolysis equals the rate of pyruvate oxidation. It has long been suggested that the rate of glucose utilization is controlled by the two phosphorylating enzymes, hexokinase and phosphofructokinase. Many cell components have been invoked as regulating factors for these two enzymes, including feedback inhibition products of mitochondrial oxidations. Clearly, when ketone bodies are oxidized by mitochondria these regulatory mechanisms do not reduce the activities of these two enzymes sufficiently to make them equal the lowered activity of pyruvate dehydrogenase. This also applies to the regulation of other enzymes in the glycolytic pathway. Since glycolysis is not synchronous with the rate of oxidation, to prevent tissue lacticacidosis a continuous rapid removal of lactate is essential.

The rates of acetyl-CoA formation from pyruvate and ketone bodies over the range of concentration of blood ketone bodies found in the animals used, are given in Table I. It can be seen that in some animals the contribution made by ketone bodies equalled that made by pyruvate. In most animals the contribution was less, with a mean value of 30%.

COMPARISON BETWEEN ESTIMATED FLUXES THROUGH ENZYMIC STEPS *IN VIVO* AND PROPERTIES OF THE ENZYMES AS DETERMINED *IN VITRO*

The first metabolic steps in the oxidation of pyruvate and ketone bodies and the individual enzymes involved are listed in Table II. The fluxes *in vivo* through each step are given, calculated from data in Table I. Values for the maximal activities of individual enzymes in the 18 day old rat brain have been taken from the literature. The value given for 3-oxoacid transferase is for the formation of acetoacetyl-CoA rather than the more usually quoted non-physiological backward direction which is about 5 times higher. It can be seen from Table II that each of the enzymes catalysing the reactions for converting ketone bodies to acetyl-CoA is present in the brain with a potential activity several-fold in excess of the estimated rate at which it functions *in vivo*. This is so even allowing for lipogenesis (see Table III). Although there may be many factors regulating these enzymes, the following considerations show substrate limitation to be a possibility. The observed blood concentration of acetoacetate and 3-hydroxybutyrate (see the legend to Table II) were similar to the Km values for 3-oxoacid CoA transferase and 3-hydroxybutyrate dehydrogenase which are about 0.29 mM (Tildon and Sevadalian, 1972) and 0.6 mM (Meinzel and Hammers, 1973) respectively. However, the rates through these enzymic steps *in vivo* are less than one-tenth of the Vmax values. One explanation could be a restricted entry of ketone bodies into the brain from the blood. Crone and Gjedde (1973) found evidence, in adult rats, for a carrier mediated facilitated diffusion of acetoacetate and 3-hydroxybutyrate across brain capillaries. Although these authors did not indicate rates of transport, a carrier mechanism may play a regulatory role in ketone body metabolism.

In contrast to the enzymes that convert acetoacetate and 3-hydroxybutyrate to acetyl-CoA, the maximal activity of pyruvate dehydrogenase, as determined *in vitro*, is very close to the estimated flux *in vivo* through this enzymic step (Table II). Properties of this enzyme will be discussed by other contributors and will be mentioned only briefly here. The enzyme is present in the brain, as in other tissues, partly as an active dephosphorylated form and partly as an inactive phosphorylated form (Seiss *et al.*, 1971; Cremer and Teal, 1974). In a study of the

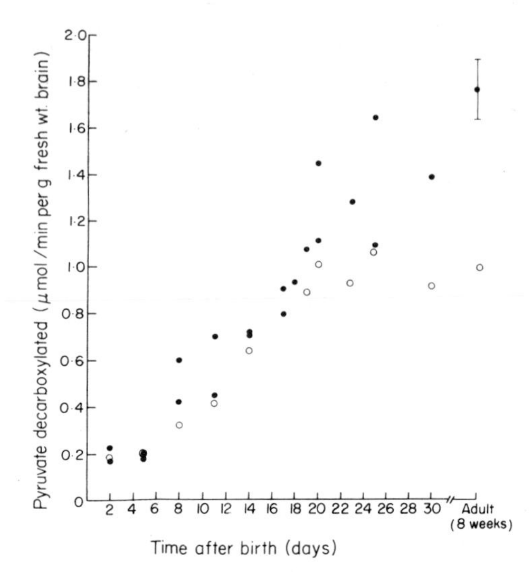

Fig.1 Pyruvate dehydrogenase activities in rat brain. ○, 'active' enzyme assayed in homogenates of rapidly frozen brains; ●, 'total activity' assayed after incubation for 30 min at 37° in 10 mM $MgCl_2$. Data are from Cremer and Teal (1974).

Table II. Estimated metabolic fluxes *in vivo* compared with maximal enzyme activities determined *in vitro* for the 18 day old rat brain.

Reaction catalysed	Enzyme	Flux *in vivo*	Vmax *in vitro*	Literature reference
D-3-hydroxybutyrate→acetoacetate	D-3-hydroxybutyrate dehydrogenase (EC 1.1.1.30)	0.034 to 0.164	1.6 (at 25°)	Krebs *et al.* (1971)
Acetoacetate→acetoacetyl-CoA	3-oxoacid CoA-transferase (EC 2.8.3.5)	0.055 to 0.265	1.1 (at 25°)	Krebs *et al.* (1971)
Acetoacetyl-CoA→2acetyl-CoA	Acetoacetyl-CoA thiolase (EC 2.3.1.9)* mitochondrial	0.055 to 0.265	4.5 (at 30°)	Middleton (1973)
Pyruvate→acetyl-CoA	Pyruvate dehydrogenase System	0.52 to 0.94	1.0 (at 37°)	Cremer & Teal (1974)

Values are given as μmol/min g of brain. Fluxes *in vivo* are from data in Table I; values for individual ketone bodies were calculated using a ratio of 3.0 for blood [3-hydroxybutyrate]/[acetoacetate] and a ratio of 1.85 for the uptake proportionality constant for k[AcAc]/k[3-OHB] applied to the bulked data in Table I. Blood concentrations of acetoacetate were from 0.11 to 0.5 mM and of D-3-hydroxybutyrate from 0.33 to 1.62 mM.

activity in whole rat brain homogenates during post-natal development an almost 10-fold increase was found between neonatal and adult values and the proportion present in the inactive form was greater in the older animals (Fig.1). The increase in activity during brain maturation is similar to that of many other enzymes, but is different from the situation with the ketone body metabolizing enzymes which reach a maximum around 25 days of age and then decline to much lower values in the adult (Krebs *et al.*, 1971). Acetoacetyl-CoA thiolase has been shown by Middleton (1973) to be present in the brain in two forms, one of them associated with

Table III. Activities of acetoacetyl-CoA thiolase and pyruvate dehydrogenase in rat brain during development.

Age	Acetoacetyl-CoA thiolase (a)	Pyruvate dehydrogenase (b)	Ratio (a)/(b)
5-day	1.8	0.3	6.0
10-day	2.5	0.58	4.3
15-day	3.5	0.70	5.0
20-day	4.8	1.05	4.6
25-day	5.5	1.30	4.2
30-day	4.8	1.60	3.0
Adult	2.0	1.80	1.1

Activities are given as μmol/min g of fresh wt. of brain. Values for (a) are from Middleton (1973) for the mitochondrial enzyme; values for (b) are the 'total activities' shown in Fig.1.

mitochondria.Only the mitochondrial enzyme shows a marked increase in activity from 0 to 30 days post-natally, followed by a fall. The other shows a slow decrease between birth and adulthood.

During the simultaneous oxidation of pyruvate, acetoacetate and 3-hydroxybutyrate, all substrates feed into a common pool of acetyl-CoA, as shown by isotopic tracer studies. Acetyl-CoA is formed in the mitochondria by the pyruvate dehydrogenase system and acetoacetyl-CoA thiolase. In the 18 day old rat brain the latter enzyme has a 4.5 times greater maximal activity than pyruvate dehydrogenase. This ratio of activities for the two enzymes is nearly constant during the first few weeks after birth, but in adult animals the activities become almost identical (Table III). It seems very probable that the difference with age in the relative activities of the two enzymes is directly connected with the greater capacity of the developing brain to oxidize preferentially ketone bodies when both they and pyruvate are available.

INCORPORATION OF LABEL INTO BRAIN LIPIDS FROM [^{14}C] GLUCOSE AND [^{14}C] KETONE BODIES

The question of the importance of ketone bodies in providing carbon for lipogenesis during brain development has been raised frequently, but quantitative estimates are lacking from the literature. We have attempted to obtain some measure of the rate of lipid synthesis from glucose and ketone bodies in the 18 day old rat. Values from typical experiments are given in Table IV.

The rates are those estimated from isotopic data and the dilution of the radioactive pyruvate formed from [^{14}C] glucose by unlabelled blood lactate has been taken into account. The rates are low compared to those for oxidation (Table I), but are not unreasonable if they represent net accumulation of lipid in the brain, which is known to occur during the developmental period. Of interest is the observation that the estimated rates of incorporation into lipids were in the ratio, pyruvate/ketone bodies, 2.25 ± 0.38 (S.E.M.), whereas for the rates of formation of the acetyl-CoA oxidized the ratio was 4.9 ± 0.3 (S.E.M.) at the mean blood concentration of ketone bodies (see Table I and allow for the formation of 2 mol of acetyl-CoA from 1 mol of ketone body).

Table IV. Labelling of brain lipids in 18 day old rats after injection of [^{14}C] glucose or [^{14}C] 3-hydroxybutyrate.

Compound injected	Min after injection	Total lipids 10^{-2} X dpm/g of brain	*Mean of lipogensis μmol/min g	Overall mean rate (± S.E.M.)
[^{14}C] glucose	3	1.05 (1)		
	5	2.32 (2)	0.011 (3 to 5 min)	
	7	5.72 (1)	0.018 (3 to 7 min)	0.018
	10	12.33 (1)	0.022 (3 to 10 min)	± 0.002
	16	19.35 (1)	0.108 (3 to 16 min)	
3-hydroxy [^{14}C] butyrate	3	6.10 (4)		0.008
	6	15.80 (4)	0.008 (3 to 6 min)	± 0.001

Suckling rats received either 3.3 μCi of [2-^{14}C] glucose or 1 μCi of D-3-hydroxy [3-^{14}C] butyrate per 45 g body wt by intraperitoneal injection. Lipogenesis was calculated from the equation $R = [q(t2) - q(t1)] / \int_{t_1}^{t_2}$ (S.A. of precursor) dt. The precursors were brain pyruvate and blood ketone bodies; the mean S.A.-t curves in 10^{-3} x dpm/μmol were, respectively:
S.A. pyruvate = $12.5e^{-0.025(t-1)} - 13.6e^{-0.275(t-1)} + 1.1e^{-3.0(t-1)}$
S.A. Ketone bodies = $59.3\ [e^{-0.084(t-0.5)} - e^{-0.80(t-0.5)}]$.

* Rates of lipogenesis are rates of incorporation of either pyruvate or ketone body moities.

If incorporation of carbon from ketone bodies into lipids occurred entirely via acetyl-CoA formed within mitochondria then no difference would be expected between these two ratios. The relatively higher rate of lipid labelling from ketone bodies can be explained by the finding that, following an injection of [^{14}C] 3-hydroxybutyrate the acetoacetate in blood rapidly becomes labelled. Within 3 minutes the ratio of specific radioactivity of blood acetoacetate/blood 3-hydroxybutyrate was found to become constant at 0.605. This means that acetoacetate entering the brain is highly labelled and it can be converted directly to acetoacetyl-CoA extramitochondrially by the enzyme acetoacetyl-CoA synthetase, described by Buckley and Williamson (1973) and thence into cholesterol and lipids by the routes outlined in the scheme of Williamson and Buckley (1973). Devivo *et al.* (1973) reported that when [^{14}C] 3-hydroxybutyrate was injected into 15 day old rats the relative labelling of brain cholesterol to proteins gave a higher ratio than the one obtained following an injection of [^{14}C] glucose. They suggested that acetoacetyl-CoA might be transported across the mitochondrial membrane to account for the high labelling of cholesterol. This postulate is unnecessary if, as seems more likely, the blood acetoacetate in their animals was labelled as well as 3-hydroxybutyrate.

The isotopic data provide evidence for the frequently suggested role of ketone bodies as sources of carbon for cholesterol synthesis and lipogenesis. However, quantitatively the contribution to total brain lipids is probably less than that made by glucose.

CONCLUDING REMARKS

Perhaps the most interesting observation from the more recent animal studies described in this chapter, which might be of relevance to young children with certain inborn errors of metabolism, is the rate-limiting activity in the brain of pyruvate dehydrogenase. Evidence from the isotopic tracer studies that the extra-mitochondrial pathway for lipid synthesis from ketone bodies is functional in developing brain, may also have some clinical significance.

The absolute quantities of glucose and ketone bodies utilized by the human infant brain may be of less importance than the flexibility to use different amounts of each class of substrate in response to available nutrients. If this flexibility is lost through defects in any of the usual regulatory mechanisms, then the brain could suffer. We still do not know which are the most important regulatory factors, but a few processes have been suggested here based on biochemical data obtained in young rats.

REFERENCES

Buckley, B.M. and Williamson, D.H. (1973). *Biochem. J.* **132**, 653-656.

Cremer, J.E. (1971). *Biochem. J.* **122**, 135-138.

Cremer, J.E. and Heath, D.F. (1974). *Biochem. J.* **142**, 527-544.

Cremer, J.E. and Teal, H.M. (1974). *FEBS Lett.* **39**, 17-20.

Crone, C. and Gjedde, A. (1973). *Acta Physiol. Scand.* **87**, 48A-49A.

Crone, C. and Sørensen, S.C. (1970). *Acta Physiol. Scand.* **80**, 47A.

Devivo, D.C., Fishman, M.A. and Agrawal, H.C. (1973). *Lipids* **8**, 649-651.

Gottstein, U., Müller, W., Berghoff, H., Gärtner, H. and Held, K. (1971). *Klin. Wschr.* **49**, 406-411.

Hawkins, R.A., Williamson, D.H. and Krebs, H.A. (1971). *Biochem. J.* **122**, 13-18.

Kraus, H., Schlenker, S. and Schwedesky, D. (1974). *Hoppe-Seyler's Z. Physiol. Chem.* **335**, 164-170.

Krebs, H.A., Williamson, D.H., Bates, M.W., Page, M.A. and Hawkins, R.A. (1971). *Adv. Enzyme Regul.* **9**, 387-409.

Meinzel, H.M. and Hammes, G.G. (1973). *J. biol. Chem.* **248**, 4885-4889.

Middleton, B. (1973). *Biochem. J.* **132**, 731-737.

Oldendorf, W.H. (1972). *European Neurol.* **6**, 49-55.

Oldendorf, W.H. (1973). *Am. J. Physiol.* **224**, 1450-1453.

Persson, B., Settergren, G. and Dahlquist, G. (1972). *Acta Paediat. Scand.* **61**, 273-278.

Seiss, E., Wittman, J. and Wieland, O. (1971). *Hoppe-Seyler's Z. Physiol. Chem.* **352**, 447-452.

Tildon, J.T. and Sevdalian, D.A. (1972). *Arch. biochem. Biophys.* **148**, 382-390.

Van den Berg, C.J. (1973). *In:* "Inborn Errors of Metabolism" (F.A. Hommes and C.J. Van den Berg, eds) pp.69-79. Academic Press, London and New York.

Williamson, D.H. and Buckley, B.M. (1973). *In:* "Inborn Errors of Metabolism" (F.A. Hommes and C.J. Van den Berg, eds) pp.81-96. Academic Press, London and New York.

DISCUSSION

Van den Berg: Why would pyruvate dehydrogenase be so close to its maximal activity? Would you think a tissue which is so dependent on its energy supply would be so critically provided with one of the essential enzymes?

Cremer: I fully sympathize with your query which agrees with our own thinking. When we first measured pyruvate dehydrogenase and did not correlate the *in vitro* and *in vivo* activities we were not so concerned. When we tried to compare the two, we did worry. We checked our assay conditions, and we are still somewhat concerned about them. The absolute values may be the minimal for the *in vitro* activity, but as far as I know one has found substantially higher activities. Even if the activity should be double, it would not make much difference to your argument.

Greengard: Did you use optimal substrate concentrations in your assay?

Cremer: Yes. The enzyme is known to be inhibited by the end product, acetyl-CoA. We added transacetylase to remove the acetyl-CoA. Another critical point in the assay is the concentration of magnesium, but this is no problem when the total activity of the enzyme is assessed rather than the proportion in the active form.

Tyson Tildon: I have one question, which grows out of the fact that maybe glucose oxidation is inhibited by ketone bodies somewhere prior to pyruvate formation. Your suggestion that pyruvate dehydrogenase was the point of difficulty would suggest that lactate should be an energy source for the brain. Or is there a single direction of flux for this substrate?

Cremer: I think the latter is true. The point is that the actual movement of lactate molecules is very rapid in both directions. Your second point, I think, is a correct one: the net movement of lactate is only outwards. Lactate is not an energy source for the brain.

Clark: You said that when pyruvate is produced in excess of requirement lactate is pushed out of the brain. Can you actually account for all of the pyruvate in that fashion? In other words, is there any evidence that pyruvate is also utilized by other routes, such as via pyruvate carboxylase?

Cremer: That is a good question. We cannot show a net output of lactate by our radioactive data. What we are able to do is to link these data with the A.V. difference measurements across the brain as it has been studied by Hawkins *et al.* (*Biochem. J.* 122, 13 (1971)). The difference between glucose uptake and lactate output accounts very well for the complete oxidation of the remaining glucose and ketone bodies. It does not mean that some of the glucose is not going to end up somewhere else, but this will be small because the overall net balance is in very good agreement. You suggested that there could be CO_2 fixation into pyruvate and net generation of four carbon units. This may well be occurring. We have not found in our data a high specific radioactivity of aspartate, which would be a consequence of this, but Van den Berg has (this volume, p.211). This may be related to the different developmental stages and species of animal used. If it was occurring in 18 day old rat, it was below the limits of accuracy of our measurements.

Snell: I am interested in what happens at an earlier stage of neonatal development. At this time the enzymes for ketone body utilization in brain are still relatively low, as also is the pyruvate dehydrogenase, while ketone body substrates are present at high concentrations by one day after birth. The questions I have are: Have you measured *in vivo* rates with your labelling techniques at earlier age, say five days, and what are the relative rates of pyruvate dehydrogenase flux and the flux through the ketone body enzymes at this age?

Cremer: I appreciate your question. I think the enzymes of ketone bodies utilization could function sufficiently to cope with energy demands at an earlier age. But we have not tested this yet in relation to fluxes *in vivo*.

Snell: Did you measure the activity of pyruvate dehydrogenase and the enzymes of ketone bodies utilization during early development? How do they compare?

Cremer: Yes, and the ratio of acetyl-CoA thiolase to pyruvate dehydrogenase is very constant, about 4 to 6, from a few days after birth until weaning. We find levels of pyruvate dehydrogenase about 1.5 times higher than you reported but I do not think that this is a significant difference.

THE EFFECTS OF HYPERKETONEMIA ON NEONATAL BRAIN METABOLISM

J.T. Tildon, P.T. Ozand and M. Cornblath

Department of Pediatrics, University of Maryland School of Medicine Baltimore, Maryland, USA
Rosewood Hospital Center Research Laboratories, Owings Mills Maryland, USA

Ketosis is known to be associated with a variety of inborn metabolic abnormalities. In most instances, however, a direct relationship between the enzymatic or hormonal defect and the increased concentrations of ketone bodies is not understood. At the present time, the role of ketone bodies in normal metabolic homeostasis remains to be defined; however, there is good evidence that these processes are different in the neonate and the adult.

In the adult, physiological ketosis occurs after prolonged fasting in both man (Owen *et al.*, 1967) and animals (Williamson *et al.*, 1971; Tildon *et al.*, 1971). In the neonate the levels of ketone bodies are higher than those found in adults and they are unrelated to blood glucose concentration (Page *et al.*, 1971; Bailey and Lockwood, 1973). More important, however, the rate of ketone body utilization by brain in the adult is directly proportional to the circulating concentration of these metabolites (Williamson *et al.*, 1971). On the other hand, the metabolism of ketone bodies by the neonatal rat brain is related both to the concentrations of these substrates and the activities of ketone utilizing enzymes (Bailey and Lockwood, 1973).

Studies using animals have shown a unique developmental pattern for the key enzymes of ketone body utilization including β-hydroxybutyrate dehydrogenase (Page *et al.*, 1971; Klee and Sokoloff, 1967), CoA transferase (Tildon *et al.*, 1971; Page *et al.*, 1971) and an isoenzyme of acetoacetyl CoA thiolase (Middleton, 1971). Furthermore, the increased activities occur at a time when there is an increased rate of utilization of ketone bodies (Hawkins *et al.*, 1971; Itoh and Quastel, 1970).

In man, the rate of ketone body utilization by the neonatal brain is also higher than in the adult brain (Persson *et al.*, 1972; Kraus *et al.*, 1974) and again, this increase appears to involve factors other than increased plasma concentrations of substrates (Kraus *et al.*, 1974). Of all the tissues examined, the human brain appears to have the highest activities of the enzymes for ketone body utilization

Table I. The blood ketone body concentrations and brain protein contents of control and hyperketonemic pups.

	Blood ketone bodies (mg %)		Brain protein content (mg/g brain weight)	
Days of Age	Control	Hyperketonemic	Control	Hyperketonemic
1	2.6 ± 0.4	11.9 ± 0.2*		-
2	1.8 ± 0.4	15.4 ± 0.4*	70.2 ± 1.5	73.2 ± 2.5
3	-	9.9 ± 0.3	-	-
5	2.5 ± 0.4	12.9 ± 0.3*	75.5 ± 3.0	72.5 ± 1.5
7	2.3 ± 0.3	15.4 ± 0.5*	-	-
9	2.1 ± 0.3	13.9 ± 0.6*	-	-
15	1.8 ± 0.3	13.4 ± 0.7*	84.0 ± 2.5	81.0 ± 1.5
18	-	11.2 ± 0.5*	-	-

Results are given as the mean ± S.E. Each value represents 6 to 10 animals.
* P value < 0.05 for hyperketonemic pups when compared to controls.

Table II. The concentrations of glucose-6-phosphate and fructose 1,6 diphosphate in the developing brain.

	G-6-P μmoles/g protein		FDP μmoles/g protein	
Days of Age	Control	Hyperketonemic	Control	Hyperketonemic
Newborn	4.82 ± 0.45	5.78 ± 0.50	1.41 ± 0.52	2.86 ± 0.92
1	4.89 ± 0.45	3.33 ± 0.18*	1.60 ± 0.73	1.21 ± 0.69
2	3.92 ± 0.23	2.10 ± 0.42*	1.82 ± 0.18	0.68 ± 0.29*
3	3.20 ± 0.48	1.98 ± 0.39*	1.92 ± 0.63	0.33 ± 0.18*
5	2.74 ± 0.30	2.44 ± 0.22	3.63 ± 1.01	0.56 ± 0.24*
7	1.83 ± 0.35	2.44 ± 0.35	3.58 ± 0.88	0.30 ± 0.04*
9	1.88 ± 0.34	1.86 ± 0.14	3.27 ± 0.73	0.20 ± 0.05*
15	0.91 ± 0.08	1.26 ± 0.08	2.19 ± 0.54	0.34 ± 0.10*
18	0.49 ± 0.12	1.02 ± 0.15	2.87 ± 0.63	0.36 ± 0.05*

Results are given as the mean ± S.E. Each value represents 8 to 10 animals for each age group.
* P < 0.05 for hyperketonemic pups as compared to controls.

(Tildon and Cornblath, 1972; Page and Williamson, 1971). These observations are in agreement with Lockwood and Bailey's (1971) conclusion that the brain is a major site of ketone body utilization during infancy.

Recently, results from our laboratory have indicated a profound influence of hyperketonemia on the neonatal pup (Dierks-Ventling, 1971; Cornblath *et al.*, 1972; Wapnir *et al.*, 1973). These experiments utilized the brains of fetal and neonatal rat pups from dams fed a 45% fat diet during the last eight days of gestation. This model takes advantage of the fact that dams fed a "high" fat diet develop ketosis, that ketone bodies cross the placenta (Scow *et al.*, 1964), and that higher concentrations of these substrates appear in fetal than in maternal circulation (Melichar *et al.*, 1967).

Among the results was the finding that following a 24 h fast, a significant decrease in brain glycogen occurred in the hyperketonemic pups as compared to the controls (Wapnir *et al.*, 1973). This observation prompted a detailed study of the effect of hyperketonemia on glycolysis and related metabolic processes. The purpose of this report is to present the results of these experiments which suggest a new "regulator" role for ketone bodies in developmental homeostasis.

The experimental design for producing hyperketonemic pups and the analytical methods for measuring the enzyme activities and intermediates have been described (Dierks-Ventling, 1971; Cornblath *et al.*, 1972; Wapnir *et al.*, 1973; Ozand *et al.*, 1974). In the present study, the metabolic intermediates, glucose-6-phosphate (G-6-P), fructose 1,6 disphosphate (FDP), lactate and pyruvate were measured. Two citric acid cycle intermediates, malate and α-ketoglutarate and two amino acids, glutamate and glutamine were also measured. In conjunction with the measurement of glutamine and glutamate, the activities of glutamine synthetase, glutaminase and glutamic dehydrogenase in these brains were determined. In addition, the phosphorylated compounds, ATP, ADP, AMP, and creatine phosphate were also measured.

The results provide evidence of the striking difference between the metabolic processes in the brains of hyperketonemic as compared to control animals. The results in Table I confirm the earlier findings that circulating ketone bodies in neonatal rat pups can be increased by feeding pregnant dams a 45% fat diet in the last trimester (8 days) of pregnancy and during lactation. Since the concentration of ketone bodies in the normal suckling rat are 3 to 5 times higher than adults (Williamson and Buckley, 1973), the further elevation in the concentration of ketone bodies in these experimental animals has been designated as hyperketonemia. Newborn pups from the dams fed the ketogenic diet had birth weights similar to those of controls (5.44 ± 0.06 versus 5.47 ± 0.08 g), and the mean number of pups in a litter was also essentially the same (10.3 ± 0.5 versus 10.6 ± 0.7). In addition, content of protein in the brain of the experimental animals was comparable to that in control animals at 2, 5 and 15 days of age (Table I).

During the first weeks of life, there was an increase in cerebral glycolysis, and this adaptation was reflected in the changes in concentrations of several glycolytic intermediates. In the brains of control animals, the concentration of FDP increased from 1.4 μmoles/g protein at birth to 2.9 μmoles/g protein at 18 days of life (Table II). In contrast, G-6-P concentrations decreased from 4.8 to 0.49 during the same period (Table II). Therefore, the ratio of FDP/G-6-P increased ten-fold during this period (Fig.1). This increase was consistent with an enhanced capacity of the brain to utilize glucose in agreement with the findings of Itoh and Quastel (1970) and Moore *et al.* (1971).

In the brains of hyperketonemic pups, the concentrations of FDP and G-6-P were significantly different from controls, and the ratio of FDP/G-6-P decreased from 0.5 at birth to 0.2 by the 9th day of life (Fig.1). These changes are similar to those observed by Newsholme *et al.* (1962) who concluded that ketone bodies inhibited glucose utilization at the site of phosphofructokinase.

The concentrations of lactate and pyruvate were also affected by elevated levels of ketone bodies (Table III). During the first week of life, the increased lactate and decreased pyruvate resulted in a change in the ratio of these two metabolites such that it rose significantly in hyperketonemic animals to values more than 70

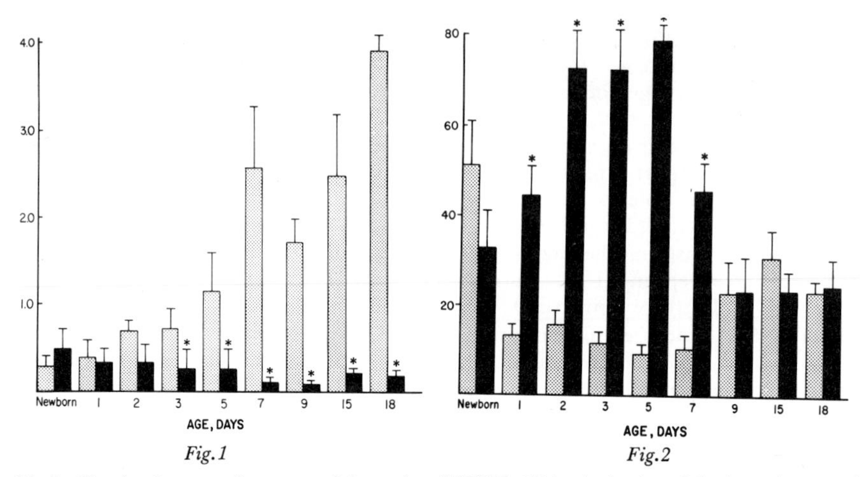

Fig.1

Fig.2

Fig.1 The developmental pattern of the ratio of FDP/G-6-P in the brains of the hyperketonemic and control rats. The solid bars represent the mean value ± S.E. for hyperketonemic animals and the speckled bars represent the mean value ± S.E. for control animals. (*) denotes a P value < 0.05.

Fig.2 The developmental pattern of the ratio of lactate/pyruvate in the brains of hyperketonemic and control rats. Symbol denotations are the same as Fig.1.

Table III. The concentrations of pyruvate and lactate in the developing rat brain.

	Pyruvate μmoles/g protein		Lactate μmoles/g protein	
Days of Age	Control	Hyperketonemic	Control	Hyperketonemic
Newborn	2.82 ± 0.24	3.23 ± 0.97	100.9 ± 19.1	87.2 ± 8.5
1	4.51 ± 0.91	1.38 ± 0.48*	37.2 ± 3.1	73.3 ± 17.7*
2	3.62 ± 0.72	1.79 ± 0.59*	35.1 ± 2.5	78.5 ± 15.5*
3	3.77 ± 0.76	1.66 ± 0.33*	37.7 ± 5.7	53.0 ± 13.0*
5	4.68 ± 0.72	1.07 ± 0.22*	32.1 ± 4.9	66.8 ± 12.0*
7	3.87 ± 0.96	1.09 ± 0.32*	46.1 ± 14.4	36.3 ± 2.5
9	3.13 ± 0.53	2.09 ± 0.37	63.1 ± 12.3	39.8 ± 8.0
15	2.93 ± 0.34	2.69 ± 0.46	47.8 ± 6.7	40.0 ± 5.5
18	3.01 ± 0.43	2.66 ± 0.32	51.3 ± 5.5	62.7 ± 7.1

Results are given as the mean ± S.E. Each value represents 8 to 10 animals for each age group.
* P < 0.05 for hyperketonemic pups when compared to controls.

(Fig.2), which was 3 to 5 times higher than the ratio found in the brain of control pups. In the controls, the ratio decreased from birth to a value of about 15 by the first day of life, and it remained relatively constant during the next 3 weeks (Fig.2). Changes in the lactate/pyruvate ratio in the presence of increased ketone bodies have been reported for other tissues, but the values observed in the brains of hyper-

Table IV. Malate and α-ketoglutarate concentrations in the developing rat brain.

	Malate µmoles/g protein		α-ketoglutarate µmoles/g protein	
Days of Age	Control	Hyperketonemic	Control	Hyperketonemic
Newborn	6.4 ± 0.9	7.1 ± 1.0	3.6 ± 0.3	5.4 ± 1.0
1	3.3 ± 0.7	3.8 ± 0.6	3.9 ± 0.5	1.3 ± 0.4*
2	6.2 ± 0.6	5.6 ± 0.9	3.9 ± 0.7	2.3 ± 0.5*
3	5.4 ± 0.2	5.8 ± 0.2	4.0 ± 0.8	2.4 ± 0.4*
5	7.2 ± 0.9	6.4 ± 1.0	5.8 ± 1.0	2.6 ± 0.6*
7	6.9 ± 1.0	5.4 ± 0.3	5.6 ± 1.0	2.9 ± 0.3*
9	6.5 ± 0.6	7.3 ± 0.3	4.0 ± 0.5	0.6 ± 0.2*
15	10.4 ± 1.1	10.7 ± 1.1	3.8 ± 0.6	0.9 ± 0.2*
18	10.5 ± 0.7	12.1 ± 0.2	4.8 ± 1.2	2.0 ± 0.3*

Results are given as the mean ± S.E. Each value represents 8 to 10 animals for each age group.
* $P < 0.05$ for hyperketonemic pups as compared to controls.

ketonemic pups were almost twice as high as the values found in the liver of alloxan diabetic rats (Krebs, 1966). These data suggest that elevated ketone bodies cause a shift toward anaerobic glycolysis resulting in an increased lactate production and an increased redox potential.

It seems reasonable to postulate that the alteration in the glycolytic pattern in hyperketonemic animals during development is the result of more than one factor since the increased lactate/pyruvate ratio was only seen during the first week of life, while the increased FDP/G-6-P ratio remained throughout the course of the three week experiment. Itoh and Quastel (1970) have shown that acetoacetate caused a decreased glucose oxidation in the brain of five day old rats. It has also been demonstrated that glycolysis was controlled by the concentrations of citrate and acetyl CoA (England and Randle, 1967; Randle *et al.*, 1968). Therefore, the alteration in glycolysis reported here may be due to an increased supply of acetyl CoA resulting from the increased concentration of ketone bodies. The effect might also be due to a delayed maturation of one or more rate-limiting enzymes of glycolysis, and experiments are now in progress to determine whether phosphofructokinase or other enzymes of glycolysis are altered in hyperketonemic pups.

Two tricarboxylic acid cycle intermediates, malate and α-ketoglutarate, were measured to determine whether the observed alteration extended into this portion of the glycolytic pathway. From the first to the eighteenth days of life, there was a three-fold increase in the concentration of malate in the brains of both hyperketonemic and control pups with no significant difference between these two groups. In contrast, the concentrations of α-ketoglutarate in the hyperketonemic pups were only 1/5 to 1/2 those of controls during the entire developmental period of 18 days (Table IV).

Since glutamate and glutamine are immediately related to α-ketoglutarate, and these amino acids have several functions in the brain, including a role as possible energy sources, they were also measured. The results demonstrate that glutamine and glutamate concentrations were significantly different in the brains of hyper-

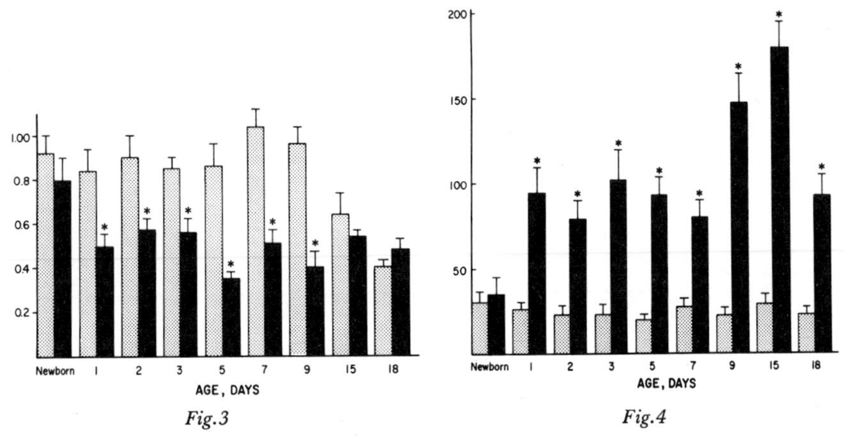

Fig.3 The developmental pattern of the ratio of glutamine/glutamate in the brains of hyperketonemic and control rats. Symbol denotations are the same as Fig.1.

Fig.4 The developmental pattern of the ratio of glutamate/α-ketoglutarate in the brains of hyperketonemic and control rats. Symbol denotations are the same as in Fig.1.

Table V. Glutamate and glutamine concentrations in the developing rat brain.

	Glutamate μmoles/g protein		Glutamine μmoles/ protein	
Days of Age	Control	Hyperketonemic	Control	Hyperketonemic
Newborn	86 ± 3	79 ± 6	78 ± 5	60 ± 4*
1	95 ± 4	99 ± 5	87 ± 6	54 ± 6*
2	111 ± 6	135 ± 6*	96 ± 6	53 ± 7*
3	101 ± 5	128 ± 3*	85 ± 7	50 ± 7*
5	119 ± 12	133 ± 11	80 ± 11	35 ± 7*
7	94 ± 7	121 ± 2*	97 ± 9	49 ± 5*
9	90 ± 8	120 ± 4*	83 ± 5	41 ± 6*
15	112 ± 8	132 ± 3*	64 ± 3	49 ± 6
18	118 ± 7	138 ± 8	44 ± 4	57 ± 5

Results are given as the mean ± S.E. Each value represents 8 to 10 animals for each age group. * $P < 0.05$ for hyperketonemic pups as compared to controls.

ketonemic animals as compared in control animals (Table V). When the ratio of glutamine to glutamate in the developing brain of hyperketonemic pups was compared to that of controls, the ratio was found to be significantly reduced during the first weeks of life (Fig.3). In contrast, the ratio of glutamate to α-ketoglutarate in the experimental animals was three to five-fold higher than the ratio in control animals during the same period of development (Fig.4). These findings indicate that glutamine and glutamate metabolism during development is altered as a consequence of increased concentrations of ketone bodies. The results also

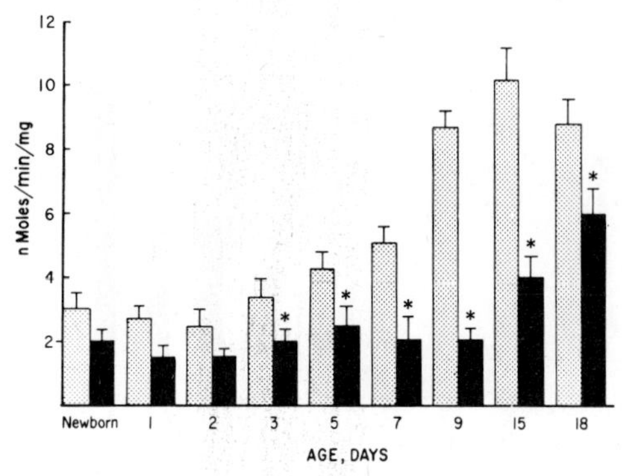

Fig.5 The developmental pattern of glutamic dehydrogenase activity in the brains of hyperketonemic and control animals. Symbol denotations are the same as in Fig.1.

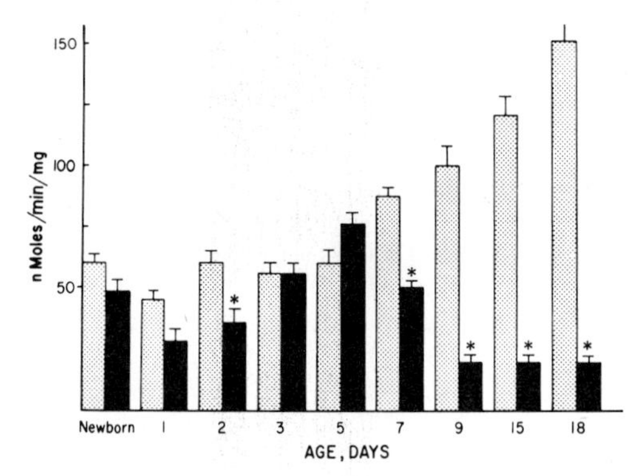

Fig.6 The developmental pattern of glutaminase activity in the brains of hyperketonemic and control animals. Symbol denotations are the same as in Fig.1.

suggest that the decreased concentation of α-ketoglutarate was related to changes in the metabolic flux through the α-ketoglutarate-glutamate axis rather than a change in the flow through the citric acid cycle.

It has been shown that the activities of glutamine synthetase, glutaminase and glutamic dehydrogenase (GDH) are directly related to the glutamate and glutamine pools in the developing rat brain during the first weeks of life (Bayer and McMurray, 1967; Prosky and O'Dell, 1972). Therefore, these enzymes were also assayed in homogenates from these brain samples.

The specific activity of glutamic dehydrogenase increased during development from a 3 nmoles/min/mg protein at birth to a level of 10 nmoles/min/mg protein at 15 days of life (Fig.5). In the brains of hyperketonemic animals this maturation was delayed (Fig.5), but the enzyme activity appeared to approach control levels by the 18th day. In the brains of control pups the specific activity of glutaminase gradually increased after the 5th day of life some three-fold until it was about 90% of the adult activity by the 18th day of life (Fig.6). In hyperketonemic pups the specific activity of glutaminase decreased after the 5th day of life, and it re-

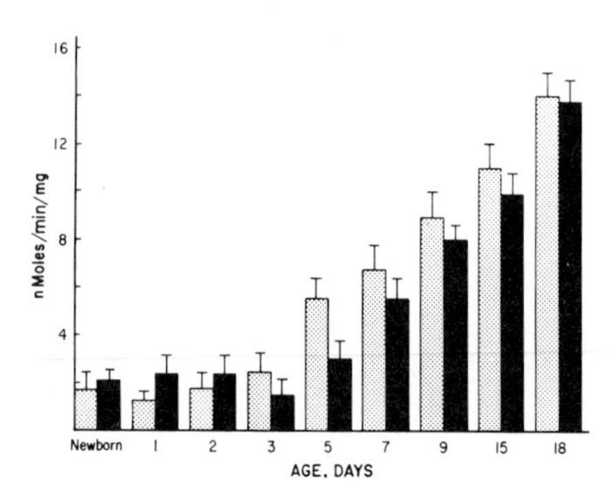

Fig. 7 The developmental pattern of glutamine synthetase activity in the brains of hyperketonemic and control animals. Symbol denotations are the same as in Fig.1.

Table VI. High and low energy phosphate concentrations in rat brain.

	ATP + CP µmoles/g protein		ADP + AMP µmoles/g protein	
Days of Age	Control	Hyperketonemic	Control	Hyperketonemic
Newborn	90.0 ± 4.2	89.0 ± 6.5	11.2 ± 1.4	12.4 ± 3.5
1	103.9 ± 10.8	97.6 ± 9.5	13.5 ± 2.1	7.5 ± 1.9*
2	102.5 ± 9.1	97.5 ± 11.1	14.2 ± 1.4	5.8 ± 1.5*
3	107.0 ± 7.7	105.7 ± 8.8	16.2 ± 1.5	12.1 ± 2.1
5	97.0 ± 13.8	88.1 ± 9.2	11.2 ± 2.9	18.6 ± 6.4
7	90.3 ± 4.8	83.3 ± 9.3	6.5 ± 1.3	15.8 ± 3.1
9	96.8 ± 11.7	87.0 ± 3.9	7.6 ± 2.0	9.8 ± 2.9
15	87.3 ± 5.6	99.8 ± 5.4	15.0 ± 1.7	13.9 ± 4.0
18	83.2 ± 5.3	79.0 ± 5.9	9.8 ± 1.9	11.2 ± 2.4

Results are given as the mean ± S.E. Each value represents 8 to 10 animals for each age group.
* $P < 0.05$ for hyperketonemic pups as compared to controls.

mained significantly lower than controls during the next two weeks (Fig.6). It would appear from these results that the change in the glutamine-glutamate pool in the brain may result at least in part from the delayed maturation of two enzymes. On the other hand, the reduced ratio of glutamine/glutamate may be responsible for the altered enzyme pattern. These changes seemed to be relatively specific since the developmental pattern of glutamine synthetase was not altered by increased exposure to ketone bodies (Fig.7).

Measurements of ATP, ADP, AMP, and creatine phosphate were also included in these experiments. For convenience the values for ATP and creatine phosphate were combined and ADP and AMP were combined (Table VI). There were no significant differences in the concentrations of high energy phosphates in the brain due to hyperketonemia, and the ADP and AMP were reduced only on the first and second days of life (Table VI). Although the data presented above indicate

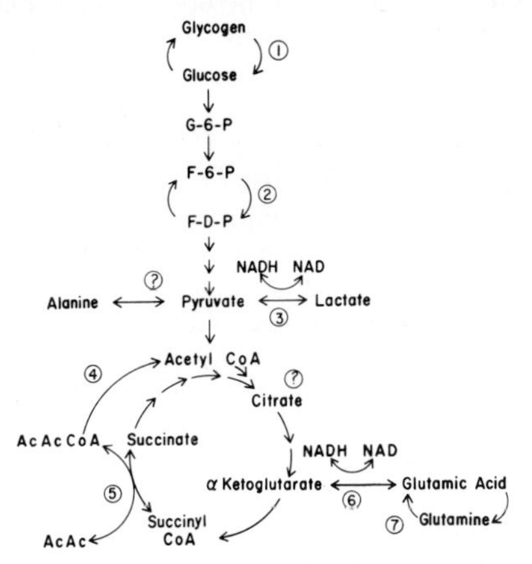

Fig. 8 Metabolic sites affected by ketone bodies in the developing rat brain. (1) Glycogen breakdown (Wapnir *et al.*, 1973). (2) Conversion of F-6-P to FDP. (3) Interconversion of pyruvate and lactate. (4) Thiolase (Dierks-Ventling, 1971). (5) CoA transferase (Dierks-Ventling, 1971). (6) Glutamic dehydrogenase. (7) Glutaminase; (?) are sites which are questionably affected by ketone bodies.

that increased levels of ketone bodies cause a significant alteration in the metabolism of both glucose and glutamine, it is apparent that these changes occur with little or no decrease in the available energy stores (Table VI).

It is apparent from these data that increased concentrations of ketone bodies cause striking changes in intracellular homeostasis. These findings suggest that ketosis as a disease condition may have broader consequences than have been considered in the past. In particular, these results may help to explain some of the puzzling aspects of a unique form of ketoacidosis due to a deficiency of CoA transferase (Tildon and Cornblath, 1972; Spence *et al.*, 1973). In those reports, it was concluded that the absence of this enzyme caused a decreased utilization of ketone bodies by peripheral tissue resulting in a persistent ketonemia and intermittent episodes of severe ketoacidosis. It was also demonstrated that in conjunction with this enzyme deficiency there was a loss of intracellular homeostasis and a reversible inhibition of glucose utilization (Tildon and Cornblath, 1972; Tildon *et al.*, 1971; and Tildon, 1973). The similarity of a decreased glycolysis resulting from hyperketonemia and the decreased glycolysis associated with the absence of a key enzyme of ketone metabolism suggest that comparable mechanisms may be involved. Since a deficiency in CoA transferase could lead to the intracellular accumulation of acetoacetate and β-hydroxybutyrate, it seemed reasonable to speculate that these substrates or related metabolites could inhibit glycolysis. Support for this hypothesis is offered by the earlier findings of Newsholme *et al.* (1962) who showed that ketone bodies caused a decrease in glucose utilization in heart tissue. Furthermore, the possibility that elevated intracellular levels of ketone bodies causes a decrease in glucose utilization was also suggested by the *in vitro* experiments of Itoh and Quastel (1970) and by the results of Mebane and Madison (1964) who showed a reduction in peripheral glucose utilization during ketonemia.

In summary, the concentration of ketone bodies in the neonatal animal is three times that of the adult. While this increase seems to be required for normal tissue development, further elevation causes pronounced alterations in the brain glycolysis and amino acid metabolism.

At least seven sites and perhaps nine (Fig.8) appear to be affected by an increased exposure to ketone bodies. Because of the different patterns in concentrations of the various metabolites, it is apparent that no single mechanism is responsible for all of these changes. The increased mobilization of brain glycogen as a result of starvation (Wapnir *et al.*, 1973) appears to reflect a loss of regulatory control. The increased activities of thiolase and CoA transferase (Dierks-Ventling, 1973) and the decreased activities of GDH and glutaminase seem to reflect changes in protein synthesis. On the other hand, the change in the ratio of lactate/pyruvate seems to result from a shift in intracellular metabolites. The decreased FDP/G-6-P may be affected by both a change in the amount of enzymes and changes in metabolites.

It is interesting to note that two of these sites affect the cytosolic and mitochondrial redox potential in the cell. These results suggest that the NAD/NADH ratio in both the cytoplasma and the mitochondria would be decreased. Since ammonia levels were not measured, an accurate determination of the mitochondrial redox potential cannot be determined. However, the values for the lactate to pyruvate ratios, which can be used to calculate cytosolic redox potential, are similar to literature values (Newsholme and Start, 1973). The shift toward a lower NAD/NADH ratio as a result of increased ketones in these experiments is similar to the decrease observed by Williamson, Lund and Krebs (1967) for the adult animal under ketotic conditions. Indeed, on several occasions the cytosolic redox potential falls much lower than the values for alloxan diabetic rats. This may reflect the difference in metabolism of ketone bodies between adult and neonatal, or it might be explained by the differences in the tissues examined in these two experiments, the brain as opposed to the liver.

Although the results of these experiments provide cogent evidence for an intracellular relationship between glucose, amino acids and ketone bodies, they give rise to more questions than they answer. When some of these answers are available, it may help us to understand better many of the inborn errors of metabolism which are associated with ketosis. These include methylmalonicacidemia and hyperglycinemia. Although the enzyme defect is known in both of these diseases, the direct connection between the defect and ketosis remains obscure; and finally, the continued exploration of these questions should provide us with some clues to the myriad of problems associated with ketotic hypoglycemia.

ACKNOWLEDGEMENTS

The collaboration of Drs. Christa Dierks-Ventling and Raul Wapnir in several of the early aspects of these studies is acknowledged. These studies were supported in part by grants from the John A. Hartford Foundation and the National Institute of Child Health and Human Development (HD-035959-06 and HD-06291-03).

REFERENCES

Bailey, E. and Lockwood, E.A. (1973). *Enzymes* **15**, 239-253.
Bayer, S.M. and McMurray, W.C. (1967). *J. Neurochem.* **14**, 695.

Cornblath, M., Gingell, R.L., Fleming, G.A., Tildon, J.T., Leffler, A.T. and Wapnir, R.A. (1971). *J. Pediat.* **79**, 413.
Cornblath, M., Tildon, J.T. and Wapnir, R.A. (1972). *Israel Med. Sci.* **8**, 453.
Dierks-Ventling, C. (1971). *Biol. Neonat.* **19**, 426-433.
England, P.J. and Randle, P.J. (1967). *Biochem. J.* **105**, 907.
Hawkins, R.A., Williamson, D.H. and Krebs, H.A. (1971). *Biochem. J.* **122**, 13-18.
Itoh, T. and Quastel, J.H. (1970). *Biochem. J.* **116**, 641-655.
Klee, C.B. and Sokoloff, L. (1967). *J. biol. Chem.* **242**, 3880-3883.
Kraus, H., Schlenker, S. and Schwedesky, D. (1974). *Hoppe-Seyler's Z. Physiol. Chem.* **355**, 164-170.
Krebs, H.A. (1966). *Adv. Enzyme Regul.* **4**, 339-353.
Lockwood, E.A. and Bailey, E. (1971). *Biochem. J.* **124**, 249-254.
Mebane, D. and Madison, L.L. (1964). *J. Lab. and Clin. Med.* **63**, 177-192.
Melichar, V., Drahota, Z. and Hahn, P. (1967). *Biol. Neonat.* **11**, 23-38.
Middleton, B. (1971). *Biochem. J.* **125**, 70P.
Moore, T.J., Lione, A.P., Regen, D.M., Tarpley, H.L. and Raines, P.L. (1971). *Amer. J. Physiol.* **221**, 1746.
Newsholme, E.A., Randle, P.J. and Manchester, K.L. (1962). *Nature* **193**, 270-271.
Newsholme, E.A. and Start, C. (1973). "Regulation in Metabolism". John Wiley & Sons, New York.
Owen, O.E., Morgan, A.P., Kemp, H.G., Sullivan, J.M., Herrera, M.G. and Cahill, G.F. (1967). *J. clin. Invest.* **46**, 1589-1595.
Ozand, P.T., Stevenson, J.H., Tildon, J.T. and Cornblath, M. Submitted to *J. Neurochem.*
Page, M.A., Krebs, H.A. and Williamson, D.H. (1971). *Biochem. J.* **121**, 49-53.
Page, M.A. and Williamson, D.H. (1971). *Lancet* **2**, 66-68.
Persson, B., Settergren, G. and Dahlquist, G. (1972). *Acta Paediat. Scand.* **61**, 273-278.
Prosky, L. and O'Dell, R.G. (1972). *J. Neurochem.* **19**, 1405.
Randle, P.J., Denton, R.M. and England, P.J. (1968). *In:* "Metabolic Roles of Citrate" (T.W. Goodwin, ed) p.87. Academic Press, London and New York.
Scow, R.D., Chernick, S.S. and Brinley, M.S. (1964). *Amer. J. Physiol.* **206**, 796-804.
Spence, M.W., Murphy, M.G., Cook, H.W., Ripley, B.A. and Embil, J.A. (1973). *Ped. Res.* **2**, 394.
Thaler, M.M. (1972). *Nature New Biol.* **236**, 140-141.
Tildon, J.T. (1973). *Proc. natn. Acad. Sic. USA* **70**, 210.
Tildon, J.T. and Cornblath, M. (1972). *J. clin. Invest.* **51**, 493-498.
Tildon, J.T., Cone, A.L. and Cornblath, M. (1971). *Biochem. biophys. Res. Commun.* **43**, 225-231.
Tildon, J.T., Leffler, A.T., Cornblath, M. and Stevenson, J. (1971). *Ped. Res.* **5**, 518.
Wapnir, R.A., Tildon, J.T. and Cornblath, M. (1973). *Amer. J. Physiol.* **224**, 489.
Williamson, D. H. and Buckley, B.M. (1973). *In:* "Inborn Errors of Metabolism" (F.A. Hommes and C.J. Van den Berg, eds) pp.81-96. Academic Press, London and New York.
Williamson, D.H., Lund, P. and Krebs, H.A. (1967). *Biochem. J.* **103**, 514.
Williamson, D.H., Bates, M.W., Page, M.A. and Krebs, H.A. (1971). *Biochem. J.* **121**, 41-47.

DISCUSSION

Gaull: I wonder whether one can equate the neonatal rat with the neonatal human. They are really at quite different stages of development. The neonatal rat is probably somewhat closer to a second trimester human in terms of the development of the brain. Did you measure γ-amino butyric acid?

Tildon: No, we did not.

Gaull: You showed that there was an increase in uric acid in your patient. Did the patient become very dehydrated during episodes of ketosis?

Tildon: I think so, yes.

Gaull: That might be the answer.

Tildon: That is a possibility.

Tager: I was particularly interested in the effects of hyperketonemia on the levels of glutamine, glutamate and α-oxoglutarate in brain and on your speculation about what the cause might be. In the very last part of your talk you mentioned the effect of hyperketonemia on the redox state of the mitochondria and the cytosol. Is it possible that the primary effect of hyperketonemia, in the mitochondria at least, might in fact be on the mitochondrial redox state, which would influence the glutamate to α-oxoglutarate ratio in the way you showed experimentally and which would, in turn, influence the glutamate/glutamine ratio?

Tildon: I think your inference is quite correct. The regrettable thing is that we did not have the ammonia assays. I think we will still have to do the ammonia, because I will not feel comfortable unless the two steady-state equilibria agree, that is to say the calculation from both, the β-hydroxybutyrate dehydrogenase as well as glutamic dehydrogenase.

Van den Berg: How was the condition of these animals?

Tildon: They were perfectly healthy, unexpectedly so. We have no indication now whether or not there are behavioural changes.

Van den Berg: It seems to be a general rule that when the level of glutamic acid in the brain changes more than 20% there is generally something very wrong in the whole brain. The changes you observed are very large. One therefore would expect these animals to have some severe brain problem.

Tildon: We are now looking in more detail at the behaviour of these animals, but as I mentioned already no severe problems were seen at day 18.

Wick: In medicine it is something like a dogma that a shift in excess lactate, a shift in the redox state, is, practically always, connected with some difficulty in maintaining a proper oxygen tension in the tissue. Did you have measurement of PO_2?

Tildon: No, we have not.

Van den Berghe: Do you have any idea of the mechanism by which ketone bodies interfere with the degradation of glycogen?

Tildon: As a matter of fact, I think, that is a puzzle for which we do not have an answer.

Tager: Did you measure citrate levels?

Tildon: We only measured citrate levels in the fibroblasts and they were somewhat reduced, but not as markedly as α-oxoglutarate.

Snell: Is there any chance that this activation of glycogenolysis by ketone bodies might be physiological, in the sense that the brain builds up stores of glycogen before birth and that after birth the glycogen levels fall off. Could this coincide with the natural rise in hyperketonemia after birth?

Tildon: When you say physiological, I am attracted to it, except that we are dealing with a pathological state in view of the levels of ketone bodies present. The normal newborns have an elevated ketone level and do not mobilize their brain glycogen as a consequence of starvation. So I think that there is an extra message given somewhere, as opposed to it being simply a physiological process.

THE CHANGING PATTERN OF BRAIN MITOCHONDRIAL SUBSTRATE UTILIZATION DURING DEVELOPMENT

J.M. Land and J.B. Clark

Department of Biochemistry, St. Bartholomew's Hospital Medical College University of London, Charterhouse Square, London

INTRODUCTION

Arteriovenous differences of the cerebral blood supply may be performed by examining the carotid arterial blood and internal jugular venous blood, samples of which may be taken easily whilst a subject is under only local anesthesia (Myerson *et al.*, 1927). Many such measurements have now been made of the adult human cerebral blood supply (for review see Sokoloff, 1973) and the consensus has been that in the normal conscious state, glucose and oxygen are the only substrates taken up in significant quantities by the brain and further that the respiratory quotient of the brain is close to unity (Table I). Attempts to illustrate other positive arteriovenous differences have invariably foundered although Adams *et al.* (1955) did suggest that glutamate has a positive arteriovenous difference, though this finding has never been confirmed by others.

Mainly from the work of Cahill's group (Owen *et al.*, 1967) it has become clear that the human adult brain may in times of hypoglyceamia caused by starvation, adapt and utilize ketone bodies as a primary substrate. More recently it has been established that the normal immature rat brain utilizes both ketone bodies and glucose as substrates (Williamson and Buckley, 1973; Page *et al.*, 1971) as does the human immature brain (Kraus *et al.*, 1974).

Studies of cerebral oxygen consumption during development (Fazekas *et al.*, 1941) have shown that in the rat oxygen consumption is relatively low at birth, shows a marked increase at approximately one week after birth and rises rapidly to values approaching adult rates by 3 weeks of age. Further studies of the changes in the total cerebral activity for glycolytic and tricarboxylic acid cycle enzymes show a similar developmental pattern, rising to a maximal level by 3 weeks of age which is retained throughout adult life (for a review see Benjamins and McKhann, 1972).

However the mitochondrial enzymes associated with ketone body utilization:- D-3-hydroxybutyrate dehydrogenase, 3-oxoacid CoA transferase and acetoacetyl CoA thiolase, have a somewhat different developmental pattern. Although maxi-

Table I. Cerebral substrate utilization by man and by rat.

	μmoles/g wet weight/min	
Man	Adult*	Adult starved (38-41d)†
Oxygen	1.56	1.32
Carbon dioxide	1.46 – 1.56	0.82
Respiratory quotient	0.93 – 1	0.62
Glucose	0.31	0.07
Ketone bodies	< 0.005	0.205
Rat	Adult+	Adult starved (4 days)+
Oxygen	2.84	N.D.
Carbon dioxide	2.70	N.D.
Respiratory quotient	0.95	N.D.
Glucose	0.51	0.43
Ketone bodies	0.026	0.16

ND = not determined. Taken from: * Sokoloff (1973); † Owen *et al.* (1967); + Hawkins *et al.* (1971) and Moore *et al.* (1971).

Table II. Incorporation of (2-^{14}C) glucose and DL-3-hydroxy(3-^{14}C) butyrate into aspartate and glutamate in the 10 day old and adult mice brain.

		Specific radioactivity of aspartate / Specific radioactivity of glutamate	
	Precursor time (min)	2-(^{14}C) glucose	DL-3-hydroxy (3-^{14}C) butyrate
10 day old	5	1.32	0.31
	10	0.83	0.83
	15	0.70	0.82
Adult	5	0.53	0.73
	10	0.69	0.94
	15	0.73	0.73

The results are taken from Van den Berg (1973).

mal activity is attained at approximately three weeks of age, in common with the enzymes of glycolysis and the tricarboxylic acid cycle, subsequently the ketone body utilizing enzymes fall by approximately 60% to give their adult levels (Page *et al.*, 1971; Middleton, 1973), a phenomenon not inconsistent with the observed differences in substrate utilization as illustrated by arteriovenous difference measurements made across the brain during maturation.

An interesting and as yet unexplained observation which should be considered in the light of the previous comments are the results of some elegant experiments by Van den Berg (1970, 1973) who has studied the incorporation of radioactivity as a function of time from 2-^{14}C glucose and DL-3hydroxy-(3-^{14}C) butyrate into asparate and glutamate by adult and immature cerebral tissue (Table II). A probable explanation of the observed differences between mature and immature animals is that pyruvate derived from the metabolism of glucose is being predominantly carboxylated to oxaloacetate and that it is 3-hydroxybutyrate which is giving rise to the major proportion of acetyl-CoA in the immature brain and not pyruvate. From a consideration of both this work and that of Williamson and Buckley (1974), Van den Berg at the previous symposia in this series (p.94) raised the question: "Why is the brain using 3-hydroxybutyrate in the young animal? Just because it is present in the blood? Well glucose is present."

In an attempt to provide some answeres to this question, it was felt necessary to investigate the rates of oxidation of pyruvate and 3-hydrobutyrate by cerebral mitochondria during maturation. Further to examine the cerebral levels during development and some of the properties of the two enzymes responsible for the incorporation of carbon skeleton from pyruvate into tricarboxylic acid cycle intermediates, whose development did not appear to have been examined before, notably: pyruvate dehydrogenase and pyruvate carboxylase.

EXPERIMENTAL

Materials

The sources of all materials are given in Clark and Land (1974) and Land and Clark (1974).

Animals

Rats of Wistar strain were used throughout. The birth dates of all litters were recorded after daily inspection and litters culled to 8-10 pups. Animals of either sex were used up to 21 days of age and after this date only male rats were used. The animals were fed *ad libitum* on Laboratory Diet Nos. 1 (Spratts, U.K.) and drinking water was freely available at all times.

Methods

Mitochondria were prepared by the method of Clark and Nicklas (1970). Homogenates of brain were prepared in 0.25 M sucrose, 10 mM tris-HCl pH 7.4 and 0.5 mM EDTA-K^+ salt, and all enzyme activities in this preparation were determined within one hour of sacrificing animals.

Enzyme Assays

Pyruvate dehydrogenase and citrate synthase were determined as described previously (Clark and Land, 1974). Pyruvate carboxylase was assayed as described by Martin and Denton (1970).

Mitochondrial Respiration

Mitochondrial respiration was determined as described by Land and Clark (1974) in high (100 mM) and low (5 mM) potassium media as described by Clark and Nicklas (1970).

RESULTS

Table III shows the respiratory activities in State 3 and respiratory control ratios of adult brain mitochondria respiring in media containing either 5 or 100

Table III. The oxidation of substrates by adult rat brain mitochondria.

Added substrate	K^+ concentration in medium			
	5 mM		100 mM	
	State 3	R.C.R.	State 3	R.C.R.
None	0	–	0	–
5 mM Pyruvate	0	–	0	–
+ 2.5 mM Malate	93	5.5	183	8.0
+ 2.5 mM Malate + 0.2 μM FCCP	106	–	196	–
2.5 mM Malate	21	1.4	25	1.6
+ 5 mM Citrate	66	3.9	92	1.9
+ 4 mM Isocitrate	40	4.0	86	3.0
+ 7.5 mM 2-oxo-glutarate	46	2.9	104	2.9
+ 10 mM DL-3-hydroxybutyrate	30	3.0	60	3.0
+ 2 mM Glutamate	65	6.0	90	3.5
+ 10 mM Glutamate	N.D.	N.D.	159	5.0
10 mM Succinate	122	3.6	158	1.8
+ 5 mM Malonate	20	–	0	–
+ 5 mM Malate	66	–	83	–
0.5 mM NADH	0	–	0	–
10 μM Rotenone	0	–	0	–
+ 10 mM DL-glycerophosphate	52	1.0	24	1.0

Abbreviations: FCCP, carbonylcyanide-p-trifluoromethoxy phenyl hydrazone; N.D., not determined; R.C.R., respiratory control ratio. All rates expressed as n atoms oxygen/min/mg of protein. Data taken from Clark and Nicklas (1970), Land and Clark (1974) and unpublished observations.

Table IV. Metabolite levels and kinetic data related to pyruvate and ketone body metabolism in the adult rat brain.

Metabolite	n moles/g wet weight		Km mitochondrial oxygen uptake (μM)	Km for primary dehydrogenase (μM)
	Fed	Starved (48 h)		
Lactate	1890	1480	–	–
Pyruvate	80	50	50 – 60	80
Acetoacetate	40	180	–	–
3-hydroxybutyrate	90	200	500 (D-isomer)	1000 (D-isomer)

Data taken from Hawkins *et al.* (1971), Clark and Land (1974) and unpublished observations.

mM K^+. The respiration rates were assessed in both these media since it has been known for some time that K^+ ions have a stimulatory effect upon the respiration of brain preparations (Dickens and Greville, 1935; Kini and Quastel, 1959) and brain mitochondrial preparations (Ozawa *et al.*, 1967; Clark and Nicklas, 1970).

Although pyruvate alone was not oxidised appreciably, in the presence of malate it gave rise to the highest rates of oxygen uptake observed in both State 3 and the uncoupled state. Indeed, it is very appropriate that pyruvate, a product of glycolysis, is oxidised rapidly in view of the known high rate of glucose utilization by the brain (Table I).

Similarly, malate alone in State 3 conditions gave rise to a low level of respiration. Succinate was oxidised at rates approaching those observed with pyruvate/malate and the oxidation was malonate sensitive. Citrate, isocitrate, 2-oxoglutarate and glutamate were also all actively oxidised in the presence of malate (Table III). Further, these mitochondria oxidise α-glycerophosphate, suggesting that α-glycerophosphate shuttle may be of importance in the cerebral tissue for the transport of reducing equivalents from the cytoplasm to the mitochondrion.

Of particular interest also is the ability of these brain mitochondria isolated from the brains of fed adult rats to oxidise 3-hydroxybutyrate at significant rates. Further, as may be seen from Table III and Fig.2, the oxidation is coupled to oxidative phosphorylation as judged by the respiratory control of 3 and the P/O ratio of 2.8. This contrasts with the report of Klee and Sokoloff (1967) who, although able to demonstrate coupling of oxidative phosphorylation to succinate and 2-oxoglutarate oxidation in mature and immature brain mitochondria, were unable to demonstrate the coupling of 3-hydroxybutyrate oxidation in mature brain mitochondria. These authors did, however, stress that their experiments were not designed to achieve optimal P/O ratios.

Finally, the apparent Km of these mitochondria for pyruvate as measured by oxygen uptake was approximately 50-60 μM whilst that for D-3-hydroxybutyrate was much higher at 0.5 mM. In both cases these parameters were independent of the K^+ ion concentration of the respiratory media, although in both cases the observed V_{max} increased with increasing K^+ ion concentration. It is of particular interest to consider these parameters in parallel with the Michaelis constants of the appropriate substrates dehydrogenases and with the known cerebral concentrations of pyruvate and 3-hydroxybutyrate in various states (Table IV).

In view of the special roles played by glucose and ketone bodies as substrates of cerebral metabolism, the development of the capacity of cerebral mitochondria to oxidise pyruvate and 3-hydroxybutyrate was investigated.

The changes in the capacity of cerebral mitochondria to utilize pyruvate as measured by respiration in State 3 respiratory conditions (Chance and Williams, 1956) is shown in Fig.1. At birth, very low levels of oxygen uptake are observable (3% of adult). However, the rates rise, particularly from 8 days onwards, attaining maximal adult rates by 4-5 weeks of age. Further examination of the rates observed in 5 and 100 mM K^+ media shows that the oxygen uptake is stimulated by K^+ ions during all stages of postnatal development and is not a property which develops at the time of inception of nervous transmission (~14 days).

The oxygen uptake supported by 3-hydroxybutyrate during development shows in contrast to that of pyruvate a markedly different pattern (Fig.2). Respiration rates at birth are again low, but rise to a maximum at about three weeks of age before falling to an adult level approximately 50% of the rate at three weeks of age (Fig.2).

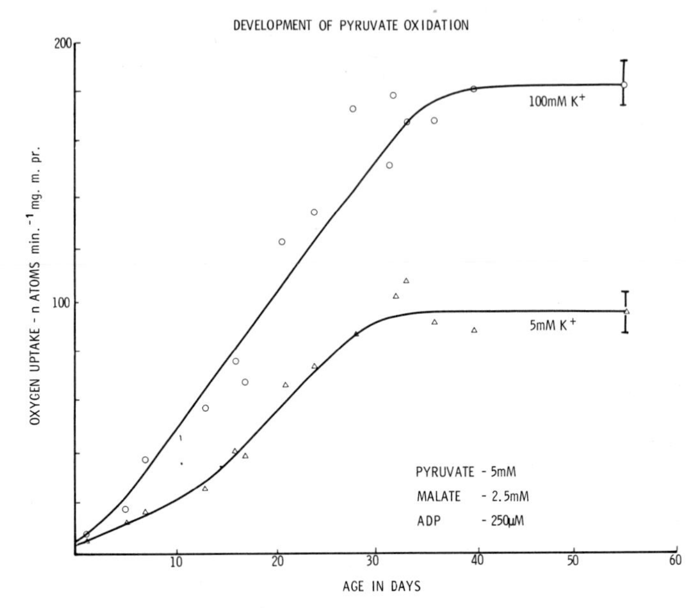

Fig.1 Mitochondria were prepared from the brains of the appropriate age rats by the method of Clark and Nicklas (1970). Respiration was measured in a medium containing 5 mM or 100 mM K^+ (see Experimental) and the substrate was 5 mM pyruvate and 2.5 mM malate. Respiration rates are expressed as n atoms of oxygen min^{-1} per mg. mitochondrial protein. State 3 was induced by the addition of 250 μM ADP and the mean respiratory control ratio was 4.5 in 5 mM K^+ and 5.4 in 100 mM K^+.

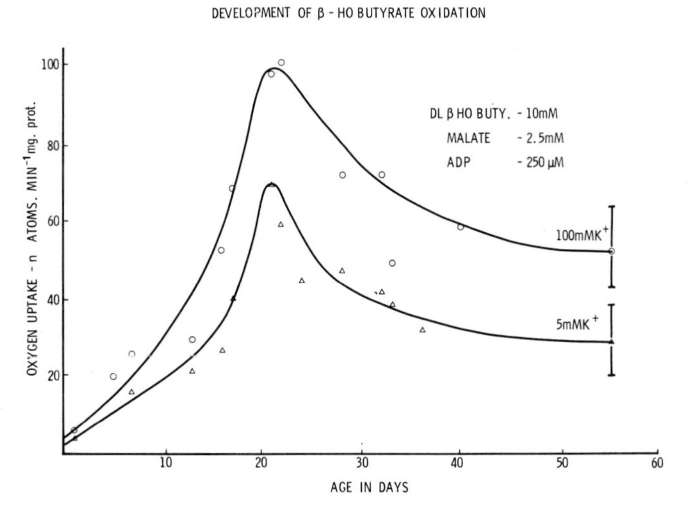

Fig.2 All details are as in the legend to Fig.1 except that 10 mM DL-3-hydroxybutyrate, 2.5 mM malate was the substrate. The mean respiratory control ratio was 3.1 in 5 mM K^+ and 3.5 in 100 mM K^+.

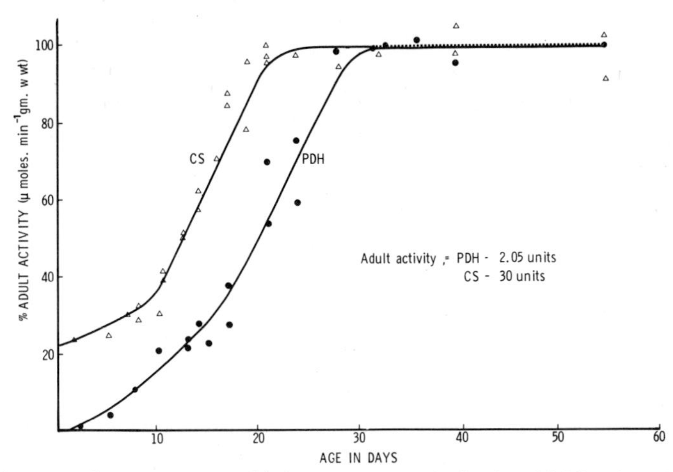

Fig.3 Citrate synthase was measured in homogenates and mitochondrial fractions derived from the brains of the appropriate age rats as described in the Experimental Section. Pyruvate dehydrogenase was assayed by the methods referenced in the Experimental Section in the mitochondrial fractions and the whole tissue level calculated on the basis of the percentage recovery of the mitochondrial fraction derived from the citrate synthase measurements.

Although the activities of the cerebral mitochondrial enzymes associated with ketone body utilization have been extensively studied during development (Klee and Sokoloff, 1967; Robinson and Hall, 1970; Pull and McIlwain, 1971; Page *et al.*, 1971; Middleton, 1973), there is a dearth of reports concerning the development of cerebral enzymes associated with pyruvate metabolism.

The development of rat cerebral pyruvate dehydrogenase during postnatal development is shown in Fig.3 together with, for comparison, that of citrate synthase, an enzyme of the tricarboxylic acid cycle, which shows a 'classical' developmental pattern as observed by Page *et al.* (1971) for glutamate dehydrogenase, Wilbur and Patel (1974) for malate dehydrogenase and aconitase and Gregson and Williams (1969) for succinic dehydrogenase.

Of particular interest is that the onset and the subsequent development of pyruvate dehydrogenase follows by approximately a week that classically observed for tricarboxylic acid cycle enzymes as illustrated here by citrate synthase (Fig.3). At one day of age pyruvate dehydrogenase activity was scarcely detectable, rising to 55% by 3 weeks of age, whilst full expression of activity is not seen until approximately 4 weeks of age, in agreement with the observed development of pyruvate supported oxygen uptake shown by cerebral mitochondria (Fig.1). Additionally, the development of pyruvic dehydrogenase as measured by 1-^{14}C pyruvate decarboxylation (Cremer and Teal, 1974) has also been shown to follow an essentially similar developmental pattern.

Additionally, it was felt necessary to investigate the effect, if any, of acetyl-CoA upon the activity of cerebral pyruvic dehydrogenase in view of this molecule's central role in ketone body and pyruvate metabolism. Blass and Lewis (1973)

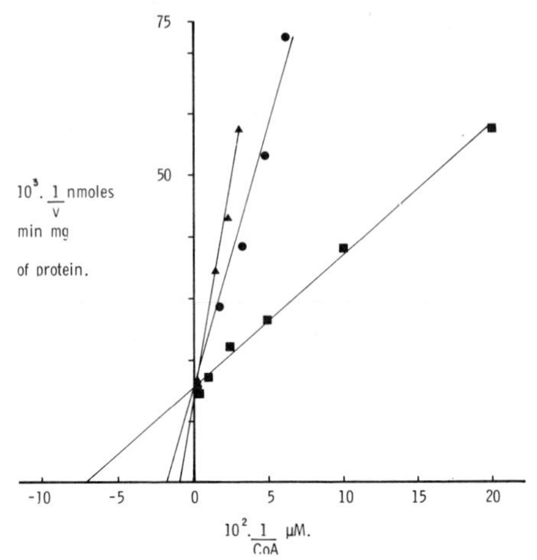

Fig.4 Pyruvic dehydrogenase activity was measured in extracts of adult brain mitochondria prepared by the method of Clark and Nicklas (1970) as described in the Experimental Section in the presence of 0 (■), 39 (●) and 78 (▲) μM acetyl-CoA at varying CoA concentrations.

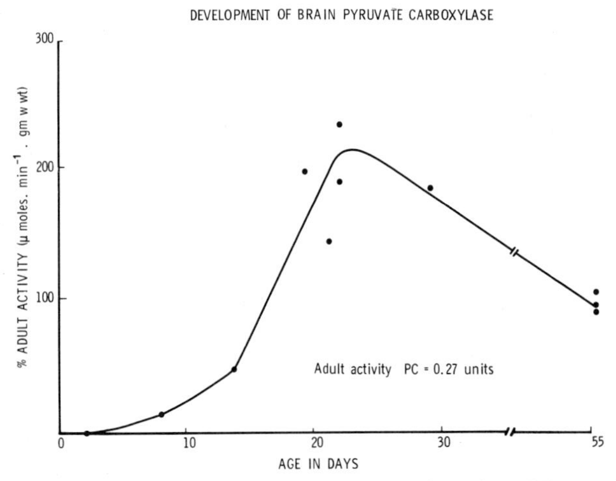

Fig.5 Pyruvate carboxylase was measured by the method of Martin and Denton (1970) within 1 h of preparing homogenates of cerebral tissue from rats of the appropriate age as described in the Experimental Section.

have reported that they were unable to observe consistent inhibition by acetyl-CoA of purified cerebral pyruvic dehydrogenase, contrasting with the report of Siess *et al.* (1971) who had found that acetyl-CoA was a classical competitive inhibitor with a K_i of 9 μM with respect to CoA.

Investigation of the effect of acetyl-CoA upon pyruvic dehydrogenase at differing CoA levels (Fig.4) illustrates that indeed acetyl-CoA acts as a classical competitive inhibitor of the multienzyme complex with a K_i of 19 μM, the Km for CoA being 15 μM, in good agreement with a value of 9 μM reported by Siess *et al.* (1971).

In contrast to the developmental patterns shown by either citrate synthase or pyruvic dehydrogenase, pyruvate carboxylase development mimics that of the mitochondrial ketone body utilizing enzymes, rising markedly in activity between 10 and 20 days, attaining a maximal level before declining in activity to its low adult level (Fig.5).

Wilbur and Patel (1974), however, estimating pyruvate carboxylase activity by (^{14}C) bicarbonate fixation in brain homogenates have found the activity to rise continuously from birth to 35 days of age. The difference between the results reported here and their results may well be a reflection of the assay systems used.

DISCUSSION

The developmental pattern of the ability of cerebral mitochondria to oxidise 3-hydroxybutyrate (Fig.2) is not inconsistent with the observed gross changes in the levels of the enzymes of ketone body utilization (Page *et al.*, 1971; Middleton, 1974). Further, the similarity of the developmental patterns does suggest that although there are striking and marked changes in the enzymic complement of the mitochondria, the proportion of total mitochondrial protein per gram wet weight does not alter significantly during development. A similar conclusion has also been arrived at by Banik and Davison (1969) who studied the development of succinic dehydrogenase and MacDowell and Greengard (1974) working on glutamate dehydrogenase and aspartate aminotransferase.

Considered in the light of the developmental pattern of the ketone body utilization, the development of the enzymes of pyruvate utilization take on an interesting meaning. The results do suggest that the glucose that the immature brain does utilize (Table I) in fact is not oxidised and utilized directly via the tricarboxylic acid cycle as in the adult rat but is preferentially guided into performing an 'Anaplerotic Role' (Kornberg, 1966) to supplement the loss of citric acid cycle intermediates used for biosynthetic purposes in the immature brain.

The probable control mechanisms involved are:

1) The development of pyruvic dehydrogenase markedly lags behind that of other citric acid cycle enzymes and the enzymes of ketone body utilization, therefore restricting the potential flux of two carbon skeleton derived from pyruvate by decarboxylation into the citric acid cycle.

2) The acetyl-CoA generated intramitochondrially from the utilization of ketone bodies further limits the potential flux of pyruvate through pyruvate dehydrogenase by inhibiting the complex, whilst at the same time stimulating pyruvate flux via pyruvate carboxylase to provide oxaloacetate.

3) The comparatively higher ratio of pyruvate carboxylase to pyruvate dehydrogenase observed in cerebral tissue during the first three weeks of life as compared to the adult state further encourages the utilization of pyruvate in an anaplerotic as opposed to a catabolic role.

A diagrammatic representation of the proposed schemes for the interaction of pyruvate and ketone body metabolism are given in Figs 6 and 7 for the immature and mature brain respectively. Within the context of these models one can now adequately explain the results of the experiments of Van den Berg (1973, Table 2).

2-^{14}C glucose will give rise to 2-^{14}C pyruvate which in the immature brain will

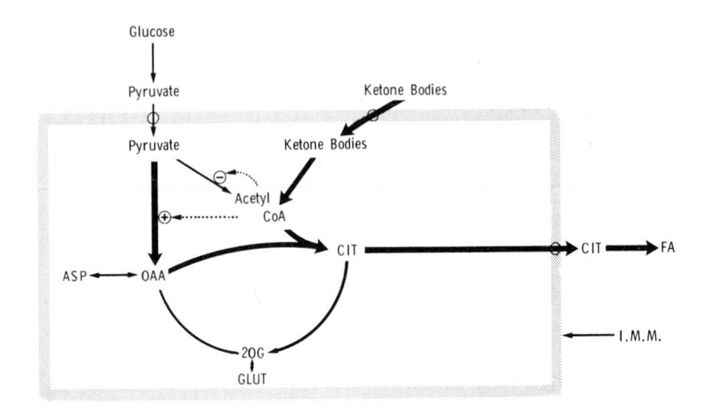

Fig.6

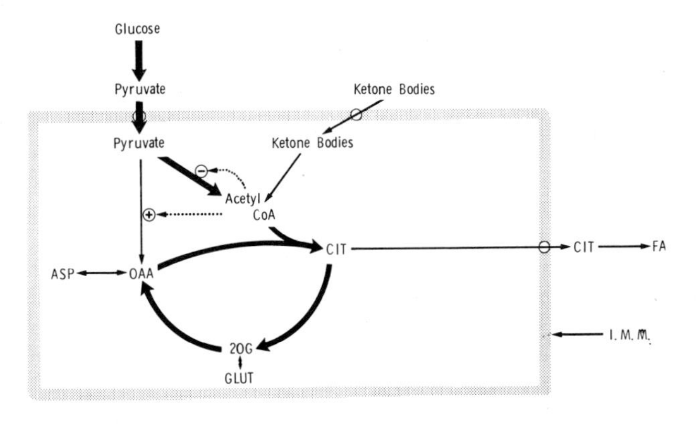

Fig.7

be rapidly incorporated directly into the oxaloacetate pool which itself is in equilibrium with the aspartate pool via aspartate aminotransferase. However, in the adult animal pyruvate will preferentially enter into the tricarboxylic acid via acetyl-CoA and although this should make little difference to the final specific activity of the oxaloacetate/aspartate pool, the time course of its incorporation into this pool will be delayed, as indeed Table II shows. Additionally, there should be no differences using 3-hydroxybutyrate as substrate between the mature and immature brain.

It now appears that ketone body utilization goes far in fulfilling both the energetic and biosynthetic requirements of the immature brain (see Cremer, 1974) whilst glucose through pyruvate, is utilized predominantly in an anaplerotic role. In the adult, on the other hand, except in extreme cases of glucose deprivation (i.e.

starvation), ketone bodies are not utilized though the potential for their utilization is ever present. Hence, in investigating an impairment of energetic and biosynthetic activities in the immature brain as when studying the possible aetiology of phenylketonuria or maple syrup urine disease, it is of importance to consider effects upon both ketone body and pyruvate metabolism (see Clark and Land, 1974).

ACKNOWLEDGEMENT

JML thanks the Science Research Council for a studentship.

REFERENCES

Adams, J.E., Harper, H.A,, Gordon, G.S., Hutchin, M. and Bentinck, R.C. (1955). *Neurol.* **5**, 100.

Banik, N.L. and Davison, A.N. (1969). *Biochem. J.* **115**, 1051-1062.

Benjamins, J.A. and McKhann, G.M. (1972). *In:* "Basic Neurochemistry" (R.W. Albers, G.J. Siegel, R. Katzman and B.W. Agranoff, eds) pp.269-298. Little Brown, Boston.

Blass, J.P. and Lewis, C.A. (1973). *Biochem. J.* **131**, 31-37.

Chance, B. and Williams, G.R. (1956). *Adv. in Enzymol.* **17**, 65-134.

Clark, J.B. and Land, J.M. (1974a). *Biochem. J.* **140**, 21-25.

Clark, J.B. and Land, J.M. (1974b). This volume.

Clark, J.B. and Nicklas, W.J. (1970). *J. biol. Chem.* **245**, 4724-4731.

Cremer, J.E. (1974). This volume.

Cremer, J.E. and Teal, H.M. (1974). *FEBS Letts.* **39**, 17-20.

Dickens, F. and Greville, E.D. (1935). *Biochem. J.* **29**, 1468-1483.

Fazekas, J.F., Alexandre, F.A.D. and Himwich, H.E. (1941). *Am. J. Physiol.* **134**, 281-287.

Gregson, N.A. and Williams, P.L. (1969). *J. Neurochem.* **16**, 617-626.

Hawkins, R.A., Williamson, D.H. and Krebs, H.A. (1971). *Biochem. J.* **122**, 13-18.

Kini, M.M.and Quastel, J.H. (1959). *Nature (Lond.)* **184**, 252-256.

Klee, C.B. and Sokoloff, L. (1967). *J. biol. Chem.* **242**, 3880-3883.

Kraus, H., Schlenker, S. and Schwedesky, D. (1974). *Hoppe-Seyler's Z. Physiol. Chem.* **355**, 164-170.

Land, J.M. and Clark, J.B. (1974). *FEBS Letts.* **44**, 348-351.

MacDowell, P.C. and Greengard, O. (1974). *Arch. biochem. Biophys.* **163**, 644-655.

Martin, D.B. and Denton, R.M. (1970). *Biochem. J.* **117**, 861-877.

Middleton, I. (1973). *Biochem. J.* **132**, 731-737.

Moore, T.J., Lione, A.P., Regen, D.M., Tarpley, H.L. and Raines, P.L. (1971). *Am. J. Physiol.* **221**, 1746-1753.

Myerson, A., Halloran, R.D. and Hirsch, H.L. (1927). *Arch. Neurol. Psych.* **17**, 807.

Owen, O.E., Morgan, A.P., Kemp, H.G., Sullivan, J.M., Herrera, M.G. and Cahill, G.F. (1967). *J. clin. Invest.* **46**, 1589-1595.

Ozawa, K., Seta, K., Araki, H. and Handa, H. (1967). *J. Biochem. (Tokyo)* **61**, 352-358.

Page, M.A., Krebs, H.A. and Williamson, D.H. (1971). *Biochem. J.* **121**, 49-53.

Pull, I. and McIlwain, H. (1971). *J. Neurochem.* **18**, 1163-1165.

Robinson, C.A. and Hall, L.M. (1970). *Fedn Proc.* **29**, **Abs. 3843**.

Siess, E., Wittman, J. and Wieland, O. (1971). *Hoppe-Seyler's Z. Physiol. Chem.* **352**, 447-452.

Sokoloff, L. (1973). *Ann, Rev. Med.* **24**, 271-280.

Van den Berg, C.J. (1970). *J. Neurochem.* **17**, 973-983.

Van den Berg, C.J. (1973). *In:* "Inborn Errors of Metabolism" (F.A. Hommes and C.J. Van den Berg, eds) pp.69-78. Academic Press, London and New York.

Wilbur, D.O. and Patel, M.S. (1974). *J. Neurochem.* **22**, 709-715.

Williamson, D.H. and Buckley, B.M. (1973). *In:* "Inborn Errors of Metabolism" (F.A. Hommes and C.J. Van den Berg, eds) pp.81-96. Academic Press, London and New York.

DISCUSSION

Blass: Could I ask you to comment on two observations. One of them is, that in the original study (Owen *et al.*, *J. clin. Invest.* **46**, 1589 (1967)) where the A.V. differences for brain ketone bodies after prolonged starvation were found, part of the glucose taken up could be accounted for by the production of lactic acid. The second observation is that several patients have been described who do not make ketone bodies on fasting at normal rates, apparently on the basis of an inherited deficiency in the carnitine acyl transferase system (Engel *et al.*, *Science* **179**, 899 (1973)). They have no particular symptoms involving the central nervous system, which is rather surprising in view of the demonstrated potential role of ketone bodies in the brain.

Land: I am afraid I cannot help you.

De Groot: I have a question from the clinical side. It is related to the last four papers. As a clinician, I see children with inborn errors and sometimes the clinical symptoms seem to be more related to special areas of the brain. In cases of pyruvate carboxylase deficiency, for example, it seems to be more related to the brain stem, the midbrain and cerebellum, and in cases of phenylketonuria more to cortical areas. My question is: do you think it is worthwhile to extend your studies to different areas of the brain?

Land: Yes, but the problem of doing biochemical studies is that when one isolates brain mitochondria, or uses brain homogenates, one is looking at a statistical population of whatever. An additional problem of using rats is that their brain is so small, but your point is well taken. The preparation of mitochondria we use, of course, reflects the whole brain of the rat.

Rutter: Has anyone looked at these metabolic phenomena in glial cells and neuroblasts or in minimal deviation glial blastomas and neuroblastomas where the function is presumably near normal? Obviously this question relates to what you were measuring in total brain, which is partially glial and partially neural cells.

Clark: We do have now a preparation of mitochondria which are derived from synaptosomes, which is not answering your question, but these are functionally active and we are in the process of studying their properties.

Greengard: I was wondering if you have estimated what percent of your total mitochondria may come from glial cells.

Land: It is very difficult to do.

Hommes: I was very happy to see the development of pyruvate carboxylase in brain. Could I persuade you to speculate on the theoretical case that pyruvate carboxylase in brain does not develop, as possibly occurs in Leigh's syndrome? What would be the results in terms of regulation of metabolism when the increased conversion of pyruvate to oxaloacetate during a certain stage of development does not take place?

Land: I would not really like to comment on that, but it seems to me that it is a rather drastic thing that you suggest.

Cremer: I just wanted to follow you up on that. What sort of proportions are we talking about in pyruvate utilization via acetyl-CoA formation and via pyruvate carboxylation? Are we talking about 30% or 2% of the pyruvate being carboxylated?

Van den Berg: I calculated the net amount of carbon added to the brain when it grows for one day in a rat as glutamic acid, aspartic acid, aspargine, glutamine and other compounds for which the formation of oxaloacetate is necessary. That comes to a few percent at the most of the total carbon turning around the citric acid cycle. All one can say is that the mechanism to make these compounds is present, but the fluxes in terms of a net addition of material to the brain is not very much.

Land: I agree in a way with Dr. Van den Berg, because effectively what the data show is that the potentiality does exist for an increased flux in a young as opposed to an adult animal.

METABOLIC SPACES IN THE DEVELOPING BRAIN: A CONTINUOUS CHANGING METABOLIC HETEROGENEITY DURING BRAIN DEVELOPMENT

C.J. Van den Berg, G.L.A. Reijnierse and D.F. Matheson

Studygroup Inborn Errors and Brain, Department of Biological Psychiatry University of Groningen, The Netherlands
Department of Biochemistry, University of Leiden, Leiden The Netherlands

Brain function and behaviour is abnormal in variable degree in many inborn errors of the metabolism of amino acids and related compounds in man. In general, the abnormality of impaired function and its associated pathology becomes evident early in the postnatal development. In most or all of these inborn errors little is known about the molecular changes in brain responsible for the pathology shown. The abnormalities in the blood due, for example, to the absence of an enzyme in the liver might lead to changes within the brain of the pattern of synthesis and degradation of a large number of small and large molecules, leading to a disruption in the organization of the brain. Is the brain composed in such a way that it is particularly sensitive to an alteration in the material reaching it? Is the brain incapable of rectifying these changes? One of the basic characteristics of the brain, making it unique to other organs, is its property of learning from experience. The future behaviour of the brain is always dependent to a large extent on the past. The interaction of external and internal events leads to a coordinated pattern of behaviour. The molecular organization of the brain must have properties allowing these adaptations to take place. Inborn errors of metabolism could result in "inadequacy" of this system to grow or mature adequately, often with dramatic consequences.

The macromolecular organization of the brain is of course of an extremely heterogeneous nature, both at the structural and dynamic biochemical level. This heterogeneity is manifest not only at the cellular level, as seen from the complexity of individual cell types such as large and small neurons and the various types of glial cells, but also presumably at the subcellular level, e.g. mitochondria, etc.. To understand the precise biochemical organization in the brain, both in time and space, the choice of biochemical parameters or units to analyse is of paramount importance. In most biochemical investigations the units chosen are: amounts, fluxes or compositions of molecules, fractions or cells. However, there is no reason why these units should necessarily be the most appropriate. The most general unit one can choose could be a "metabolic space". A metabolic space can be defined as a metabolic region in time or space, or both, showing some biochemical property by which it shows itself relative to other spaces. As very likely the metabolic patterns are present in brain to serve some useful purpose, one can add to the definition of metabolic space given that it possesses a function or is a discernible part of a functional complex. The sum of the metabolic spaces present in a tissue is certainly much larger than the total physical space. One part may belong to more than one metabolic space.

A metabolic space is not necessarily confined within a spatial border of individual cell types and can be shared by many different cell types. The problem will be to find those metabolic spaces the properties of which are closely related to some of the brain functions at a higher level. One can expect that as the composition of one or more networks of metabolic spaces is related to the capacity of the brain to learn from the past, small changes in some of the properties of a number of metabolic spaces could well be related to the "memory function" of the brain, or to the apparent ease by which some types of brain damage seem to become permanent.

In the remainder of this chapter we will describe some aspects of the analysis of the brain in metabolic spaces, and provide evidence for the continuous changing composition and properties of some of these spaces during development. The examples chosen are far removed from the theoretical discussion given in this section; nevertheless we feel that this gap may be closed by more experimentation.

MITOCHONDRIAL POPULATIONS DURING DEVELOPMENT

There is abundant evidence that the crude mitochondrial fraction from adult rat brain can be separated into more than two subpopulations (Van Kempen *et al.*, 1965; Neidle *et al.*, 1969; Blokhuis and Veldstra, 1970; Reijnierse, 1973; Van den Berg *et al.*, 1975). Glutamate dehydrogenase, for example, sediments on a sucrose-density gradient towards a higher sucrose-density than NAD-isocitrate dehydrogenase. γ-aminobutyrate transaminase sediments in roughly the same way as glutamate dehydrogenase, but there is no complete overlap between these two enzymes. Enzymes, like citrate synthase, glutaminase and the mitochondrial NADP-isocitrate dehydrogenase have also been found to differ from each other in their sedimentation profiles (Blokhuis and Veldstra, 1970). This extreme heterogeneity of brain mitochondria is very striking. The absence of detailed comparative studies with mitochondria from other sources makes it impossible to say to what extent this extreme heterogeneity is unique for the brain. Some data suggest that this might be the case (Van Kempen *et al.*, 1965).

The study of possible changes in the degree of heterogeneity can possibly shed some light on the significance of this heterogeneity for brain function and, of course, on some of the processes taking place in the brain when it develops from an undifferentiated mass of cells to the highly complex organization of many different cells in the adult tissue. In Fig.1 we have summarized data on the sedimentation behaviour of mitochondrial bound NAD-isocitrate dehydrogenase and glutamate dehydrogenase and the synaptosomally localized lactate dehydrogenase from crude mitochondrial fractions from 4-day-old, 14-day-old and adult rat brain. It can be seen that the distribution patterns for each of the enzymes is a function

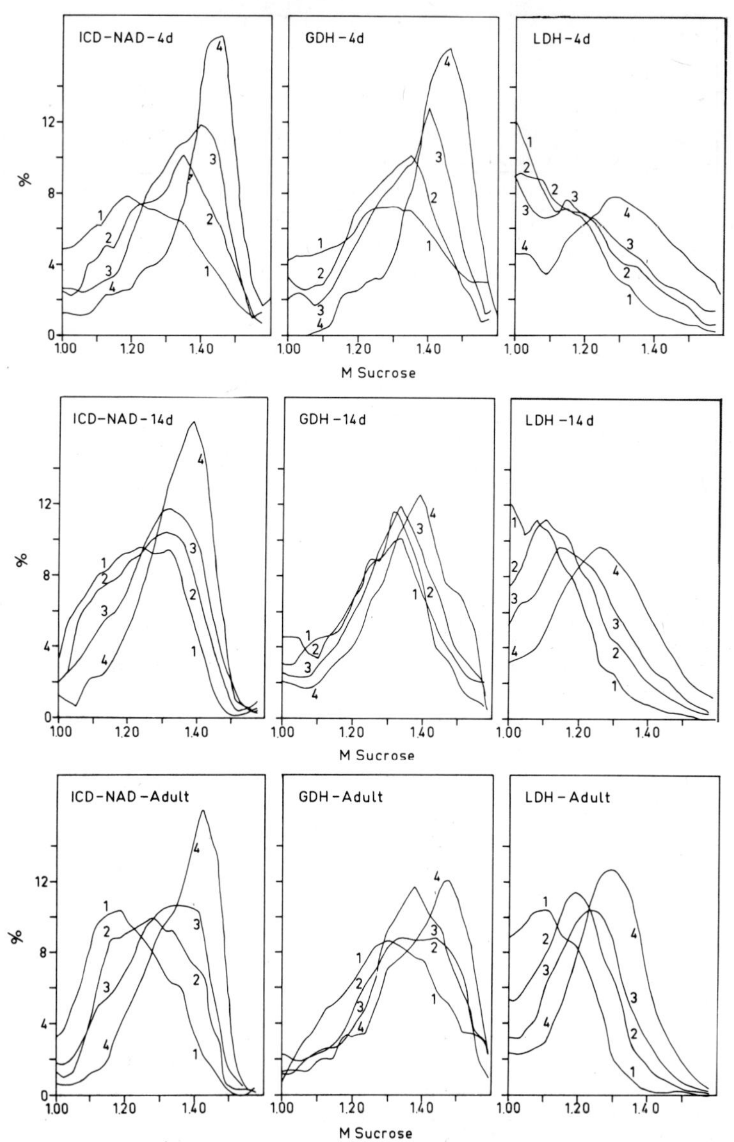

Fig.1 Distribution of NAD-isocitrate dehydrogenase, glutamate dehydrogenase and lactate dehydrogenase in sucrose-density gradients of mitochondrial fractions from the immature and mature rat brain. Crude mitochondrial fractions were layered on sucrose-density gradients and centrifuged for various times at 53,000 g_{av} in a Spinco ultracentrifuge in a SW-27 rotor. After the centrifugation 18 fractions were collected and the three enzymes assayed. The results are expressed as percentages of the activity recovered in the gradient. The points were connected by a smooth line. For a full description see Reijnierse (1973) and Reijnierse, Veldstra and Van den Berg, in preparation. 1 = 1 h, 2 = 5 h, 3 = 4 h and 4 = 20 h centrifugation.

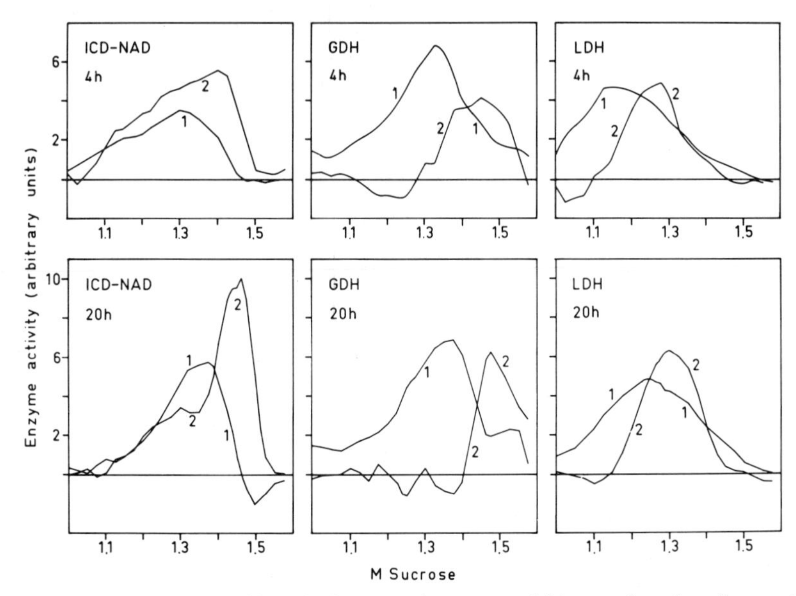

Fig.2 Curves showing the differential increase of enzyme activities as a function of age and of the density. The results were obtained from the data presented in Fig.1, and of the absolute changes in activities of the three enzymes from 4-days to the adult stage. The data are presented as absolute amounts of the three enzymes, in relative units, as a function of the density. The data are given for 4-14-days and for 14- to the adult stage. For further details see Fig.1 and the reference given there. 1: differential increase in enzyme activity from 4 to 14-day and 2: differential increase from 14-day to the adult stage.

of the age of the animal. The quantitative interpretation of sedimentation profiles, especially as a function of age, is very difficult. In order to facilitate a semi-quantitative interpretation we have constructed curves showing the differential changes of each of the enzymes between two subsequent ages, that is from 4-14 day and from 14-day-old to adult. These curves are presented in Fig.2. There is a major increase of glutamate dehydrogenase in the lower density region of the gradient between 4 and 14-day, while there is an increase in the glutamate dehydrogenase in the heavier density region from 14-day onwards. The pattern observed for NAD-isocitrate dehydrogenase is more complex.

Whatever the exact interpretation of these data is, for example, in terms of those properties of the mitochondria which determine their position in a density gradient after an exposure to a gravitational field, one can conclude from the data presented that the composition of the glutamate dehydrogenase and NAD-isocitrate dehydrogenase containing mitochondria changes during development. The changes observed for both particles are not identical. There is also a qualitative different change of the glutamate dehydrogenase containing mitochondria from 4-14 and from 14-the adult stage. Very likely these changes in composition will reflect differences in, for example, the enzyme patterns contained in these mitochondria; this again might indicate that there is during development a continuous change in metabolic patterns in which these enzymes have a function, for example, in the metabolism of glutamate and related amino acids.

METABOLIC PATTERNS *IN VIVO* OF GLUTAMATE METABOLISM

When labelled glucose and acetate are used as precursors of glutamate and related amino acids in adult brain, it is observed that there is a different product-precursor relationship of glutamate and glutamine, glucose producing a normal relationship in which the labelling of the precursor is higher than that of the product glutamine, whilst acetate labelling shows a higher labelling of glutamine with respect to glutamate. This metabolic heterogeneity can be shown to be related to the existence in brain of two or more populations of mitochondria (Neidle *et al.*, 1969; Van den Berg, 1973). Thc mctabolic heterogeneity of glutamate metabolism has been shown to be a function of the age of the animals, both with glucose and acetate as precursors as with leucine as a precursor (Van den Berg, 1970; Patel and Balázs, 1970).

In the adult brain, both acetate and propionate show with respect to their incorporation pattern into glutamate and glutamine a very great similarity (O'Neal *et al.*, 1966). In a study of the developmental changes of acetyl-CoA synthase and of propionyl-CoA synthase, we found that these two enzymes did not increase at the same rate (Reijnierse, 1973; Reijnierse, Veldstra and Van den Berg, in preparation). In view of these data and those presented in the second section of this chapter, we compared the incorporation patterns of acetate and propionate with each other in 10-day-old and adult mice. The results are shown in Table I. It can be seen that the relative specific radioactivity of glutamine is about 4 for the two precursors at 5 min after the injection in the adult animal. In the 10-day-old mice, acetate results in a relative specific radioactivity of glutamine of lower than 2, while that observed for propionate is still about 4, as in the adult brain. Clearly, therefore, these data show that the acetate and propionate metabolic spaces are not identical in the young brain, while they might be identical in the adult brain.

We have argued earlier that glutamate dehydrogenase is involved in those reactions in mice brain which result in a high labelling of glutamine (Neidle *et al.*, 1969). We therefore have investigated the effect of acute ammonia injection on the incorporation of acetate and propionate into glutamate and glutamine. Presumably, ammonia will increase the flux through glutamate dehydrogenase in the direction of glutamate synthesis. We did not observe an increased incorporation of acetate into glutamine in the adult animal, but we did observe an increased incorporation of propionate into glutamine. In the 10-day-old animals, the opposite pattern was found (Matheson and Van den Berg, in preparation).

These data suggest that the metabolic space involved in the metabolism of acetate in the adult brain reacts in another way to ammonia than the metabolic space involved in the metabolism of propionate. Further, the characteristics of these two metabolic spaces, as shown in the experiments with ammonia is a function of the age of the animal. One can expect that when more and more criteria are applied more and more metabolic spaces will appear (see Van den Berg *et al.*, 1975).

GENERAL DISCUSSION

The data presented make it abundantly clear that there are in brain many different metabolic spaces involved in the metabolism of such basic molecules as glutamate and glutamine. Although we do not know much about the function or role of these metabolic spaces, one can expect that these functions do exist. We have

Table I. Incorporation of acetate and propionate into glutamate and glutamine of adult and 10-day-old mice brain.

Relative specific radioactivity of glutamine (specific radioactivity of glutamine/specific radioactivity of glutamate).

Time (min)	10-day-old	Adult
(^{3}H) acetate as precursor		
2	1.92 (0.28)	4.50 (0.41)
5	1.70 (0.44)	3.91 (0.13)
10	1.87 (0.52)	3.38 (0.64)
(1-^{14}C) propionate as precursor		
2	1.8 (0.19)	–
5	4.4 (0.36)	3.47 (0.92)
10	2.68 (0.17)	2.63 (0.05)
15	2.04 (0.36)	1.98 (0.32)

Ten-day-old and adult mice were injected i.p. with the labelled precursors. The animals were killed by immersion in liquid N_2 at the time intervals of the injection indicated in the table. The isolation and quantification was done as described earlier (Van den Berg *et al.*, 1969). Between brackets: SD.3 experiments for the propionate series and from 3-8 for the acetate experiments. Data from Matheson and Van den Berg, in preparation; see also Van den Berg *et al.* (1975).

elsewhere argued the reasons why it is unlikely that these spaces are present in different cell types (Van den Berg *et al.*, 1975). It seems more likely to suppose that they might exist in variable proportion within the same cells. If this interpretation is indeed correct, the existence of these different metabolic spaces indicated a high degree of metabolic heterogeneity within each cell.

What, if any, is the significance in this metabolic heterogeneity? It will be evident from the action of ammonia that the precise mechanism of action can be followed by which the ammonia ion can act on the metabolism of glutamate and the related amino acids. Moreover, if ammonia (or any other compound for that matter) is present over long periods of time, it could be expected that this would have long-lasting effects on the brain. Such effects might well be dependent on the development of heterogeneity, or even on the particular stage of the heterogeneity that has been reached. Such a conclusion, if correct, could well be of profound importance to an adequate understanding of defects occurring in inborn errors of metabolism.

Although this enormous complexity might at first seem confusing and disturbing, or to have only a limited and theoretical meaning, its ramifications when united into a coherent whole may have profound importance. If indeed patterns are more important than simple quantities, such as an amount of flux in a certain relatively large space, it will not be worthwhile to study only these quantities, as is commonly done, but one will have to study patterns by all sorts of means. A more general aspect, which has been mentioned already in the introduction, is that of linking this enormous heterogeneity of the molecular organization of the brain to its role in "memory" in the widest sense possible, which includes perpetuated damage of one kind or another. Reversal of the brain damage in one of the inborn errors does not seem to be possible, certainly not by biochemical means. But one can argue that the improved social-psychological condition of many sufferers from one of the inborn errors is partly the result of "improvements" in the biochemistry of the brain, effected by "reorganization of the molecules" due to a better care of these patients. There is no doubt that many inborn errors impose a severe restriction on the development of the brain and behaviour, but external influences on the biochemistry of the brain are not only the result of changing the biochemistry directly but also by changing the whole interaction of the system with the environment.

REFERENCES

Blokhuis, G.G.D. and Veldstra, H. (1970). *FEBS Letts.* 11, 197-199.
Neidle, A., Van den Berg, C.J. and Grynbaum, A. (1969). *J. Neurochem.* **16**, 225-234.
O'Neal, R.M., Koeppe, R.E. and Williams, E.L. (1966). *Biochem. J.* **101**, 591-597.
Patel, A.J. and Balázs, R. (1970). *J. Neurochem.* 17, 955-971.
Reijnierse, G.L.A. (1973). Ph.D. Thesis, University of Leiden, The Netherlands.
Van den Berg, C.J. (1970). *J. Neurochem.* 17, 973-983.
Van den Berg, C.J. (1973). *In:* "Metabolic Compartmentation in the Brain" (R. Balázs and J.E. Cremer, eds) pp.129-166. MacMillan, London.
Van den Berg, C.J., Matheson, D.F., Reijnierse, G.L.A., Blokhuis, G.G.D., Kroon, M.C., Ronda, G., Clarke, D.D. and Garfinkel, D. (1975). *In:* "Metabolic Compartmentation" (S. Berl and D.M. Schneider, eds) *(in press)*. Plenum Press, New York.
Van Kempen, C.M.J., Van den Berg, C.J., Van der Helm, H.J. and Veldstra, H. (1965). *J. Neurochem.* 12, 581-588.

DISCUSSION

Tager: Most of us will agree with Dr. Van den Berg that there are metabolic spaces in a heterogeneous tissue like brain. We would also agree that these metabolic spaces change during development. The question is, where are these metabolic spaces with regard to mitochondria? Are they mitochondria derived from different areas of the brain or are they mitochondria derived from the same cell?

Van den Berg: There is no definite answer, but there are indications. Each mitochondrial enzyme is present everywhere, for instance you will find glutamic acid dehydrogenase everywhere, but there are different patterns as shown by gradient centrifugations.

Tager: You have then mitochondria from different areas of the brain, mitochondria containing different levels of glutamate dehydrogenase. One area of the brain has a mitochondrial population with predominantly one pattern and another area with another pattern.

Van den Berg: I think one should not use the term 'areas' so easily, but rather 'space'.

Tager: It depends on whether you mean a glial cell or a neuron.

Van den Berg: But glutamate dehydrogenase is found in both glial cells and neurons.

Tager: The ratio of glutamate dehydrogenase relative to another mitochondrial enzyme is the basis for your definition of heterogeneity.

Van den Berg: Yes. The present level of anatomical analysis is still rather crude. There are almost no areas in the brain which contain only glial cells or only neurons. Electronmicroscopy of the central nervous system has shown a regional -

and regional is now applied to very small spaces - specialization of mitochondria. So I think it is a phenomenon occurring within cells.

Greengard: Are you assuming then that there are three or four different types of mitochondria within a single neuron?

Van den Berg: Or maybe many more.

Gaull: I think that the discussion brings up in a specially valuable way the virtues of an interdisciplinary symposium like this. Earlier there was some discussion about ammonia levels being high or low. Dr. De Groot raised what I thought was a very salient question about the fact that inborn errors of metabolism seem to affect functionally different parts of the brain.Now the enormous complexity of the brain in terms of metabolic compartmentation is emphasized. In the field of inborn errors there has been a tendency to simplify, even to the point of obscuring the questions. The blood level of phenylalanine is such and such, and we must lower the blood level of phenylalanine to such and such a concentration so that it will not intoxicate the brain: and these, somehow, are explanations. I think that Dr. Van den Berg has served a very useful function to point out the enormous complexity of even beginning to study the mechanism of inborn errors. I would like to raise one more area of complexity that has been lost sight of sometimes. We speak of inborn errors of metabolism as if they were a single disease. In fact, virtually none of them is. Most patients are really double heterozygotes. For instance, the case of cystathione synthase deficiency, an enzyme with a molecular weight of about 150,000 with about 1000 amino acid residues: even taking into account the fact that you cannot substitute each amino acid in a random way, there are many possibilities of gene mutation and therefore many different alleles, which in turn is likely to result in two different mutant alleles, i.e. double heterozygotes. The finding of the mechanism of inborn errors, or even any one of them, in the present day seems to be many years off. I think we are fooling ourselves if we try to think of it in less complex terms than Dr. Van den Berg has now.

PHENYLKETONURIA AND MAPLE SYRUP DISEASE AND THEIR ASSOCIATION WITH BRAIN MITOCHONDRIAL SUBSTRATE UTILIZATION

J.B. Clark and J.M. Land

Department of Biochemistry, St. Bartholomew's Hospital Medical College University of London, Charterhouse Square, London, UK

Phenylketonuria (PKU), the most commonly occurring inborn error of amino acid metabolism, and Maple Syrup urine disease (MSUD), another example of a genetically linked inborn error of amino acid metabolism, have both clinically and pathologically many similar manifestations. Individuals suffering from either of these conditions show, in the absence of therapy, a number of symptoms, the most evident of which are the impaired synthesis and deposition of myelin and a progressive mental retardation as the individual gets older (Dancis and Levitz, 1972).

It is now reasonably well established that the genetic defect in PKU is the absence of hepatic phenylalanine hydroxylase (EC 1.14.3.1.) (Knox, 1972) which leads to phenylalanine metabolism occurring by transamination and decarboxylation pathways, which although always present, are normally of only minor significance. This leads to the accumulation in both tissues and plasma of abnormally high concentrations of a number of phenylalanine derivatives including phenylpyruvate (2-oxo-3-phenylpropionate) (see Fig.1). Maple Syrup urine disease (MSUD), however, is somewhat less well delineated in its genetic lesion. This condition is characterized by a block in the metabolism of the branched chain amino acids, leucine, isoleucine and valine and is characterized by high urine, plasma and tissue concentrations of those amino acids and their respective analogue 2-keto acids, 2-oxo-4-methylpentanoate (α-keto isocaproate), 2-oxo-3-methylpentanoate (α-keto β-methylvalerate) and 2-oxo-3-methylbutanoate (α-keto isovalerate) (Dancis and Levitz, 1972). This originally led to the suggestion that it was the branched chain 2-oxo-dehydrogenase which was genetically absent (Dancis *et al.*, 1963) but recent evidence indicates that it is probably only the enzyme responsible for decarboxylating 2-oxo-4-methylpentanoate or 2-oxo-3-methylpentanoate that is actually absent and that the accumulation of those 2-oxo acids causes the inhibition of 2-oxo-3-methylbutanoate decarboxylase (Bowden and Connelly, 1968; Connelly *et al.*, 1968) (Fig.2).

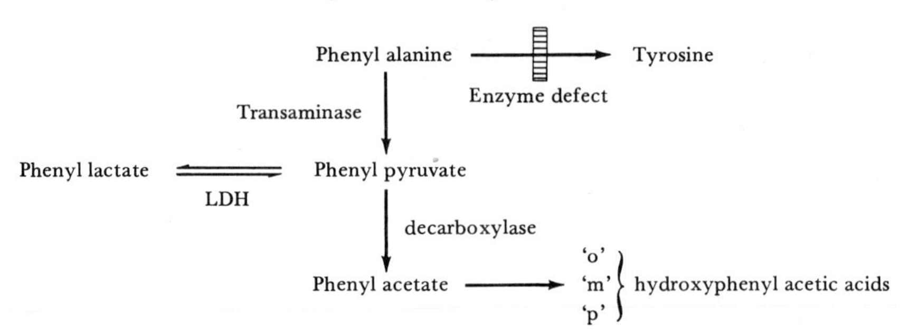

Fig. 1 Pathways of phenyl alanine metabolism in phenylketonuria.

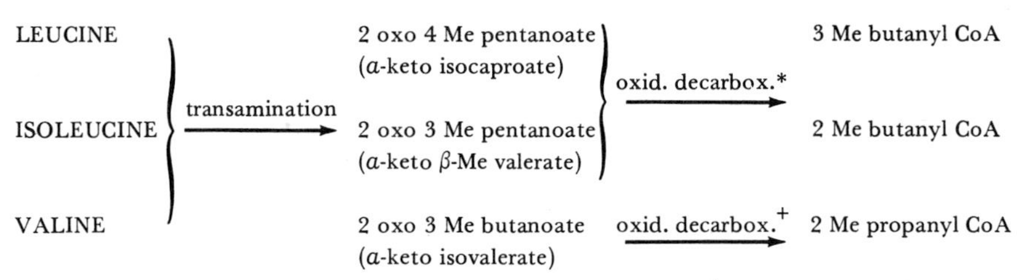

Fig. 2 Branched chain amino acid metabolism. * Absent in MSUD. + Inhibited by high 2 oxo acid levels.

In both of these syndromes two basic hypotheses have been proposed to account for the progressive and severe mental retardation that ensues in the absence of treatment. Firstly, that transmitter production is impaired, e.g. Sandler and Davison (1958), showed that 5-hydroxytryptamine and epinephrine production was inhibited, Wolley and Van der Hoeven (1966) that serotonin production was down and recently Small *et al.* (1970) have indicated that GABA production was impaired, in all cases in model PKU systems. Also Dreyfus and Prensky (1967) have shown that glutamate decarboxylase is inhibited by the 2-oxo-acids produced in MSUD. The second main hypothesis has been that the impairment of myelin formation is the prime cause of the mental retardation. Evidence for the latter has come from studies on myelinating cultures of rat cerebellum (Silberger, 1969), in which both phenylpyruvate and 2-oxo-4-methylpentanoate (α-keto isocaproate) inhibited markedly the myelination process but none of the other 2-oxo acids had any significant effect. Further support for this hypothesis came from the work of Shah *et al.* (1970) and Weber (1970) who showed that incorporation of precursors into cholesterol and lipid in brain slices was significantly impaired by phenylpyruvate. Agrawal, Bone and Davison (1970) and Oldendorf *et al.* (1971) have also proposed that myelination is deficient because general amino acid uptake in the brain is impaired by the high plasma phenylalanine levels found in phenylketonuria. From these latter-mentioned studies, two points became reasonably clear: a) that phenylpyruvate, in the case of PKU, and 2-oxo-4-Me pentanoate (α-keto isocaproate), in the case of MSUD, were probably the metabolites causing the detrimental effects and b) that although myelination and transmitter production were inhibited by these metabolites, the concentration of these compounds required to effect an inhibition were far higher than could be reasonably expected in the clinical state. We were therefore interested to locate more specifically where these metabolites might be acting and we decided initially to study some of the enzymes involved in fatty acid synthesis from precursors such as glucose or pyruvate, viz. citrate synthase (EC. 4.1.37.), acetyl CoA carboxylase (EC. 6.4.1.2.) and fatty acid synthetase. The main outcome of these experiments was the observation that both phenylpyruvate and 2-oxo-4-methylpentanoate (α-keto isocaproate) were potent inhibitors of the brain fatty acid synthetase with Ki's 250 and 930 μM respectively (Land and Clark, 1973b; Clark and Land, 1974). It is important to point out that such concentrations of phenylpyruvate and 2-oxo-4-methylpentanoate are well within reported plasma values for these compounds found in the clinical condition - for phenylpyruvate in PKU plasma values up to 500 μM (Knox, 1972; Patel, 1972) - for MSUD 2-oxo acid levels of 2-4 mM (Synderman *et al.*, 1964). However, at the same time we had been interested in observations that phenylpyruvate of 2-oxo-4-methylpentanoate markedly inhibited [1-^{14}C] pyruvate decarboxylation by brain slices or homogenates (Dreyfus and Prensky, 1967; Bowden *et al.*, 1970, 1971). The latter authors attributed this to an inhibition of the pyruvate dehydrogenase complex activity but subsequent work by ourselves suggested that this was most unlikely (Land and Clark, 1973a; Clark and Land, 1974). This has also been confirmed by Blass and Lewis (1973) and Hoffman and Hucho (1974). We therefore embarked upon a study of the effects of phenylpyruvate and 2-oxo-4-methylpentanoate on intact brain mitochondria, to see how their metabolism was affected.

METHODS

Mitochondria were prepared by the method of Clark and Nicklas (1970) from either 22-day old or adult male Wistar rats. Mitochondrial respiration was studied using an oxygen electrode in a medium containing the following (final concentrations) 100 mM KCl; 75 mM mannitol: 25 mM sucrose: 10 mM phosphate-Tris: 10 mM Tris-HCl; 0.05 mM EDTA: final pH 7.4. 1-2 mg brain mitochondrial protein was added followed by 2.5 mM malate and either 5 mM pyruvate or 10 mM DL-3-hydroxybutyrate.

Mitochondrial transport studies were carried out essentially according to the technique of Harris and Manger (1968). For uptake experiments 6-10 mg mitochondrial protein were added to 2 ml of a medium at 25° containing 125 mM KCl; 20 mM Tris-HCl, pH 7.4; 0.5 mM EDTA-K^+; 12.5 mM succinate-Tris pH 7.4; 5 μM rotenone; 17.6 mM oxamate; 0.75 mM sodium arsenite together with an appropriate quantity of titrated water. For uptake experiments, aliquots of radioactive (C^{14}) substrate were then added or, for exchange studies, a preincubation with radioactive substrate was carried out to 'load' the mitochondria followed by several additions of unlabelled substrate. In all cases samples were taken from the incubation medium and spun through a silicone oil layer into 1.5 M perchloric acid in a microcentrifuge. Then samples were taken from a solution above and below the oil layer and counted for H^3 and C^{14} on a scintillation counter. These data were then processed by computer to give details of the intramitochondrial content of radioactive substrate.

RESULTS AND DISCUSSION

Initially we studied the effects of all the amino acids and their analogue 2-oxo acids which accumulate in either PKU or MSUD on the ability of adult rat brain

Table I. Effect of some amino acids and their analogue 2-oxo-acids on pyruvate oxidation by adult rat brain mitochondira.

Addition (fc. 2 mM)	State 3 (+ADP) respiration		State 4 (−ADP) respiration	
	n atoms O min^{-1} mg.pr.	% control	n atoms O min^{-1} mg.pr.	% control
None	181	100	25	100
DL-valine	174	96	24.5	98
L-leucine	183	101	26	104
DL-isoleucine	169	93	23.5	94
L-phenylalanine	187	103	25	100
2-oxo-3-Me butanoate	181	100	31	124
2-oxo-4-Me pentanoate	114	63	32	128
2-oxo-3-Me pentanoate	178	98	28	112
Phenyl pyruvate (2-oxo-3 Ph. propionate)	82	45	23.5	94

Mitochondria were prepared from adult (150-160 g) rat brains by the method of Clark and Nicklas (1970). Respiration was measured in a medium containing 100 mM K^+ (see Methods) and the substrate was 5 mM pyruvate + 2.5 mM malate. The respiration rates are expressed as n atoms oxygen min^{-1} per mg mitochondrial protein. State 3 was induced by the addition of 250 μM ADP and the inhibitor under investigation was added during state 4 after one ADP cycle had been completed.

mitochondria to oxidise pyruvate and malate (Table I). It is evident from here that a) only phenylpyruvate (2-oxo-3-phenylpropionate) and 2-oxo-4-methyl-pantanoate (α-keto isocaproate) have any inhibitory effects, and b) that it is only the stimulated rate of oxidation (i.e. state 3 or + ADP) that is inhibited, the resting rate of oxidation (state 4) being essentially not affected. Similar results were found in 16 day old rat brain mitochondria (Clark and Land, 1974). However, it is now well accepted that the brain, particularly in the young animal uses ketone bodies (3-hydroxybutyrate) rather than pyruvate (i.e. glucose) as its main energy fuel (Hawkins *et al.*, 1971; Williamson and Buckley, 1974; Cremer, this volume; Land and Clark, this volume). Furthermore, it is during the period of ketone body utilization that the main part of the myelin is laid down in the developing brain (10-21 days in the rat). In view of the association of these inborn error diseases and the brain myelination state, we also studied the effects of the amino acids and their analogue 2-oxo-acids on 3-hydrobutyrate oxidation by brain mitochondria derived from 22 day old rats (Table II). A similar pattern emerges as with pyruvate, namely that of the compounds tested, only phenylpyruvate (2-oxo-3-phenylpropionate) and 2-oxo-4-Me-pentanoate (α-keto isocaproate) show significant inhibition and that only the stimulated rate of oxidation (state 3) and not the resting state (state 4) is affected. Previously, both Blass and Lewis (1973) and ourselves (Land and Clark, 1973a and Clark and Land, 1974) had observed that neither phenylpyruvate (2-oxo-3-phenylpropionate) nor 2-oxo-4-Me-pentanoate inhibited the semipurified brain pyruvate dehydrogenase complex significantly, and a similar situation was found for the brain 3-hydroxybutyrate dehydrogenase (Land and Clark, unpublished observations).

Table II. Effect of some amino acids and their analogue 2-oxo acids on 3-hydroxybutyrate oxidation by 22 day old rat brain mitochondria.

Addition (fc. 2 mM)	State 3 (+ADP) respiration		State 3 (−ADP) respiration	
	n atoms O min^{-1} mg.pr.	% control	n atoms O min^{-1} mg.pr.	% control
None	100	100	25	100
DL-valine	95	95	26	104
L-leucine	101	101	25	100
DL-isoleucine	93	93	23	92
L-phenylalanine	98	98	24	96
2-oxo-3-Me butanoate	82	82	24	96
2-oxo-4-Me pentanoate	71	71	28	112
2-oxo-3-Me pentanoate	80	80	23	92
Phenyl pyruvate (2-oxo-3 Ph. propionate)	64	64	28	112

Mitochondria were prepared from 22 day old rat brains by the method of Clark and Nicklas (1970). Respiration was measured in a medium containing 100 mM K^+ (see methods) and the substrate was 10 mM DL-3-hydroxybutyrate + 2.5 mM malate. The respiration rates are expressed as n atoms oxygen min^{-1} per mg mitochondrial protein. State 3 was induced by the addition of 250 μM ADP and the inhibitor under investigation was added during state 4 after one ADP cycle had been completed.

Table III. Effect of 2-oxo acids (+ derivs) on brain mitochondrial respiration.

Substrate \ Inhibitor	Control	Phenyl pyruvate		2-oxo-4 Me pentanoate		2-cyano cinnamate	
	State 3 respn.	State 3 respn.	% control	State 3 respn.	% control	State 3 respn.	% control
5 mM PYR/2.5 mM MAL	181	82	45	114	63	68*	38
10 mM DL-3 HB/2.5 mM MAL	100	64	64	71	71	32	32
10 mM Glut./2.5 mM MAL	167	139	83	159	95	160†	96
10 mM SUCCINATE	165	163	99	164	99	149†	90

Mitochondria were prepared from adult rat brains by the method of Clark and Nicklas (1970), with the exception of the 3-hydroxybutyrate studies for which 22 day old animals were used. Respiration was measured polorographically in 10 mM K^+ containing medium (see Methods section) and is expressed in n atoms O min^{-1} mg mitochondrial protein. State 3 was induced and the inhibitor added as in the legends of Tables I and II. The final concentration of the inhibitor was in all cases 2 mM except in the cases designated. *inhibitor concentration - 1 μM and † inhibitor concentration - 1 mM.

We next attempted to see whether other citric acid cycle substrates other than pyruvate and 3-hydroxybutyrate were affected (Table III) by phenylpuryvate and 2-oxo-4-Me pentanoate. From this table it can be seen that neither glutamate nor succinate oxidation were significantly affected by either of these 2-oxo acids, although phenylpyruvate did have a slight effect on glutamate utilization, possibly by acting as a transaminase acceptor at high concentrations (Benuck *et al.*, 1971). Also the effects of α-cyano cinnamate on the respiratory activities of brain mitochondria were compared. α-cyano cinnamate has recently been proposed as a

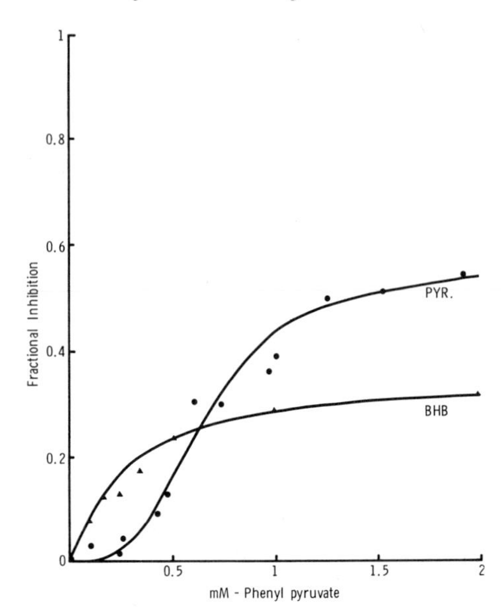

Fig.3 Effect of phenylpyruvate on pyruvate (●) and 3-hydroxybutyrate (▲) oxidation by rat brain mitochondria. Rat brain mitochondria were incubated as described in the Methods section, together with 5 mM pyruvate + 2.5 mM malate (●) or 10 mM DL-3-hydroxybutyrate + 2.5 mM malate (▲). The stimulated (state 3 - Chance and Williams, 1956) respiration rate was induced by the addition of 250 μM ADP and the inhibitor was added during the ensuing state 4. Further aliquots of ADP were added and the state 3 rate in the presence of inhibitor estimated. The results are the averages of at least 2 estimations of the state 3 rate expressed as a fractional inhibition, i.e.

$$\frac{\text{CONTROL STATE 3 - INHIBITED STATE 3}}{\text{CONTROL STATE 3}}$$

The control state 3 rates for pyruvate and malate were 183 ± 13 (n = 15) n atoms O min^{-1} mg protein (respiratory control ratio - 8) and for 3-hydroxybutyrate and malate were 106 ± 7 (n = 15) n atoms O min^{-1} mg protein (respiratory control ratio - 4). The lines drawn through the points are best fit curves derived from the Hill equation.

specific pyruvate transport inhibitor in liver mitochondria (Halestrap and Denton, 1974). The results in Table III show that α-cyano cinnamate also inhibits pyruvate oxidation in brain mitochondria very effectively (> 60% decrease of state 3 respiratory activity of 1 μM concentration) and also that it inhibits 3-hydroxybutyrate oxidation to a similar extent, although a much higher concentration is required (2 mM). However, in common with phenylpyruvate and 2-oxo-4-methylpentanoate, α-cyano cinnamate did not affect the oxidation of glutamate or succinate by rat brain mitochondria. It seemed possible, therefore, that a) both phenylpyruvate and 2-oxo-4-methylpentanoate were acting in a similar way to α-cyano cinnamate, i.e. on the mitochondrial transport system for pyruvate and b) that this transporter either also carried 3-hydroxybutyrate or that there was a distinct system but with very similar properties to the pyruvate transport system. Closer inspection of the effects of phenylpyruvate and 2-oxo-4-methylpentanoate on pyruvate and 3-hydroxybutyrate supported respiration by rat brain mitochondria are shown in

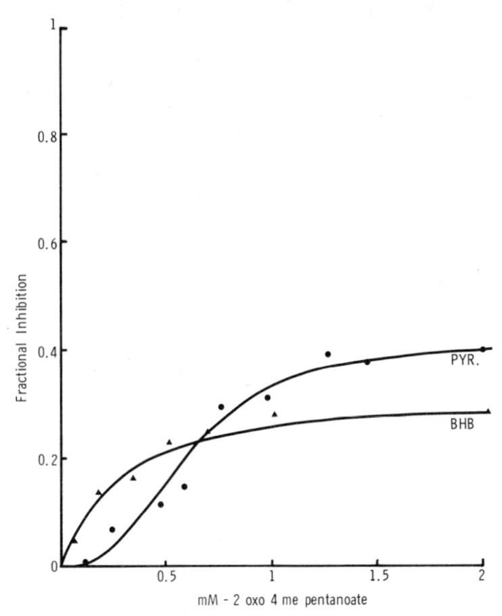

Fig.4 Effect of 2-oxo-4-methylpentanoate concentration on pyruvate (●) and 3-hydroxybutyrate (▲) oxidation by rat brain mitochondria. All conditions were as in Fig.3 except that 2-oxo-4-methylpentanoate was used instead of phenylpyruvate.

Figs 3 and 4. The effects of increasing concentrations of the inhibitors on the stimulated respiration rate (i.e. +ADP) were observed for brain mitochondria oxidizing either pyruvate and malate or 3-hydroxybutyrate and malate. The results are plotted as fractional inhibition

$$\frac{\text{(control respiration rate} - \text{inhibited rate)}}{\text{control respiration rate}}$$

against final inhibitor concentration. At the same time theoretical lines which give a best fit to the points were computed and have been drawn in. A number of points are worth noticing. For both inhibitors the inhibition of pyruvate supported respiration followed a sigmoidal relationship with increasing inhibitor concentration, whereas the inhibition of 3-hydroxybutyrate supported respiration followed a hyperbolic relationship. This may be related to the fact that the pyruvate dehydrogenase complex activity is activated by 2-keto acids by inhibition of the PDH kinase (Linn *et al.*, 1969; Hoffman and Hucho, 1974). However, the observed differences in the inhibitor response curves for pyruvate and 3-hydroxybutyrate respiration (Figs 3 and 4) may equally be a reflection of the relative activities of the translocase and enzyme for each substrate. These relationships, however, may have important consequences for PKU and MSUD since it means that at low concentrations (up to 200-300 μM) of phenylpyruvate or 2-oxo-4-Me pentanoate there will be considerably more inhibition of 3-hydroxybutyrate oxidation than of pyruvate which in the young brain, would have marked detrimental effects on the normal development of the brain. It is, however, also worth noticing that at

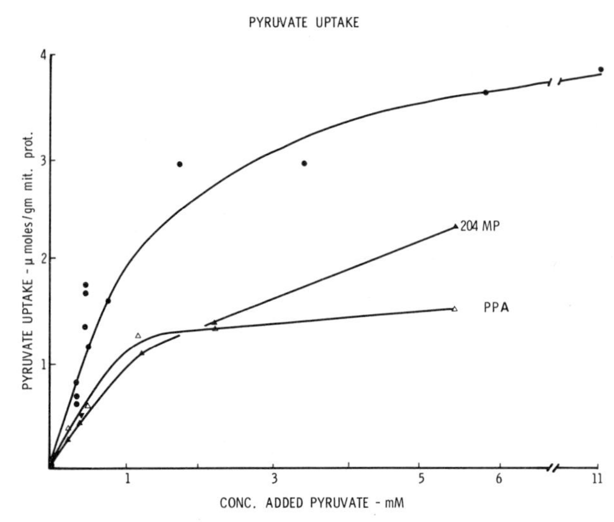

Fig.5 Effect of phenylpyruvate and 2-oxo-4-methylpentanoate on [2^{14}C] pyruvate uptake by rat brain mitochondria. Experiments were carried out as outlined in the Methods section. Both inhibitors - phenylpyruvate (PPA - △—△) and 2-oxo-4-methylpentanoate (204 MP - ▲—▲) were added at final concentrations of 4.2 mM. Pyruvate uptake in the absence of inhibitor is plotted as ●—●. The internal matrix space of the mitochondria was the same in the presence or absence of inhibitor - 1.43 ± 0.4 μl per mg mitochondrial protein.

Table IV. Characteristics of uptake process for pyruvate+3-hydroxybutyrate in rat brain mitochondria.

Substrate	Maximum substrate concentration μmoles/g mit. prot.	External substrate concentration required ½ maximal substrate content — Ks — mM
Pyruvate	5	1.72
DL-3-HO butyrate	4.2	0.90

Data derived from control curves of Figs 3 and 4 plotted as double reciprocal plots according to Lineweaver Burke.

higher inhibitor concentrations, pyruvate oxidation is considerably more inhibited than 3-hydroxybutyrate, particularly by phenylpyruvate. This may be related to the relative proportions of pyruvate and 3-hydroxybutyrate which obligatorily enter the mitochondria by the transport mechanism as distinct from simple diffusion. In this connection the fact that at the pH of these experiments, one would expect to find two orders of magnitude more 3-hydroxybutyrate (pK 4.4) in the undissociated form than pyruvate (pK 2.5) may have some relevance.

We then decided to look at the effects of these inhibitors on the uptake and exchange of pyruvate and 3-hydroxybutyrate across the mitochondrial membrane.

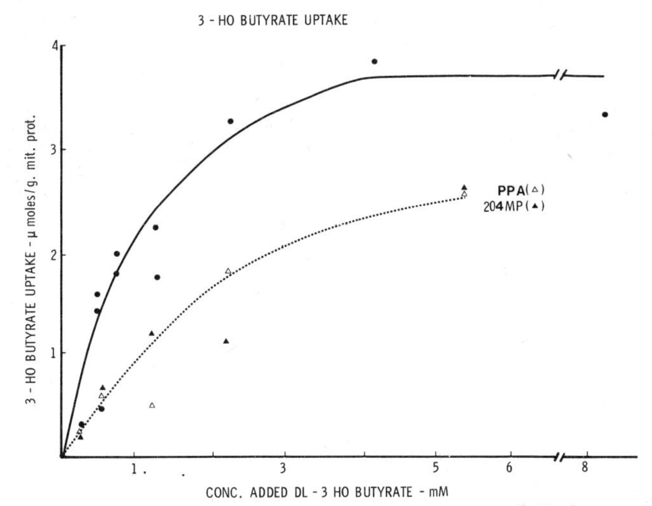

Fig.6 Effect of phenylpyruvate and 2-oxo-4-methylpentanoate on [3^{14}C] DL-3-hydroxybutyrate uptake by rat brain mitochondria. All conditions were the same as in Fig.5 except that [3^{14}C] DL-3-hydroxybutyrate was used instead of pyruvate.

The uptake experiments were performed by adding increasing concentrations of ^{14}C pyruvate or ^{14}C 3-hydroxybutyrate to mitochondria whose metabolism had been blocked with rotenone and arsenite. These mitochondria were then separated from the medium by spinning them through a silicone oil layer into an acid medium. It may be seen from Figs 5 and 6 that in the case of both pyruvate and 3-hydroxybutyrate, phenylpyruvate and 2-oxo-4-methylpentanoate have a marked inhibitory effect on the uptake of these substrates, e.g. at 1 mM concentrations of these inhibitors there is less than 50% of the control content of either substrate. It is also worth noting the uptakes in the absence of inhibitors. If these data are treated in the classical 'double reciprocal plot' fashion, values which approximate to the maximal substrate concentration attainable and the external substrate concentration required to attain 50% of this value (Ks) may be derived (Table IV). It is interesting that brain mitochondria appear to be capable of taking up considerable quantities of both pyruvate and 3-hydroxybutyrate, in fact considerably more than, for example, liver mitochondria under similar circumstances ~ 1 μmole/g mitochondrial protein (Mowbray, 1974). However, the concentration of pyruvate required to achieve this level of saturation is at least double and if it is assumed that the transporter only carries the D isomer, some four times greater than that of 3-hydroxybutyrate. This suggests that the brain mitochondria has, in fact, a distinct preference for 3-hydroxybutyrate as compared with pyruvate. An alternative, and perhaps more rigorous way of studying mitochondrial transport is to preload the mitochondria to saturation with a radioactive substrate and then to add excess non-radioactive substrate and to study the efflux of radioactively labelled substrate from the mitochondria. Figures 7 and 8 show the results of such experiments. Figure 7 represents rat brain mitochondria which have been preloaded with saturating amounts of 3 C^{14}C 3-hydroxybutyrate and then aliquots of non-radioactive 3-hydroxybutyrate

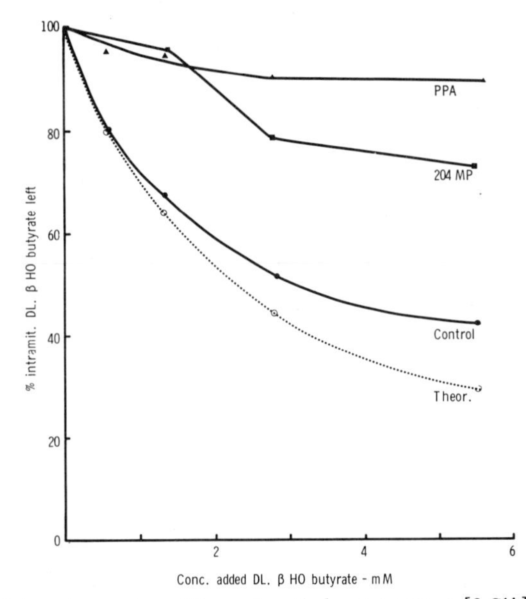

Fig.7 Effect of phenylpyruvate and 2-oxo-4-methylpentanoate on [3 C^{14}] DL-3-hydroxybutyrate exchange by rat brain mitochondria. Mitochondria were preloaded with radioactive 3-hydroxybutyrate and then aliquots of cold 3-hydroxybutyrate were added in the presence and absence of inhibitors and the efflux of C^{14}-3-hydroxybutyrate measured. The final concentration of the inhibitors was 11.6 mM - PPA (▲—▲); 2-oxo-4-methylpentanoate (■—■); no inhibitor (●—●). Also shown is the calculated decrease (○---○) in specific activity of the radioactive 3-hydroxybutyrate that would occur assuming complete equilibration between the added cold substrate and the radioactive substrate. The results are expressed as a % of the original control content of 3-hydroxybutyrate: these were: 3.98 ± 0.25 μmoles/g mitochondrial protein in the absence of inhibitor; 3.09 ± 0.27 μmoles/g mitochondrial protein in the presence of phenylpyruvate; 3.42 ± 0.3 μmoles/g mitochondrial protein in the presence of 2-oxo-4-Me pentanoate. The mitochondrial internal matrical space did not change significantly in the presence of inhibitors (1.49 μl per mg mitochondrial protein).

have been added either in the presence or absence of inhibitors. From the control line it may be seen that there is a rapid efflux of ^{14}C 3-hydroxybutyrate as cold 3-hydroxybutyrate is added which corresponds fairly well with the calculated decrease in the specific activity of the radioactive 3-hydroxybutyrate that would be expected if complete equilibration occurred. However, in the presence of inhibitors, although there is an initial immediate efflux of labelled substrate on adding the inhibitor to 86% of the control in the case of 2-oxo-4-methylpentanoate and to 75% of the control for phenylpyruvate, thereafter as cold substrate is added there is very little exchange occurring across the mitochondrial membrane. A similar situation is shown in Fig.8 where pyruvate instead of 3-hydroxybutyrate is titrated into a preparation of mitochondria preloaded with ^{14}C 3-hydroxybutyrate in the presence and absence of inhibitor. Again in the absence of inhibitor the efflux of radioactive 3-hydroxybutyrate approximates closely to the theoretical decline in specific activity of the 3-hydroxybutyrate which one would expect if the cold pyruvate was equilibrating directly with the hot 3-hydroxybutyrate as

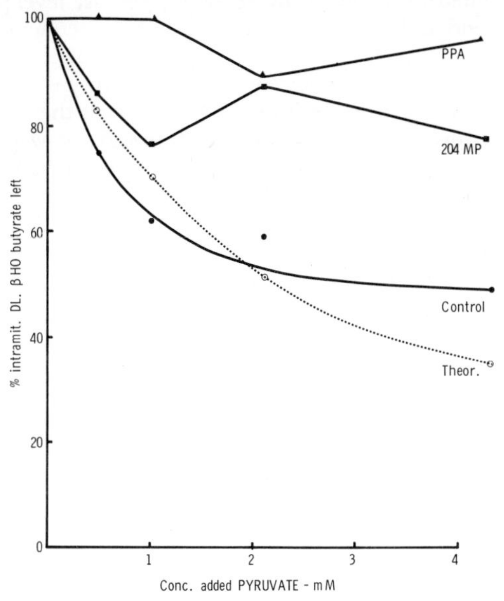

Fig.8 Effect of phenylpyruvate and 2-oxo-4-methylpentanoate on pyruvate and [3-C[14]] DL-3-hydroxybutyrate exchange by rat brain mitochondria. Mitochondria were preloaded with [3-C[14]] DL-3-hydroxybutyrate and aliquots of cold pyruvate were added in the presence and absence of inhibitors and the efflux of C^{14}-3-hydroxybutyrate measured. Inhibitors were present at 11.6 mM (f.c) - PPA (▲–▲); 2-oxo-4-methylpentanoate (■–■); no inhibitor (●–●). The calculated decrease in specific activity (○---○) of the ^{14}C-3-hydroxybutyrate is also shown assuming that the non-radioactive pyruvate added equilibrates completely with the 3-hydroxybutyrate pool. The results are expressed as a % of the original control content of 3-hydroxybutyrate - these and other parameters were as in Fig.7.

if they were one pool. This suggests that both the pyruvate and 3-hydroxybutyrate are involved on the same carrier. Furthermore, it is also found that both inhibitors cause an immediate loss of ^{14}C 3-hydroxybutyrate but that subsequently there is very little efflux of ^{14}C-3-hydroxybutyrate as pyruvate is added. Both these experiments support strongly the hypothesis that phenylpyruvate and 2-oxo-4-methylpentanoate are inhibiting the transport of both pyruvate and 3-hydroxybutyrate across the mitochondrial membrane.

In conclusion, therefore, we would like to propose the following scheme which we believe may in part contribute to the etiology of PKU and MSUD and in particular explain the defective myelination and depressed energy metabolism which appears as a 'hallmark' of these conditions. Firstly, that brain mitochondria possess a transport system to allow pyruvate or 3-hydroxybutyrate to enter the mitochondria at rates in excess of the diffusion rate. This is a property held in common with other mitochondria (Papa *et al.*, 1971; Brouwer *et al.*, 1973; Halestrap and Denton, 1974; Mowbray, 1974). This transport system is inhibited by phenylpyruvate or 2-oxo-4-methylpentanoate which in the case of the developing brain means that the brain is deprived of adequate substrate for both synthetic and general energy purposes since at this stage of brain development it is almost entirely dependent on ketone bodies for these functions, unlike other tissues. This is probably in part due to the low pyruvate dehydrogenase level present in the developing brain (Land and Clark, this volume). As a consequence there will be an impaired production of ATP, an impaired availability of extra mitochondrial citrate for fatty acid synthesis, and an impaired availability of precursors for transmitter synthesis, e.g. glutamate for GABA. Although there may be other metabolic lesions involved in the overall phenylketonuria or Maple Syrup urine disease syndrome, it is not difficult to assess that the lesion we have discussed here will have extensive malevolent implications for the normal development of brain function.

ACKNOWLEDGEMENT

J.M.L. thanks the SRC for a studentship held during this work.

REFERENCES

Agrawal, H.C., Bone, A.H. and Davison, A.N. (1970). *Biochem. J.* **117**, 325-331.
Benuck, M., Stern, F. and Lajtha, A. (1971). *J. Neurochem.* **18**, 1555.
Blass, J.P. and Lewis, C.A. (1973). *Biochem. J.* **131**, 31-37.
Bowden, J.A., Brestel, E.P., Cope, W.T., McArthur, C.L., Westfall, D.M. and Fried, M. (1970). *Biochem. Med.* **4**, 69-76.
Bowden, J.A. and Connelly, J.L. (1968). *J. biol. Chem.* **243**, 3526-3531.
Bowden, J.A., McArthur, C.L. and Fried, M. (1971). *Biochem. Med.* **5**, 101-108.
Brouwer, A., Smits, G.G., Tas, J., Meyer, A.J. and Tager, J.M. (1973). *Biochimie* **55**, 717-725.
Clark, J.B. and Land, J.M. (1974). *Biochem. J.* **140**, 25-29.
Clark, J.B. and Nicklas, W.J. (1970). *J. biol. Chem.* **245**, 4724-4731.
Connelly, J.L., Danner, D.L. and Bowden, J.A. (1968). *J. biol. Chem.* **243**, 1198-1203.
Dancis, J. and Levitz, M. (1972). *In:* "Metabolic Basis of Inherited Disease" (J.B. Stanbury, J.B. Wyngaarden and J.S. Fredrickson, eds) 3rd edn., pp.426-439. McGraw Hill, N.Y.
Dreyfus, P.M. and Prensky, A.L. (1967). *Nature* **214**, 276.
Halestrap, A.P. and Denton, R.M. (1974). *Biochem. J.* **138**, 313-316.
Harris, E.J. and Manger, J. (1948). *Biochem. J.* **109**, 239.
Hawkins, R.A., Williamson, D.H. and Krebs, H.A. (1971). *Biochem. J.* **122**, 13-18.
Hoffman, B.T. and Hucho, F. (1974). *FEBS Letts.* **43**, 116-119.
Knox, W.E. (1971). *In:* "Metabolic Basis of Inherited Disease" (J.B. Stanbury, J.B. Wyngaarden and D.S. Fredrickson, eds) 3rd edn., pp.266-295. McGraw Hill, N.Y.
Land, J.M. and Clark, J.B. (1973a). *Biochem. J.* **134**, 539-544.
Land, J.M. and Clark, J.B. (1973b). *Biochem. J.* **134**, 545-555.
Linn, T.C., Pettit, F.M., Hucho, F. and Reed, L.J. (1969). *PNAS* **64**, 227-234.
Mowbray, J. (1974). *FEBS Letts.* **44**, 344-347.
Oldendorf, W.H., Sisson, W.B., Mehter, A.C. and Treciokas, L. (1971). *Arch. Neurol.* **24**, 423-430.
Patel, M.S. (1972). *Biochem. J.* **128**, 677-684.
Sandler, M. and Davison, A.N. (1958). *Nature* **181**, 186-187.
Shah, S.N., Peterson, M.A. and McKean, C.M. (1970). *J. Neurochem.* **17**, 279-284.
Silberberg, D.H. (1969). *J. Neurochem.* **16**, 1141-1146.
Small, N.A., Holton, J.B. and Ancill, R.J. (1970). *Brain Res.* **21**, 55-62.
Synderman, S.E., Morton, P.M., Roitman, E. and Holt, L.E. (1964). *Pediatrics* **34**, 454-472.
Weber, G., Glazer, R.I. and Ross, R.A. (1970). *Adv. Enz. Regul.* **8**, 13-36.
Wolley, D.W. and Van der Hoeven, Th. (1966). *Science* **144**, 883-884.
Williamson, D.H. and Buckley, B.M. (1973). *In:* "Inborn Errors of Metabolism" (F.A. Hommes and C.J. Van den Berg, eds) pp.81-92. Academic Press, London and New York.

DISCUSSION

Tager: I think two important points have emerged. The first is the evidence you have brought forward for an inhibition of pyruvate transport by these two compounds which accumulate and the second is the demonstration of an exchange of pyruvate for β-hydroxybutyrate in mitochondria.

Snell: Have you looked at acetoacetate in your system?

Clark: No, we have not.

Snell: So you have no idea whether pyruvate also exchanges with acetoacetate as well as with β-hydroxybutyrate.

Clark: We do not know. Papa (*Eur. J. Biochem.* 49, 265 (1974)) has produced evidence indicating that acetoacetate does exchange for pyruvate in liver mitochondria.

Cremer: You have stressed that ketone bodies provide the major portion of energy during development. I showed (p.134) that they can contribute, but that they at no time overtake glucose. That leads on to your oxygen uptake rates. β-hydroxybutyrate gave a lower rate of respiration than pyruvate plus malate. The question is, does the rate of β-hydroxybutyrate oxidation ever reach the same rate as that for pyruvate plus malate?

Land: If one considers mitochondria derived from 22 day old animals the pyruvate plus malate state 3 oxidation rate is effectively equivalent to the β-hydroxybutyrate plus malate oxidation rate. But with adult animals the 3 oxidation rate with β-hydroxybutyrate is about one third of that seen with pyruvate.

Greengard: Have your compared your results using mitochondria from animals of different ages?

Clark: No.

Gaull: I would like to discuss some of the assumptions you made about the relevance of your experimnets. I wonder whether the percentage of inhibition is as important as the actual flux through the pathway. There was a point that came up in work similar to yours (Miller *et al., Science* 179, 904 (1973)).

Clark: One of the problems is what relevance these experiments have to the *in vivo* situation and this is very difficult to evaluate.

Tager: May I make a comment on this point? I think that if one tries to ascertain what the pathogenesis of some particular lesion is, one starts with an artificial system and one deliberately chooses unphysiological conditions in order to be able to pinpoint the effect. The point you made about the flux rates is, of course, a good one. The flux rate depends not only on the concentration of the enzyme, but also on the concentration of metabolites. The inhibition of a process is in many cases strongly dependent on the concentration of the metabolites of the enzyme. In this case, too, I would suspect that the inhibition by something like phenylpyruvate at a concentration of 1 millimolar would be very effective at the pyruvate concentration one normally finds in a cell.

Gaull: Thank you. You mentioned also that millimolar concentrations of phenylpyruvate had been found in tissues. I take it that you refer to animal experiments?

Clark: I think that there have been reports of millimolar concentrations of these ketoacids in affected infants. But I must add that the whole problem of concentration of these metabolites in clinical cases, to my mind, is very sparse and poorly documented.

Tager: What is the level of phenylpyruvate found in the blood? What is the range?

Wadman: It depends on the phenylalanine load, but you certainly can find phenylpyruvic acid in plasma and in cerebrospinal fluid.

Tager: But would you expect a concentration of 1 mM which has been mentioned?

Wadman: In non-fasting plasma, perhaps. M.S. Patel (*Biochem. J.* 128, 677 (1972)) referring to *Jervis Proc. Soc. exp. Biol. Med.* 82, 514 (1952) mentioned fasting serum phenylpyruvate levels of 0.1 mM in patients with PKU, increasing by about four-fold after ingestion of phenylalanine. Coburn *et al.* (*Clin. Chem.* 17, 378 (1971)) give for plasma phenylpyruvate in 10 fasting phenylketornuria subjects 0.03 - 0.15 mM. Some incidental fasting values from our laboratory: *(gas chromatography of trimethyldilyl derivates): concentrations are given in mM.

J.V.D.	phenyl-alanine	phenyl-pyruvate*	phenyl-lactate*	0-OH phenyl acetate*
serum 9-12-'71	2.21	0.073	0.126	0.033
serum 11-12-'71	1.60	0.024	0.054	0.033
T.v.D.				
serum 3-2-'73	1.56	0.067	0.049	—
cerebrospinal fluid 3-2-'73	—	0.006	—	—
K.v.d.B.				
serum 28-8-'71	2.79	0.200	0.205	—
cerebrospinal fluid 28-8-'71	1.89	0.098	0.054	—
M.P.				
serum 6-7-'71	2.60	0.037	0.024	—
serum 7-7-'71	2.71	nd	nd	nd
cerebrospinal fluid 7-7-'71	0.81	0.671	0.223	0.099
F.K.				
serum 12-5-'72	1.96	0.152	0.157	—
cerebrospinal fluid	0.82	—	0.018	—
D.P.				
serum 3-9-'72	2.04	0.195	0.139	trace

Gaull: I would likc to makc just onc general point. The entire world's literature on the actual concentrations of amino acids and their ketoacids in these diseases are based on a total of four post-mortem brains in phenylkeronuria, and a total of two post-mortem brains in maple syrup urine disease (Gaull *et al., In:* "Biology of Brain Dysfunction", G. Gaull, ed., Vol.III, 1975). And certainly the determination of α-ketoacids in post-mortem tissues is highly unreliable.

Clark: I think that the argument does not rest on the actual levels. The point is that there is a qualitative inhibition and that it may have some relevance.

Gaull: This is not a criticism of your experiments. This is a general statement about the difficulty of determining the relevance of these experiments to the clinical problem. The fact is (I think that Dr. De Groot made this point the other day) that maple syrup urine disease and phenylketonuria present clinically in very different ways. MSUD classically presents as an acute fulminating condition in the neonate, resulting in coma and death if untreated. Phenylketonuria presents with a very gradual onset without any of the dramatic symptoms. There is an impairment of higher cortical functions which appears at a time scale measured in months and years, rather than hours and days.

Clark: I am not suggesting that the inhibition we have shown is the only lesion which matters. I am adding it to the list of ones already reported, but I do think that it could have in the long term quite a number of deleterious effects.

Hommes: The question arises when the change-over from ketone body utilization by the human brain to glucose utilization really takes place. Is this in any way related to the five to eight year period during which PKU patients are treated?

Clark: I do not know, and I must say that this is a very interesting point, particularly from the point of view of treating phenylketonuria as to when they can be safely put back on to a normal diet.

Wick: I just wonder where you got the information that the decarboxylation system of valine is different from that of leucine and isoleucine. I once had the same idea, and tried to give a high dose of valine to a MSUD patient. We had to stop it, because valine was rising very high, and I think that contradicts to a certain extent this statement.

Clark: Connelly and Bowden (*J. biol. Chem.* 243, 3526 (1968)) reported this. They came to the conclusion that there were two decarboxylases.

Cremer: It may be of interest to mention, and perhaps of some relevance, that leucine is degraded in the brain in a different "metabolic space" from the substrates I was talking about yesterday, i.e. glucose and ketone bodies.

Van den Berg: That is again a difference between leucine and phenylalanine. The degradation of phenylalanine in brain is quite low or even absent. The inhibitions by the ketoacids were never more than about 40%, despite the presence of very high concentrations. Why is that?

Clark: I think it comes back to a point Professor Tager mentioned; that it is quite possible that a certain amount of pyruvate and indeed β-hydroxybutyrate may diffuse across the mitochondrial membranes as opposed to being actively translocated. The oxidation of pyruvate is also more inhibited at the higher inhibition concentrations than the oxidation of β-hydroxybutyrate. This can possibly be explained on the basis of the pH at which experiments were done, since the β-hydroxybutyrate will be more undissociated than pyruvate.

CLINICAL AND METABOLIC ABNORMALITIES ACCOMPANYING DEFICIENCIES IN PYRUVATE OXIDATION

J.P. Blass, S.D. Cederbaum and G.E. Gibson

Mental Retardation Research Center
and
Departments of Psychiatry, Pediatrics and Biological Chemistry
University of California at Los Angeles Medical School, USA

Reducing the supply of oxygen to the brain promptly impairs brain function. For decades, there have been attempts to find patients with neuropsychiatric disorders and defective metabolism of glucose, the main normal substrate of the brain, or of other carbohydrates including lactic acid. The abnormalities that have been reported were in general nonspecific, except that brain metabolism was low in old people with generalized mental deterioration (McIlwain and Bachelard, 1971).

In the 1960's, several children were described who had severe and often fatal neurological disease and whose blood contained elevated levels of pyruvic and lactic acids (Hartmann *et al.*, 1962; Israels *et al.*, 1964; Erickson, 1965; Worsley *et al.*, 1965; Haworth *et al.*, 1967). Clinical studies suggested that these children might have inherited deficiencies in carbohydrate utilization. In the last five years, it has been possible to show that a number of similar children had inherited deficiencies affecting the pyruvate dehydrogenase complex (PDH) (Blass *et al.*, 1970, 1971a, 1971b, 1972; Cederbaum and Minkoff, 1972; Cederbaum *et al.*, 1973; Farmer *et al.*, 1973; Falk *et al.*, 1974). This multienzyme complex catalyzes the conversion of pyruvic acid to acetylCoA, a central reaction in the main pathway of glucose oxidation by the brain.

In the following discussion we review the clinical abnormalities in patients with PDH deficiencies, the identification of their enzymatic defects, studies of the pathophysiology using experimental animals, and some implications of these observations.

1. CLINICAL ABNORMALITIES

A. *History, Signs and Symptoms*

We know of thirteen patients in whom there is substantial biochemical evidence for abnormalities of PDH (Table I). The presenting and most prominent clinical

Table I. Patients with abnormalities of PDH

Patient	Sex	Abnormal Enzyme	Onset	Current Age or Age at Death	Presenting Sign or Symptom	Prominent Neurological Findings	Growth Failure	Acute Acidosis
1	M	PDC†	infancy	15	Intermittent ataxia	Intermittent ataxia	no	no
2	M	PDC†	2 years	12	Intermittent ataxia	Intermittent ataxia, optic atrophy	no	no
3	M	PDC†	5 years	10	Intermittent ataxia	NA	no	no
4	F	LAT††	infancy	5 (dead)	Poor wt. gain, psychomotor retardation, neonatal tachypnia	Psychomotor retardation, intermittent lethargy, microcephaly	no	yes
5	M	LAT††	infancy	10	Psychomotor retardation, hypotonia	Psychomotor retardation, small stature, microcephaly	yes	yes
6	F	LAT†† (inference)	infancy	7½ (dead)	Psychomotor retardation, hypotonia	Psychomotor retardation, small stature, microcephaly	yes	yes
7	F	LAT†† (inference)	infancy	3½ (dead)	Psychomotor retardation, hypertonia	Psychomotor retardation, small stature, microcephaly	yes	yes
8	M	PDH*	infancy	11	Psychomotor retardation, hypotonia	Psychomotor retardation, seizures	moderate	no
9	M	PDH*	infancy	2	Hypotonia	Hypotonia, intermittent bulbar palsy, seizures	moderate	no
10	M	PDC†	infancy	6 m.(dead)	Acidosis	Acidosis, psychomotor retardation, seizures	NA	yes
11	M	PDC†	1 year	NA	Ataxia	Developmental regression, hypotonia, global CNS deterioration	NA	no
12	F	PDC† (probable)	2 years	NA	Intermittent ataxia	Hypotonia during acute episode	no	no
13	M	PDC† (probable)	3 years	NA	Intermittent ataxia	NA	no	no

Patient	Response to High Carbohydrate	Family History	Blood Lactate (mg/d)** Pyruvate (mg/d)***	L:P Ratio	Alaninemia or Alaninuria	Reference
1	no	no	L:normal - 4.0 P:normal - 4.0	normal	intermittent	Blass *et al.*, 1970; Blass *et al.*, 1971a, 1971b
2	no	Brother of 3	L: 21.0 - 30.0 P: 1.8 - 4.2	normal	yes	Lonsdale *et al.*, 1969; Blass *et al.*, 1971b
3	NA	Brother of 2	L: NA P: NA	NA	NA	Blass *et al.*, 1971b
4	Maintained by parents on low fat diet	Consanguinity	L: 30 - 100 P: 1.8 - 4.2	normal	yes	Blass *et al.*, 1972
5	Acute acidosis	Brother of 6 & 7	L:normal - 180 P:normal - 8	normal	yes	Cederbaum and Minkoff, 1972; Cederbaum, Blass and Cotton, 1973; Cederbaum and Blass, 1974
6	NA	Sister of 5 & 7	L:normal - 150 P:normal - 2.8	NA	NA	Cederbaum, Minkoff, Brown and Blass, in preparation
7	NA	Sister of 5 & 6	L:normal - 120 P:normal - NA	NA	NA	Cederbaum, Minkoff, Brown and Blass, in preparation
8	Increased blood pyruvate	Brother of 9	L:normal - 30 P:normal - 4.5	normal	intermittent	Falk, Cederbaum and Carrel, 1974
9	Increased blood pyruvate	Brother of 8	L:normal - 30.0 P:normal - 3.7	normal	borderline	Falk, Cederbaum and Carrel, 1974
10	Normal lactate on high fat diet	no	L:greatly elevated P:greatly elevated	normal	intermittent	Farrell *et al.*, 1974 C.R. Scott, personal communication
11	NA	NA	L: 17 P: 1.2 - 2.1	normal	NA	Farmer *et al.*, 1973
12	NA	Sister of 13	L:elevated P:elevated	normal	NA	P.S. Gerald, personal communication
13	NA	Sister of 12	L:elevated P:elevated	normal	NA	P.S. Gerald, personal communication

† PDC, thiamin-dependent first enzyme of the PDH complex; †† LAT, second enzyme of the complex; * site of defect under study; NA, information not available; ** normal = 5 to 18: *** normal = 0.4 to 1.0.

abnormality in those less severely ill has been cerebellar ataxia. It was the only clinical abnormality in one (no.3) and the presenting complaint in six (nos 1-3, 11-13). Of course, physicians interested in these disorders may have looked for children with ataxia. The movement disorder is typically intermittent and brought out by febrile illnesses such as colds or by other stresses which increase metabolic rate.

Other abnormalities are prominent in patients with more severe disease (Table I). Optic atrophy of varying degree is common. Mental and physical development is often slow. Six patients had microcephaly (nos 4-9), and in three siblings development was too low for meaningful measurement (nos 5-7). In contrast, a patient with mild disease who has been followed for six years is doing well in an advanced class in junior high school (no.1). Mixtures of increased and decreased muscle tone are common. In one family, two affected children were hypotonic (nos 5,6) and one spastic (no.7). Other movement disorders occurred in some of these patients and have sometimes been hard to classify. One patient (no.1) had unique eye movements (Podos, 1970). Seizures occurred in three (8-10), including two brothers with petit-mal episodes.

In ten patients, symptoms were apparent by the age of one. Three have been minimally ill (nos 1-3). One child died in infancy (no.10) and three more before the age of 10 (nos 4,6,7). Nine patients had affected siblings and the parents of a tenth were consanguinous (no.4). The absence of more families with single patients suggests a relative failure to diagnose this condition until two children have been affected. Five of the patients with pyruvate decarboxylase deficiency are male and two female. Three girls and one boy have been deficient in the transacetylase component of the PDH complex. The data suggest that both types of deficiency are inherited in an autosomal recessive manner (see below).

We cannot diagnose PDH deficiencies, even presumptively, by clinical observations alone. Ataxia that varies in severity raises our suspicions, as does psychomotor retardation accompanying low serum bicarbonate.

B. *Clinical Laboratory Tests*

The blood and/or urine of these patients usually contains elevated levels of pyruvate and lactate or alanine (the α-amino isomer). Urinary alanine elevation seems a more sensitive indicator than blood alanine. Urinary pyruvate and lactate are less reliable indicators, perhaps because they must cross the renal threshold. Occasional samples of blood or urine have been normal even from patients whose pyruvate and lactate levels are usually elevated. Normal concentrations of lactate and pyruvate in a single sample do not rule out this disorder. Similarly, three patients with severe disease (nos 5,6,7) had persistently near normal blood pyruvate and lactate and only an occasional elevation of alanine in the urine. Lactate to pyruvate ratios have been normal in all our patients. Blood bicarbonate is generally low when lactate levels are high.

After oral or intravenous glucose loads, the levels of pyruvate especially and of lactate rose excessively in the blood of the three patients so studied (nos 5,8,9). In two patients (nos 1,2), oral glucose loads led, reproducibly, to only slight increases in blood sugar (Lonsdale *et al.*, 1969; Blass *et al.*, 1971b).

Feeding the patients a diet high in carbohydrates can bring about severe lactic acidosis. One patient (no.5), whose levels of lactate and pyruvate were near normal on a 20% protein, 40% carbohydrate, 40% fat diet, developed life-threatening acidosis when the carbohydrates were increased to 60%. During the next 36 hours, his blood pH dropped to 7.0. He excreted 30 grams of lactate into his urine, and he received 650 meq of $NaHCO_3$, twice his calculated total body bicarbonate.

A muscle biopsy from one of the patients (no.1) contained unusual fat droplets in ATP-ase type I fibers (Blass *et al.*, 1971b). In another, only neurogenic changes were seen (no.8).

C. *Treatment*

An infant with severe, chronic lactic acidosis and severe muscle weakness (no. 10) became stronger and more alert on a high-fat, low carbohydrate diet (Scott, 1974). Lactate levels fell to normal and treatment with bicarbonate was no longer necessary. He worsened when challenged again with a normal diet. Although fed a high fat diet, he died in his first year. Four other patients (nos 4,5,8,9) became weaker and more acidotic when the proportion of carbohydrate in their diet increased. The brain can utilize ketone bodies when their concentrations in the blood are high. It then by-passes the PDH reaction. In our opinion, the more seriously ill patients with deficiencies of PDH deserve a trial of a high fat, low carbohydrate, ketogenic diet; but, they should be watched carefully for hypoglycemia and clinical deterioration.

Care of symptoms and close attention to nursing are important. These patients, like many others with metabolic disorders, tolerate mild infections and similar stresses poorly. Five patients (nos 4-7,10) have been treated with bicarbonate for persistent metabolic acidosis.

Treatment with glucocorticoids lysed the symptoms in a patient with intermittent ataxia (Blass *et al.*, 1971) and probably did so in another (no.8) with periodic exacerbations (Falk *et al.*, 1974). It also led to an increase in strength in a patient with "ragged-red myopathy" and lactic acidosis associated with an unknown mitochondrial defect (Shapira *et al.*, 1973, 1974). It is not known whether steroids simply reduced the patients' reactions to intercurrent stresses or had a more specific action.

We have not observed benefits in our patients of doses of the vitamins thiamin and nicotinamide as high as 600 mg a day.

D. *Pathology*

The brain of a PDH-deficient child who died in the first year of life had multiple anomalies including no corpus collosum (Scott, 1974). In the brains of two sisters who died with lactic acidosis, and whose brother has a demonstrated deficiency of PDH, myelin was reduced to less than 10% of normal and neurones had been lost, diffusely. Detailed studies of their brain pathology are in progress. The changes of Leigh's subacute necrotizing encephalomyelopathy did not occur in the brains of these children.

E. *Related Disorders*

Elevations of blood lactate and pyruvate can occur secondary to a wide variety of conditions including hypoxia, poor perfusion, anemia, and with certain poisons (Oliva, 1970). They occur with inherited errors in enzymes of a gluconeogenesis, notably pyruvate carboxylase, EC 6.4.1.1 (Hommes *et al.*, 1968; Grover *et al.*, 1972; Tang *et al.*, 1972; Tada *et al.*, 1973; Delvin *et al.*, 1974) and fructose-1,6-diphosphatase, EC 3.1.3.11 (Baker and Winegrad, 1970; Hulsman and Fernandes, 1971; Pagliara *et al.*, 1972; Greene *et al.*, 1972; Melancon *et al.*, 1973), and with a deficiency of glucose-6-phosphatase, EC 3.1.3.10 (Howell, 1972).

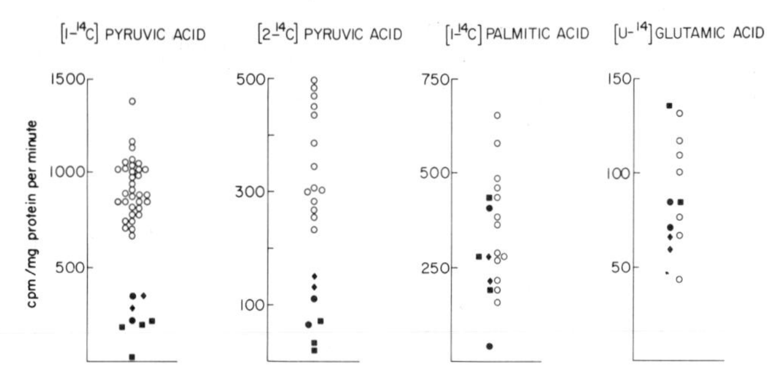

Fig.1 Oxidation of several substrates by intact fibroblasts. Intact cells were incubated with [1-^{14}C] pyruvic acid, [2-^{14}C] pyruvic acid, [1-^{14}C] palmitic acid on albumin, and [U-^{14}C] glutamic acid, as described in detail elsewhere (Blass *et al.*, 1970, 1972; Cederbaum and Blass, 1974). Each point represents the mean of duplicate or triplicate determinations with at least 3 cultures. ○, controls; ■, PDC deficient; ●, deficiency of transacetylase; ◆, site of enzymatic deficiency not yet determined.

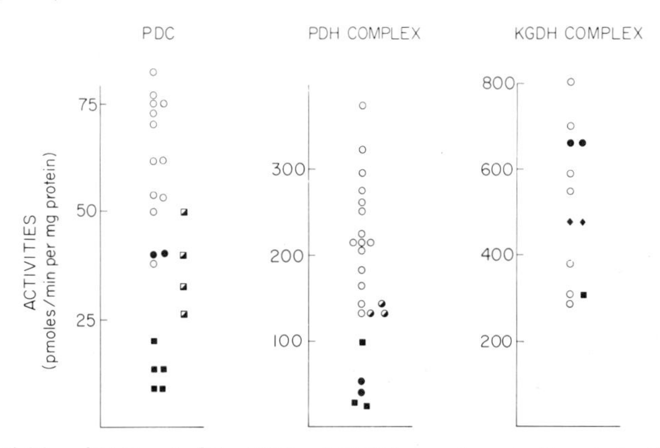

Fig.2 Activities of PDC and of the PDH and KGDH complexes in disrupted cells. Activities of PDC (the thiamin-dependent first enzyme of the PDH complex) or of the overall pyruvate dehydrogenase (PDH) or 2-oxoglutarate dehydrogenase (KGDH) complexes were measured as described elsewhere (Blass *et al.*, 1970, 1972; Cederbaum and Blass, 1974). Each point represents the mean of duplicate or triplicate determinations with extracts from at least 3 cultures for each individual. ◪, parents of patients with PDC deficiency; ◑, parents of patients with abnormalities affecting the transacetylase; other symbols as for Fig.1. Note that for the two patients with defects of the second enzyme (LAT) activities for PDC (the first enzyme) were within the range for controls. For patients with deficiencies of PDC, activities of the whole PDH complex were low, as expected (Blass *et al.*, 1974; Cederbaum and Blass, 1974).

An entity which might prove to be related to inherited deficiencies of pyruvate metabolism is subacute necrotizing leukoencephalopathy (SNE) described by Leigh (1951). The clinical features of this condition vary, but psychomotor retardation, ataxia, and signs of dysfunction of the brain stem often occur (Pincus, 1972). Pathologically, these patients have lesions resembling but not identical to those of Wernicke-Korsakoff syndrome. Pincus *et al.* (1974) and Murphy (1973) studied an inhibitor of TTP-ATP phosphoryl transferase in these patients and have proposed its use as a diagnostic test for SNE. Others have reported deficiencies of pyruvate carboxylase (Hommes *et al.*, 1968; Tada *et al.*, 1973) and of PDC (Farmer *et al.*, 1973) in occasional patients. Pincus *et al.* (1973) have reported that many of their patients responded to high doses of thiamin. Others have found that occasional patients responded to biotin (Keen, 1974). Further studies are needed to allow more precise statements about SNE.

2. ENZYME DEFECTS

The abnormalities in PDH in these patients have been defined by applying standard biochemical techniques *in vitro* to samples of their tissues. Cultured skin fibroblasts have been particularly useful, although defects have also been demonstrated in white blood cells, biopsied muscle and liver and brain from autopsy (Blass *et al.*, 1970; Farrel *et al.*, 1974). The cultured cell lines allowed repeated experiments without added risk or inconvenience to the patients. Furthermore, after 10 serial passages the original material biopsied from the patient has been diluted about a million-fold. Anomalies which persist in culture are coded into the cells and cannot be attributed to environmental factors or to the nonspecific effects of disease.

Defects in the conversion of radioactive pyruvate to $^{14}CO_2$ have been detected in intact cells from 8 patients (Fig.1). Oxidation of [U-^{14}C] glutamate has been normal in all of these lines and of [1-^{14}C] palmitate in all but one. As discussed in detail elsewhere, we cannot explain the low oxidation of palmitate by that patient's cells simply on the basis of their well-documented defect in PDH (Blass *et al.*, 1972).

Low activities of the overall PDH complex were found in cell-free extracts of lines from the five patients (nos 1-5) studied in detail so far (Fig.2). Activities of the 2-oxoglutarate dehydrogenase complex (KGDH) and of a number of other enzymes examined were normal (Blass *et al.*, 1972), including aconitate hydratase (EC 4.2.1.3) and NAD$^+$-linked isocitrate dehydrogenase (EC 1.1.1.41). The enzymes were assayed by radiochemical modifications of standard techniques. Studies of their specificity have been presented elsewhere (Blass *et al.*, 1970, 1972, 1974; Cederbaum and Blass, 1974). Activities for the parents were around the lower end of the control range, as expected if the primary inherited defects are in the PDH complex. The biochemical data, like the pedigrees, suggest that the patients receive abnormal genes from both parents. Mixing experiments did not show a soluble inhibitor of PDH, and addition of excess NAD$^+$, CoA, thiamin pyrophosphate, or oxidized or reduced lipoic acid did not significantly ameliorate the defect in any of the lines studied so far (Blass *et al.*, 1970, 1972; Cederbaum and Blass, 1974).

It has been possible to specify the site of the anomalies in the PDH complex more closely. This complex consists of three catalytic and two regulatory enzymes (Reed and Cox, 1970). The first enzyme (PDC) uses thiamin pyrophosphate as a cofactor, and is correctly called pyruvate dehydrogenase [pyruvate:lipoate oxidoreductase (decarboxylating and acceptor acetylating), EC 1.2.4.1]. The second (LAT) uses lipoic acid and is named lipoate acetyltransferase (Acetyl-CoA: dehydrolipoate S-acetyltransferase, EC 2.3.1.12). The third, which uses NAD$^+$ and FAD, appears to be the same protein in the PDH and KGDH complexes (Sakurai, 1970). It is called lipoamide dehydrogenase (NADH:lipoamide oxidoreductase, EC 1.6.4.3). A specific kinase inactivates the first enzyme by phosphorylating it and a specific phosphatase reactivates it by removing the phosphate (Linn *et al.*,

1969; Walajtys *et al.*, 1974). The complex has been fully activated in extracts of the 6 lines of fibroblasts tested so far, but we have not yet examined this point in cells from most of the patients (Cederbaum and Blass, 1974).

As discussed in detail elsewhere (Blass *et al.*, 1970, 1972, 1974; Cederbaum and Blass, 1974), it has been possible to study the activity of each of the catalytic enzymes, by varying the pH, substrates, and acceptors in the assays. The four patients (nos 4-8) with deficiencies affecting the second, core enzyme - the transacetylase - had severe neurological disease and lactic acidosis (Table I). Three are dead. The first three patients (nos 1-3) with deficiencies in PDC, the thiamin-dependent first enzyme, were relatively mildly ill. Their only prominent abnormality was intermittent ataxia. The patients with PDC deficiency identified since then have been sicker. One (no.10) was the infant mentioned above who died in the first year of life. Activity in his tissues was very low. We have not been able to relate the severity of the symptoms in the other patients to the magnitude of the enzymatic deficiency as measured in their cells by our current techniques.

Similar patients have been identified in a numer of centers, but detailed clinical and enzymatic studies are not yet available.

3. PATHOPHYSIOLOGY

The relationship between the clinical and the biochemical anomalies in these patients has been of great interest to us, particularly since the first patients we encountered with PDC deficiency appeared to have relatively limited and intermittent neurological abnormalities, along with a constant and general enzyme defect (Blass, 1972; Blass *et al.*, 1974). We have attempted to gain a better understanding of the effects on the brain of partial deficiencies in the PDH complex, by posing discrete questions in animal experiments.

A. *Isoenzymes*

No evidence for qualitatively different forms of PDH was found in extensive studies of the purified proteins from a variety of pig and beef tissues, including brain, by Reed's group (Burgett, 1972; Barbera *et al.*, 1972), nor in extensive kinetic studies of PDH from beef brain and kidney by us (Blass and Lewis, 1973a, b) or from rat tissues by Siess *et al.* (1971). Brain and liver as well as cultured fibroblasts were deficient in PDH and PDC in a patient (no.10) who died with lactic acidosis (Farrel *et al.*, 1974). The available evidence suggests that one can draw conclusions about the enzyme in the brain by studying more accessible tissues such as cultured fibroblasts. There is no evidence that cell specific isoenzymes could explain relatively selective effects of PDC deficiency on certain cells in the nervous system.

B. *PDH Activity and Pyruvate Flux*

The next question concerned the quantitative effects of partial deficiencies of the PDH complex on pyruvate metabolism in the brain. Studies in various laboratories indicate that PDH activity in rat brain is about 120-220 μmoles/g wet weight per hour (McCandless *et al.*, 1968; Siess *et al.*, 1972; Cremer and Teal, 1974; Jope and Blass, unpublished). The flux of pyruvate is about 70-140 μmoles/g per hour, whether calculated from O_2 uptake (McIlwain and Bachelard, 1971; Norberg and Siesjö, 1974) or from the flux of radioactive glucose (Gaitonde, 1965) There are fewer values for PDH activities in human brain. We found 90-300 μmoles g per hour in samples from 4 autopsies. Pyruvate flux in human brain *in vivo* is

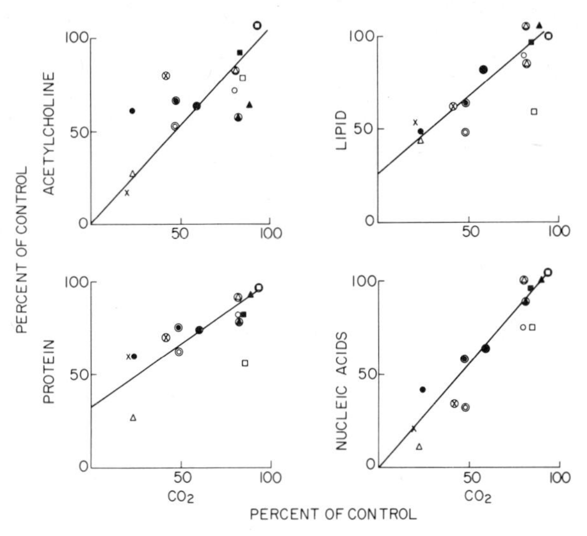

Fig.3 Effects of various compounds on the utilization of [U-^{14}C] glucose. Minces of adult rat brain were incubated in a modified Krebs-Ringer solution containing 31 mM-K^+ and 5 mM-[U-^{14}C] glucose for 1 h, and radioactivity in CO_2 and in other fractions was determined. The experimental details are described elsewhere (Gibson *et al.*, 1974; Gibson and Blass, 1974a,b). Each point represents the mean of triplicate determinations with at least two preparations. SEM was smaller than the magnitude of the symbols. ●, incubated under N_2 ; △, 1.0 mM-KCN; ■, 1 mM-malonate; ○, 5 mM-malonate; □, 1.0 mM-NaF; ×, 0.5 mM-pentobarbitol; ⊗, 0.5 mM-amobarbitol; ⊖, 2 mM-leucine; ⊘, 2 mM-isoleucine, ▲, 2 mM-valine; ⊛, 2-oxo-4-methylpentanoate; ◉, 2 mM-2-oxo-3-methylpentanoate; ◉, 2 mM-2-oxo-3-methylbutanoate; ◎, amino and keto acids combined.

about 85-135 μmoles/g per hour, calculated from O_2 uptake (McIlwain and Bachelard, 1971). Thus, defects which reduced PDH activity to less than a third of normal might be expected to reduce the flux of pyruvate to acetylCoA in human or rat brain.

C. *Metabolic Effects*

A third question concerned the effects on brain metabolism of mild to moderate impairments in pyruvate oxidation. Quantitatively, the major fate of the acetyl-CoA produced is oxidation to CO_2 by the tricarboxylic acid cycle, but studies in at least four laboratories have demonstrated that impairments of carbohydrate oxidation can permanently abolish electrical activity without reducing brain ATP (King *et al.*, 1967, MacMillan and Siesjö, 1971; Siesjö and Plum, 1971; Drewes and Gilboe, 1973; Ljungren *et al.*, 1974; Yatsu and Liao, 1974). To examine directly the effects of impaired pyruvate oxidation, a variety of compounds which reduced pyruvate oxidation were added to minces of rat brain respiring in buffered saline containing either 5 mM-glucose [U-^{14}C] or 5 mM-sodium pyruvate [1-^{14}C] or [2-^{14}C]. With glucose as substrate, (Gibson and Blass, 1974a,b) agents which impaired the production of $^{14}CO_2$ also impaired the incorporation of radioactivity into acetylcholine (ACh) and into lipids, proteins and nucleic acids (Fig.3). Effects were roughly proportional, even though less than 2% of the radioactivity converted to $^{14}CO_2$ was incorporated into the synthetic products examined and less than

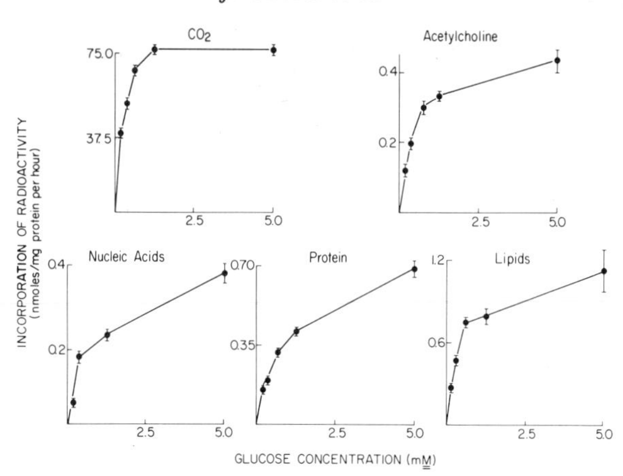

Fig.4 Glucose utilization with varying concentrations of glucose. Rat brain minces were incubated as described for Fig.3, except that the concentrations of glucose were varied. The points indicate the means of triplicate determinations for at least two preparations; the bars show SEM. For details, see Gibson and Blass (1974a).

Table II. Activities of the pyruvate dehydrogenase complex in regions of cat brain grey matter.

	PDH (nmoles/min per mg wet wt.)	PDH/SDH ($\times 10^2$)	PDH/Cytochrome Oxidase	PDH/QO_2
Cerebellar vermis	0.51 ± 0.06	0.64	0.11	0.45
Sensory cortex	0.91 ± 0.05	2.04	0.38	1.13
Upper medulla	0.91 ± 0.20	1.98	0.30	1.17
Temporal cortex	0.98 ± 0.23	ND	ND	0.84
Motor cortex	1.11 ± 0.20	2.02	0.30	1.40
Thalamus	1.20 ± 0.18	1.98	0.40	1.20
Caudate nucleus	1.66 ± 0.11	2.60	0.48	2.06

Reynolds *et al.* (1974) discuss in detail the methods used to measure the relative activity of PDH and of succinate dehydrogenase (SDH), cytochrome oxidase, and O_2 uptake (QO_2) in pieces from various regions of cat brain grey matter. Values for PDH are means ± SEM for triplicate determinations on 6-18 samples from each region. The QO_2 was determined in duplicate in 6-8 samples for each region, and SEM was 6-15%. Cytochrome oxidase and SDH were determined in duplicate in 4 samples from each area; SEM was less than 7%. ND, not determined.

0.5% into ACh. With radioactive pyruvate as substrate the reduction in ACh synthesis was proportional to the reduction in $^{14}CO_2$ produced, but effects on proteins, lipids, and nucleic acids were less marked (Gibson *et al.*, 1974). Reducing the concentration of glucose in the media also reduced ACh synthesis (Fig.4). Effects were particularly sharp below 1 mM-glucose, the normal level of glucose in rat brain. LeFresne *et al.* (1973) have shown that the synthesis of ACh by rat brain slices depended on the concentration of pyruvic acid in the incubation media. Others have shown that inhibiting carbohydrate metabolism with flurocitrate reduces the synthesis of amino acids and of another neurohormone, GABA

Table III. Disorders where abnormal proteins predispose to the development of clinical disease.

Organ System	Disorder	Enzyme	External Stress
Nervous system	PDC deficiency	PDC	Fever, hypermetabolism
Nervous system	Intermittent maple syrup urine disease	Branched chain dehydrogenase	Fever, protein load
Nervous system	Intermittent isovaleric acidemia	Isovaleryl-CoA dehydrogenase	Fever, protein load
Nervous system	Acute intermittent porphyria	Uroporphyrin 1-synthetase	Barbiturates, other drugs
Nervous system	Serum pseudocholinesterase deficiency	Pseudocholinesterase	Succinylcholine
Red blood cells	G6PD deficiency	Glucose-6-phosphate dehydrogenase	Drugs, fava beans, etc.
Red blood cells	Methemoglobinemia	Heterozygous methemoglobin reductase deficiency	Antimalarials, other drugs
Red blood cells	Hemolytic anemia	Certain hemoglobin abnormalities	Hypoxia, certain drugs
Clotting system	Mild hemophilia	Factor VIII or IX deficiency	Trauma, surgery
Lungs	Emphysema	Heterozygous α-antitrypsin deficiency	Cigarette smoking
Skin	Xeroderma pigmentosum, several forms	DNA excision-repair enzymes	Sunlight
Skeletal muscle	Starvation myoglobinuria	Carnitine acyltransferase deficiency	Fasting
G.I. Tract	Fermentative diarrhea	Lactase deficiency	Milk products, lactose
Liver	Fructose interolerance	Fructose-1-phosphate aldolase	Sweet fruits, sucrose

Only selected examples are given. For references, see McKusick (1974). Disorders in which loss of dietary control leads to symptoms, such as galactosemia, phenylketonuria, or maple syrup urine disease, are not listed.

Clarke *et al.*, 1970; Patel and Koenig, 1971). Even partial reductions in pyruvate oxidation can impair a number of biosynthetic pathways in brain, including the synthesis of neurohormones and particularly of ACh, *in vitro.* Recent experiments suggest that impairing carbohydrate metabolism does impair the synthesis of a physiologically important pool of ACh (Gibson and Blass, 1974a). Furthermore, a variety of studies in the literature indicate that ACh synthesis and cholinergic transmission are relatively sensitive to impairments of carbohydrate metabolism *in vivo* and *in vitro*, in various species, and by physiological and biochemical tests (LeFresne *et al.*, 1973; Browning and Schulman, 1968; Grewall and Quastel, 1973; Guyenet *et al.*, 1973; Dolivo, 1974; Blass *et al.*, 1974).

D. *Regional Susceptibility*

The activity of PDH appears to be relatively low in some areas of brain compared to others (Reynolds *et al.*, 1974; Blass *et al.*, 1974). Among several areas of cat brain grey matter, PDH activity was lowest in an area of cerebellar vermis, whether measured per unit tissue or as a ratio of the activity of succinate dehydrogenase (EC 1.3.99.1), cytochrome oxidase (EC 1.9.3.1), or O_2 uptake (Table II). Others have found pyruvate oxidation in preparations of rat cerebella to be lower and more sensitive to impairment from thiamin deficiency or acetaldehyde than in preparations from other parts of rat brain (McCandless *et al.*, 1968; Kiessling, 1962). The area of vermis studied was chosen to be analogous to an area often damaged in thiamin-deficient human alcoholics (Victor *et al.*, 1971). As noted, cerebellar ataxia was the striking symptom in the patients we studied with inherited deficiencies of PDC. These observations suggest that certain structures in the cerebellar vermis might have relatively little excess of PDH compared to pyruvate flux and therefore, differential susceptibility to deficiencies in PDH (Blass *et al.*, 1974). A firmer conclusion would require much more extensive measurements on more discrete pieces of brain as well as physiological studies.

4. IMPLICATIONS

Despite the many unanswered questions about the inherited defects of pyruvate oxidation, they do have some implications. Firstly, they document that mutations affecting major pathways of carbohydrate metabolism need not be lethal. They occurred in people who were sick but not dead. Inherited deficiencies have also been demonstrated to affect enzymes of glycolysis (Valentine and Tanaka, 1972), gluconeogenesis (see above), glycogenolysis (Howell, 1972), and probably in oxidative phosphorylation (Luft *et al.*, 1962; Haydar *et al.*, 1971; Shapira *et al.*, 1973, 1974). Many - but not all - of these defects are specific to certain tissues. There are a group of disorders of glucose metabolism analogous to the aminoacidopathies and to the lipidoses.

Secondly, the mild form of PDC deficiency with intermittent ataxia is one of several conditions in which an abnormality in an enzyme predisposes to brain disease, when the affected individual is under the appropriate stress. Other examples include intermittent porphyria, Leber's optic atrophy (Wilson, 1965), and the intermittent forms of maple syrup urine disease (Dancis *et al.*, 1967), of isovaleric acidemia (Budd *et al.*, 1967), and of propionic acidemia (Glaser, 1974). Many diseases of other organs are associated with abnormalities of particular proteins which predispose to the development of clinical symptoms (Table III). Garrod (1928) suggested, Childs (1970) reemphasized, and the known genetic heterogeneity of proteins (Harris, 1971) would support the notion that mild anomalies in specific enzymes could underlie genetic predispositions to diseases. It is interesting that genetic predisposition to the development of neuropsychiatric disorders are well documented (Heston, 1970), and that the brain, which has a minute-to-minute dependence on oxidative metabolism, might be expected to be affected by anomalies which altered its ability to use carbohydrates.

Finally, exploiting genetic defects has provided unique information about the chemical physiology of many organisms. To do such studies with man, precise and detailed clinical observations are as necessary as discrete experiments. We hope that such studies of inherited disorders of energy metabolism may not only apply basic knowledge but also contribute to it.

ACKNOWLEDGEMENTS

We thank the physicians who cared for the patients for samples of skin or fibroblasts and for making their records available to us. We also wish to thank Ms. E. Hom, M. Cotton, S. Harris, C. Lewis Pollito, and L. Graul who provided expert technical help. These studies were supported by grants HD-06051, HD-06576, HD-34504, HD-461205, GM-00364 and by the California State Department of Mental Hygiene.

REFERENCES

Baker, L. and Winegrad, A.I. (1970). *Lancet* **II**, 13-16.

Barbera, C.R., Namahira, G., Hamilton, L., Munk, P., Eley, M.H., Linn, T.C. and Reed, L.J. (1972). *Arch. biochem. Biophys.* **148**, 343-352.

Blass, J.P. (1972). *Int. J. Neurosci.* **4**, 65-69.

Blass, J.P., Avigan, J. and Uhlendorf, B.W. (1970). *J. clin. Invest.* **49**, 423-432.

Blass, J.P., Gibson, G. and Kark, R.A.P. (1974). *Proc. US-Japan Conference on Thiamin (in press).*

Blass, J.P. and Lewis, C.A. (1973a). *Biochem. J.* **131**, 31-37.

Blass, J.P. and Lewis, C.A. (1973b). *Biochem. J.* **131**, 415-417.

Blass, J.P., Kark, R.A.P. and Engel, W.K. (1971a). *Arch. Neurol.* **25**, 449-460.

Blass, J.P., Lonsdale, D., Uhlendorf, B.W. and Hom, E. (1971b). *Lancet* **I**, 1302.

Blass, J.P., Lonsdale, D., Uhlendorf, B.W. and Hom, E. (1972). *J. clin. Invest.* **51**, 1845-1851.

Browning, E.T. and Schulman, M.P. (1968). *J. Neurochem.* **15**, 1391-1405.

Budd, M.A., Tanaka, K., Holmes, L.B., Efron, M.L., Crawford, J.D. and Isselbacher, K.J. (1967). *New Eng. J. Med.* **277**, 321-327.

Burgett, M. (1972). Ph.D. Thesis, U. of Texas at Austin, University Microfilms 73-7519. Ann. Arbor, Michigan.

Cederbaum, S.D., Blass, J.P. and Cotton, H.E, (1973). *Clin. Res.* **21**, 530.

Cederbaum, S.D. and Minkoff, N. (1972). *Amer. J. Hum. Gen.* **24**, 23a.

Cederbaum, S.D. and Blass, J.P. (1974). *(In preparation).*

Childs, B. (1970). *New Eng. J. Med.* **282**, 71-77.

Clarke, D.D., Nicklas, W.J. and Berl, S. (1970). *Biochem. J.* **120**, 345-351.

Cremer, J.E. and Teal, H.M. (1974). *FEBS Letts.* **39**, 17-20.

Dancis, J., Hutzler, J. and Rokkones, T. (1967). *New Eng. J. Med.* **276**, 84-89.

Delvin, E., Scriver, C.R. and Neal, J.L. (1974). *Biochem. Med.* **10**, 97-106.

Dolivo, M. (1974). *Fedn Proc.* **33**, 1043-1051.

Drewes, L.R. and Gilboe, D.D. (1973). *J. biol. Chem.* **248**, 2489-2496.

Erickson, R.J. (1965). *J. Pediatr.* **66**, 1004-1016.

Falk, R.E., Cederbaum, S.D. and Carrel, R.E. (1974). *Amer. J. Hum. Gen.* **26** *(in press)* **Abs.**

Farmer, T.W., Veath, L., Miller, A.L., O'Brien, J.S. and Rosenberg, R.N. (1973). *Neurology* **23**, 429.

Farrell, D.F., Clark, A.F., Scott, C.R. and Wennberg, R.P. (1974). *Science (in press).*

Gaitonde, M.K. (1965). *Biochem. J.* **95**, 803.

Garrod, A. (1928). *Brit. Med. J.* **1**, 1099-1101.

Gibson, G.E. and Blass, J.P. (1974a). *(In preparation).*

Gibson, G.E. and Blass, J.P. (1974b). *(In preparation).*

Gibson, G.E., Jope, R. and Blass, J.P. (1974). *(Submitted for publication).*

Glaser, D. (1974). Personal communication.

Greene, H.L., Stifel, F.B. and Herman, R.H. (1972). *Amer. J. Dis. Child.* **124**, 415-418.

Grewaal, D.S. and Quastel, J.H. (1973). *Biochem. J.* **132**, 1-14.

Grover, W.D,, Auerbach, V.H. and Patel, M.S. (1972). *J. Pediatr.* **81**, 39-44.

Guyenet, P., LeFresne, P., Rossier, J., Beaujouan, J.C. and Glowinski, J. (1973). *Molec. Pharmacol.* **9**, 630-639.

Harris, H. (1971). "The Principles of Human Biochemical Genetics". American Elsevier, New York.

Hartmann, A.F., Wohltmann, H.J., Purkerson, M.L. and Wesley, M.E. (1962). *J. Pediatr.* **61**, 165-180.

Haworth, J.C., Ford, J.D. and Younoszai, M.K. (1967). *Canad. Med. Assoc. J.* **97**, 773-779.
Haydar, N.A., Conn, H.L., Afifi, A., Nakid, N., Ballas, S. and Fawaz, K. (1971). *Ann. Int. Med.* **74**, 548-558.
Heston, L.L. (1970). *Science* **167**, 249-256.
Hommes, F.A., Polman, H.A. and Reerink, J.D. (1968). *Arch. Dis. Childh.* **43**, 423-426.
Howell, R.R. (1972). *In:* "The Metabolic Basis of Inherited Disease" (J.B. Stanbury, J.B. Wyngaarden and D.S. Fredrickson, eds) pp.149-173. Blakiston-McGraw Hill, New York and London.
Hulsmann, W.C. and Fernandes, J. (1971). *Pediat. Res.* **5**, 633-637.
Israels, S., Haworth, J.C., Gowley, B. and Ford, J.D. (1964). *Pediatrics* **34**, 346-356.
Keen, J.S., personal communication.
Kiessling, K.H. (1962). *Exptl. cell. Res.* **27**, 367-372.
King, L.J., Lowry, O.H., Passonneau, J. and Venson, V. (1967). *J. Neurochem.* **14**, 599-611.
LeFresne, P., Guyunet, P. and Glowinski, J. (1973). *J. Neurochem.* **20**, 1083-1097.
Leigh, D. (1951). *J. Neurol. Neurosurg. Psychiat.* **14**, 216-221.
Linn, T.C., Pettit, F.H. and Reed, L.J. (1969). *Proc. natn. Acad. Sci. U.S.A.* **62**, 234-241.
Ljunggren, B., Ratcheson, R.A. and Siesjö, B.K. (1974). *Brain Res.* **73**, 291-307.
Lonsdale, D., Faulkner, W.R., Price, J.W. and Smeby, R.R. (1969). *Pediatrics* **43**, 1025-1034.
Luft, R., Ikkos, D., Palmieri, G., Ernster, L. and Afzelius, B. (1962). *J. clin. Invest.* **41**, 1776-1804.
McCandless, D.W., Schenker, S. and Cook, M. (1968). *J. clin. Invest.* **47**, 2268-2280.
McIlwain, H. and Bachelard, H.S. (1971). "Biochemistry and the Central Nervous System", pp.8-31. Churchill Livingstone, Edinburgh and London.
McKusick, V.A. (1974). "Mendelian Inheritance in Man", 4th edn. Johns Hopkins Press, Baltimore.
MacMillan, V. and Siesjö, B.K. (1971). *Europ. Neurol.* **6**, 66-72.
Melancon, S.B., Khachadurian, A.K., Nadler, H.L. and Brown, B.I. (1973). *J. Pediatr.* **82**, 650-657.
Murphy, J.V. (1973). *Pediatrics* **51**, 710-715.
Norberg, K. and Siesjö, B.K. (1974). *J. Neurochem.* **22**, 1127-1129.
Oliva, P.B. (1970). *Amer. J. Med.* **48**, 208-225.
Pagliara, A.S., Karl, I.E., Keating, J.P., Brown, B.I. and Kipnis, D.M. (1972). *J. clin. Invest.* **51**, 2115-2123.
Patel, A. and Koenig, H. (1971). *J. Neurochem.* **18**, 621-628.
Pincus, J.H. (1972). *Develop. Med. Child. Neurol.* **14**, 87-101.
Pincus, J.H., Cooper, J.R., Murphy, J.V., Rabe, E.F., Lonsdale D. and Dunn, H.G. (1973). *Pediatrics* **51**, 716-721.
Pincus, J.H., Cooper, J.R., Piros, K. and Turner, V. (1974). *Neurology* **24**, 885-890.
Podos, S.M. (1970). *Arch. Opthal.* **83**, 504-505.
Reed, L.J. and Cox, D. (1970). *In:* "The Enzymes" (P.D. Boyer, ed) 3rd edn., pp.213-240. Academic Press, New York and London.
Reynolds, S.F., Jope, R. and Blass, J.P. (1974). *Submitted for publication.*
Sakurai, Y. (1970). *Seikagaku* **42**, 726-736.
Scott, C.R. (1974). Personal communication.
Shapira, Y., Cederbaum, S.D., Lippe, B.M. and Verity, M.A. (1973). *Ped. Res.* **7**, 195a.
Shapira, Y., Cederbaum, S.D., Cancilla, P.A., Nielsen, D. and Lippe, B.M. (1974). *Submitted for publication.*
Siesjö, B.K. and Plum, F. (1971). *Acta Anaesth. Scand., Suppl.* **45**, 81-101.
Siess, E., Wittman, J. and Wieland, O. (1971). *Z. Physiol. Chem.* **352**, 447-452.
Tada, K., Sugita, K., Fujitoni, K., Uesakai, T., Takado, G. and Omura, K. (1973). *Tohoku J. Exp. Med.* **109**, 13-18.
Tang, T.T., Good, T.A., Dykeu, P.R., Johnson, S.D., McCreadie, S.R., Sy, S.T., Lardy, H.A. and Rudolph, F.B. (1972). *J. Pediatr.* **81**, 189-190.
Valentine, W.N. and Tanaka, K.R. (1972). *In:* "The Metabolic Basis of Inherited Disease" (J.B. Stanbury, J.B. Wyngaarden and D.S. Fredrickson, eds) pp.1338-1357. Blakiston-McGraw Hill, New York and London.
Victor, M., Adams, R.D. and Collins, G.H. (1971). "The Wernicke-Korsakoff Syndrome". F.A. Davis Co., Philadelphia.
Walajtys, E.I., Gottesman, D.P. and Williamson, R.J. (1974). *J. biol. Chem.* **249**, 1857-1865.

Wilson, J. (1965). *Clin. Sci.* 29, 505-515.
Worsley, H.E., Brookfield, R.W., Elwood, J.S., Noble, R.L. and Taylor, W.H. (1965). *Arch. Dis. Childh.* 40, 492-501.
Yatsu, F.M. and Liao, C.L. (1974). *Trans. Amer. Soc. Neurochem.* 5, 88.

DISCUSSION

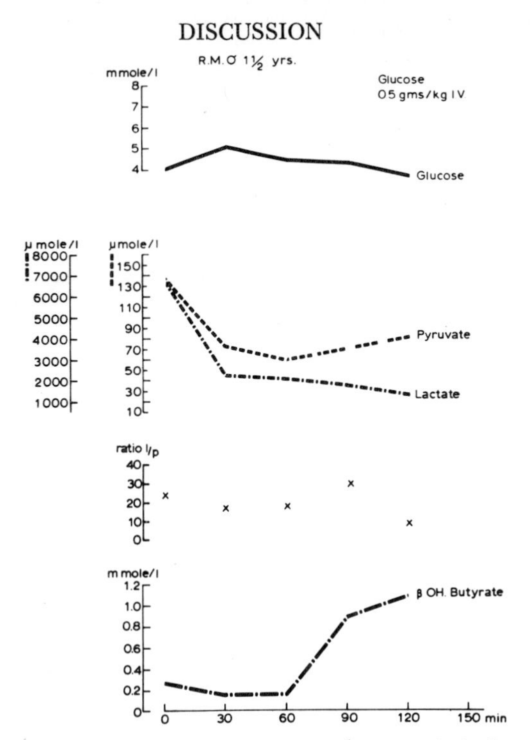

Fig.1 Intravenous glucose loading test of patient R.M. (0.5 g per kg body weight).

Koepp: Can you comment on the relationship between pyruvate dehydrogenase and Leigh's encephalopathy?

Blass: The data available right now do not allow a simple comment. In principle it certainly seems possible that more than one inherited biochemical abnormality could lead to a similar pathological pattern. The lipidosis may provide an analogy. To understand these patients better and to be able to do more for them we are, I think, going to have to do more chemistry - including unequivocal definition of the nature of the "inhibitor" described by Pincus and co-workers (*Neurology* 19, 842 (1969)) and a careful analysis of the reaction which it is inhibiting.

Hommes: How do you visualize a normal lactate to pyruvate ratio in blood of a patient with pyruvate dehydrogenase deficiency?

Blass: I assume that the deficiency is in the entrance of carbohydrate to the tricarboxylic acid cycle. This need not alter necessarily the intramitochondrial redox potential.

Hommes: The lactate to pyruvate ratio is not completely independent of the

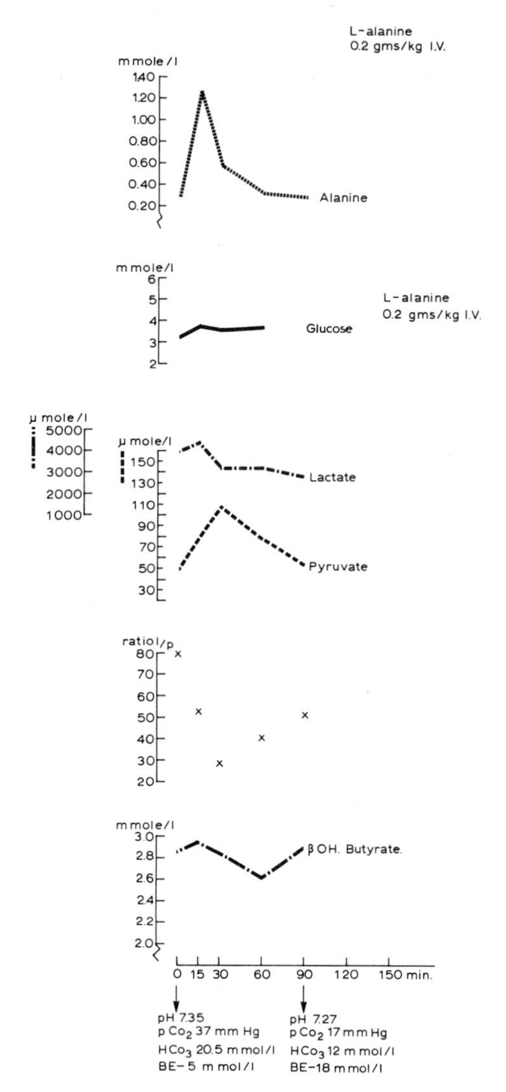

Fig. 2 **Intravenous alanine loading test of patient R.M. (0.2 g per kg body weight, given after a 23 h fast).**

mitochondrial NAD to NADH ratio. In every condition where you have some interference with oxidative metabolism you do get an increase in the lactate to pyruvate ratio.

Tager: I think that in this particular case, where the lesion is at the level of pyruvate dehydrogenase, this need not necessarily be the case. If you have an efficient means of transporting reducing equivalents from the cytosol into the mitochondrion, and of oxidizing these reducing equivalents by a pathway which need not and probably does not involve pyruvate dehydrogenase, you would have no effect on the lactate to pyruvate ratio.

Barth: To elaborate on what Dr. Hommes just said, I want to show some data on a patient with pyruvate dehydrogenase deficiency. It is a male patient of 1½ years old, who is mentally retarded at the 5 month level, ataxic not parethic. He had one episode of deterioration at the age of four months when he had a gastrointestinal infection. We found increased levels of lactate and pyruvate (up to 8 mM and 0.13 mM respectively) with an increased lactate to pyruvate ratio. An intravenous glucose loading test resulted in a decrease of the blood lactate and pyruvate concentrations (Fig.1) and in a decrease of the lactate to pyruvate ratio. β-Hydroxybutyrate showed an increase, but only after 60 min. An intravenous alanine loading test (Fig.2) resulted likewise in a decrease of the lactate to pyruvate ratio. We have no explanation for these phenomena. The patient was demonstrated to be deficient in liver pyruvate dehydrogenase but had a normal activity of liver pyruvate carboxylase and a normal activity of muscle pyruvate dehydrogenase, as well as a normal cytochrome system in muscle mitochondria (assays carried out in Dr. Hommes' laboratory).

Koepp: Was the clinical status of the patient affected by the glucose load?

Barth: No.

Tager: Were there changes in blood ketone bodies in your patients with pyruvate dehydrogenase deficiency, Dr. Blass?

Blass: We do not know, unfortunately.

Monnens: We studied a patient with pyruvate dehydrogenase deficiency and found this patient always had low levels of ketone bodies in the blood.

Tager: The reason why I bring up this point is that I am not quite sure what the mechanism of ketone body formation on glucose feeding would be in these patients with pyruvate dehydrogenase deficiency.

Hommes: There is a lag before the ketone bodies start to accumulate. I am not sure whether they are derived from glucose.

Tildon: The injection of alanine causes a decrease in ketone bodies, which would suggest just the opposite. In fact it is now accepted that alanine has an antiketogenic effect. Whether this defect in pyruvate dehydrogenase contributes to a fault in the mechanism of the antiketogenic effect of alanine is certainly worth exploring.

Wick: There could be some hormonal control involved, because how will these patients get rid of the high glucose levels if the pyruvate dehydrogenase complex is not sufficiently active? I think that they have to synthesize glycogen although they are in need for energy. This could be the cause of ketone body formation.

Monnens: There is some danger in giving alanine to patients with an increase of pyruvate and lactic acid in the blood. A patient with a deficiency of cytochrome oxidase in muscle tissue, when given alanine, had a severe metabolic acidosis with a bicarbonate level of 4.

Koster: Were the fibroblasts harvested at the same subculture?

Blass: Yes. They were grown in the same media, grown in parallel and assayed in parallel with the same substrate and at the same time.

Tager: You mentioned that pyruvate carboxylase could be assayed in fibroblasts. What is the activity of this enzyme in fibroblasts?

Blass: We measured CO_2 fixation by intact fibroblasts which is very low, but is measurable.

Willems: We measured the enzyme in fibroblasts. The activity of pyruvate carboxylase is 2 to 5% of that of the liver.

Tager: A deficiency of pyruvate carboxylase would therefore be difficult to detect in fibroblasts.

Willems: Yes.

BIOCHEMICAL ASPECTS OF HEREDITARY FRUCTOSE INTOLERANCE *

G. Van den Berghe

Laboratoire de Chimie Physiologique
International Institute of Cellular Pathology, Université de Louvain, UCL 7539
Avenue Hippocrate 75, B-1200 Brussels, Belgium

INTRODUCTION

Hereditary Fructose Intolerance was first reported by Chambers and Pratt (1956) and recognized as an inborn error of metabolism by Froesch *et al.* (1957). It was shown by Hers and Joassin (1961) that in patients with this disorder the ability of liver aldolase to split fructose 1-phosphate is almost completely lost. This finding adequately explained the accumulation of fructose that occurs in the blood of the patients after the ingestion of the hexose. The mechanism whereby the enzyme defect provokes the profound hypoglycemia that dominates the clinical picture was, however, not readily evident and became the subject of numerous clinical and biochemical investigations. The decrease of blood glucose after the ingestion of fructose was shown by Dubois *et al.* (1961) to be due to a marked inhibition of the glucose output by the liver. Cornblath *et al.* (1963) reported that the hypoglycemia could not be relieved by the administration of glucagon, indicating a block of the degradation of glycogen.

In this chapter, pertinent aspects of fructose metabolism, the primary enzyme defect and the hypotheses that have been put forward to explain the hypoglycemia of hereditary fructose intolerance will first be briefly reviewed. In the following paragraphs, the effects of the parenteral administration of fructose to normal animals, that have added a lot to the understanding of the pathogenesis of hereditary fructose intolerance, and the glycogenolytic mechanism of glucagon, which has been thoroughly worked out in recent years, will be described. Finally experimental, mostly biochemical data will be presented, concerning the mechanism by which fructose metabolism interferes with the glycogenolytic action of glucagon.

1. THE METABOLISM OF FRUCTOSE

Fructose is metabolized mainly in the liver (Levine and Huddlestun, 1947; Mendeloff and Weichselbaum, 1953), the kidney (Reinecke, 1944) and the small intestine (Bollman and Mann, 1931). The pathway, described in Fig.1, includes a number of specialized enzymes for the utilization of fructose.

* Bevoegdverklaard Navorser van het "Nationaal Fonds voor Wetenschappelijk Onderzoek".

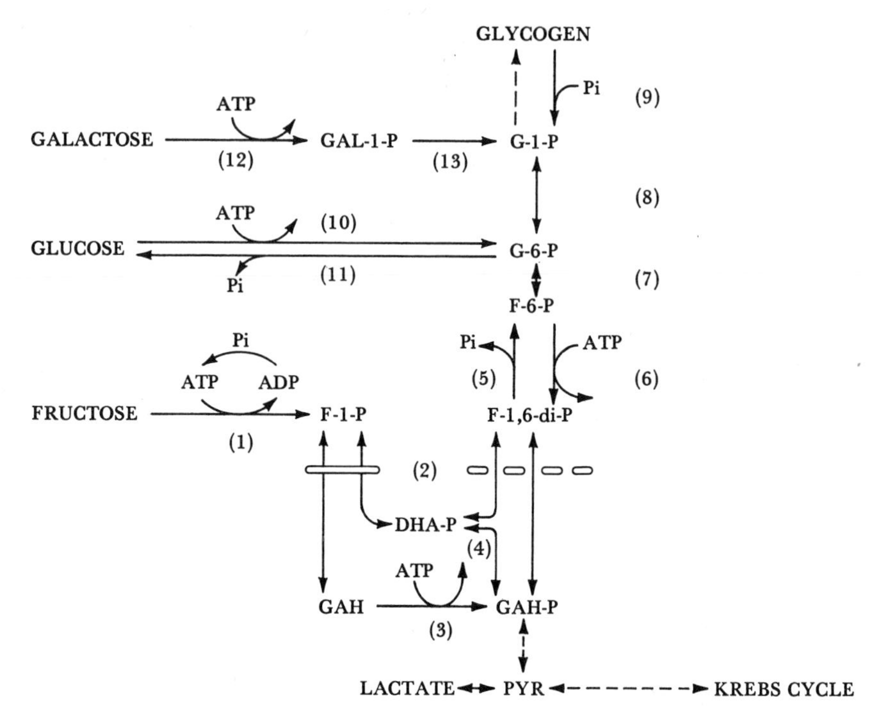

Fig.1 The pathway of fructose metabolism in liver, kidney and small intestine. (1) Ketohexokinase; (2) aldolase; (3) triokinase; (4) triose phosphate isomerase; (5) hexosediphosphatase; (6) phosphofructokinase; (7) phosphohexoseisomerase; (8) phosphoglucomutase; (9) phosphorylase; (10) hexokinase (glucokinase); (11) glucose 6-phosphatase; (12) galactokinase; (13) galactose 1-phosphate uridylyl transferase.

Fructose is first phosphorylated into fructose 1-phosphate by ketohexokinase (previously called fructokinase) (Leuthardt and Testa, 1951; Cori *et al.* , 1951; Hers, 1952a). The phosphoryl-donor is ATP (Hers, 1952b). This enzyme has a very high affinity for fructose with a Km lower than 0.5 mM (Hers, 1952a) and an appreciable Vmax of 1-3 U/g at 25° (Heinz *et al.*, 1968; Woods *et al.*, 1970). The phosphorylation of fructose is therefore a rapid process. Ketohexokinase is not specific for fructose, since it catalyses also the phosphorylation of L-sorbose, D-tagatose, D-xylulose and L-galactoheptulose (Adelman *et al.*, 1967; Sanchez *et al.*, 1971).

Fructose 1-phosphate is split by liver aldolase or aldolase B into D-glyceraldehyde and dihydroxyaceton phosphate (Hers and Kusaka, 1953; Leuthardt *et al.*, 1953). The same enzyme catalyses the splitting of fructose 1,6-diphosphate into D-glyceraldehyde 3-phosphate and dihydroxyaceton phosphate, as well as the condensation of these triose phosphates into fructose 1,6-diphosphate. The maximal activity of the liver aldolase on fructose 1,6-diphosphate is the same as on fructose 1-phosphate (7-15 U/g at 37°): this is expressed by a cleavage ratio equal to 1 (Hers and Joassin, 1961).

The metabolic fate of D-glyceraldehyde originating from the splitting of fructose 1-phosphate has been the subject of controversies (for a review, see Hue, 1974). There are three mechanisms by which D-glyceraldehyde can be converted into a triose phosphate: (1) direct phosphorylation by ATP and triokinase (Hers and Kusaka, 1953); (b) reduction to glycerol by NADH and alcohol dehydrogenase (Wolf and Leuthardt, 1953) or by NADPH and aldose reductase (Hers, 1960), followed by phosphorylation into D-glycerol 1-phosphate (Bublitz and Kennedy, 1954) and subsequent oxidation into dihydroxyaceton phosphate; (c) oxidation to glyceric acid (Leuthardt *et al.*, 1953) and conversion to 2-phosphoglycerate (Holzer and Holldorf, 1957; Ichihara and Greenberg, 1957). Experiments with [4-^{3}H,6-^{14}C] fructose (Hue and Hers, 1972) have shown conclusively that, *in vivo*, D-glyceraldehyde is phosphorylated by triokinase according to the first hypothesis. The maximal activity of the enzyme is 1.5-6 U/g at 37° (Hers, 1962; Heinz *et al.*, 1968) and the Km for D-glyceraldehyde is equal to about 0.01 mM (Sillero *et al.*, 1969). Dihydroxyaceton phosphate being another intermediate of the glycolytic-gluconeogenic pathway, fructose is thus in part broken down to lactic acid and oxidized and in part converted into glucose and glycogen.

The utilization of fructose by peripheral tissues, with the possible exception of adipose tissue (Froesch and Ginsberg, 1962) seems to be negligible (Wick *et al.*, 1953; Weichselbaum *et al.*, 1953), because these tissues do not possess the specialized enzymes ketohexokinase and triokinase. Although fructose can be phosphorylated by hexokinase, the level of glucose in the blood prevents this process, as the enzyme has 20 times more affinity for glucose than for fructose (Sols and Crane, 1954). Glucokinase on the other hand is inactive on fructose (Vinuela *et al.*, 1963).

2. THE ENZYME DEFECT IN HEREDITARY FRUCTOSE INTOLERANCE

From the finding that the administration of fructose provoked a prolonged accumulation of the ketose in the blood of patients with hereditary fructose intolerance (Froesch *et al.*, 1957) postulated that this disorder was caused by the congenital absence of an enzyme of the metabolic pathway of fructose, Hers and Joassin (1961) showed a modification of the properties of liver aldolase. In biopsies of the liver of patients with hereditary fructose intolerance the activity of aldolase towards fructose 1-phosphate was decreased to around 4% of the normal value, whereas about 25% of the activity towards fructose 1,6-diphosphate remained (Fig.1). This resulted in an increased fructose 1,6-diphosphate:fructose 1-phosphate ratio of about 6. The characteristics of the enzyme thus come closer to the ones of the muscle (aldolase A) and brain enzyme (aldolase C) that have a fructose 1,6-diphosphate:fructose 1-phosphate ratio equal to 50 and 10 respectively. Immunologic studies by Nordmann *et al.* (1968) have shown that the liver of the patients contains a protein sharing the immunological properties of purified liver aldolase. The enzyme defect has also been described in the small intestine (Nisell and Linden, 1968) and in the renal cortex of the patients (Kranhold *et al.*, 1969). It is interesting to mention that recently, correction of the fructose 1,6-diphosphate:fructose 1-phosphate ratio has been obtained in liver extracts of 3 out of 5 patients with hereditary fructose intolerance, upon addition of antibodies prepared against normal human liver aldolase (Gitzelmann *et al.*, 1974).

From the enzyme defect it can be reasonably assumed that fructose 1-phosphate accumulates in the tissues that possess ketohexokinase: liver, kidney, and small intestine, although direct evidence of this accumulation is scarce (Pitkänen and Perheentupa, 1962; Hue, unpublished observations), due to methodological difficulties. The intracellular accumulation of the phosphate ester explains the accumulation of fructose in the blood: fructose 1-phosphate has indeed been shown to inhibit ketohexokinase once its concentration reaches 10 mM (Froesch *et al.*, 1959). The fact that the major toxic effects of fructose are localized in the organs that possess ketohexokinase points towards fructose 1-phosphate as the toxic metabolite. The problem of the toxicity of fructose has been studied most extensively in experimental models that will be discussed in section 4.

3. THE MECHANISM OF THE HYPOGLYCEMIA IN HEREDITARY FRUCTOSE

It has been shown conclusively that the fructose-induced hypoglycemia of hereditary fructose intolerance is not due to a stimulation of the secretion of insulin. This mechanism would result in an increased utilization of glucose by peripheral tissues. Dubois *et al.* (1961) demonstrated that, on the contrary, the coefficient of glucose assimilation decreased during fructose-induced hypoglycemia. Since then, it has been shown repeatedly (Cornblath *et al.*, 1963; Froesch *et al.*, 1963) that the administration of fructose to patients with hereditary fructose intolerance does not modify or even lower the level of insulin in the serum.

The studies of Dubois *et al.* (1961) have also pointed out towards a block of the release of glucose by the liver as the mechanism of fructose-induced hypoglycemia. After the injection of a tracer dose of [1-^{14}C] glucose to a normal child, the specific radioactivity of the blood glucose decreased exponentially as a result of the dilution of the label by unlabelled glucose released from the liver and this decrease was not influenced by the administration of fructose. When the same study was performed in a patient with hereditary fructose intolerance, the administration of fructose resulted in a complete arrest of the decrease of the specific radioactivity, that paralleled the development of the hypoglycemia.

This block of the release of glucose by the liver could be due to an inhibition of glycogenolysis, an inhibition of gluconeogenesis or to both. Several lines of evidence suggest an impairment of gluconeogenesis after the administration of fructose in hereditary fructose intolerance. Clinical studies have shown that the administration of fructose to patients with hereditary fructose intolerance decreases the metabolisation of an oral load of dihydroxyaceton, although a certain conversion of this compound into blood glucose still appears possible (Gentil *et al.*, 1964). *In vitro* experiments have demonstrated that two gluconeogenic enzymes, glucose 6-phosphate isomerase (Zalitis and Oliver, 1967) and aldolase assayed in the direction of fructose 1,6-diphosphate formation (Bally and Leuthardt, cited by Froesch, 1972), are inhibited by fructose 1-phosphate. Even a complete block of gluconeogenesis cannot, however, by itself account for the rapid fall of blood glucose. Indeed, it takes several hours of fasting for patients with the more recently described profound deficiency of the gluconeogenic enzyme hexosediphosphatase to become hypoglycemic (Baker and Winegrad, 1970; Baerlocher *et al.*, 1971; Pagliara *et al.*, 1972; Saudubray *et al.*, 1973). Furthermore, although a fructose load can provoke hypoglycemia in these patients, a higher dose of the ketose is required than in patients with hereditary fructose intolerance.

A block of glycogenolysis was demonstrated by the finding that the fructose-induced hypoglycemia in hereditary fructose intolerance could not be corrected by the administration of glucagon (Perheentupa *et al.*, 1962; Cornblath *et al.*, 1963). An inhibition of phosphoglucomutase by fructose 1-phosphate *in vitro* has been reported (Sidbury, 1959). The fact that in patients with hereditary fructose intolerance blood glucose can be increased by the administration of galactose during a fructose-induced hypoglycemia indicates, however, that the conversion of glucose 1-phosphate into glucose 6-phosphate and free glucose is not impaired (Cornblath *et al.*, 1963). Therefore, the inhibition of the hyperglycemic action of glucagon indicates a deficient phosphorolytic breakdown of liver glycogen by the enzyme phosphorylase. This could be due to various mechanisms, since this enzyme is subjected to a complex regulation that will be described in section 5.

4. THE EFFECTS OF FRUCTOSE ON NORMAL ANIMALS

When large doses of fructose (1 g/kg body weight or more) are injected to experimental animals, fructose 1-phosphate, which is barely detectable in control liver, may reach concentrations as high as 10 mM within a few minutes (Günther *et al.*, 1967; Heinz and Junghänel, 1969; Burch *et al.*, 1969). This accumulation of fructose 1-phosphate is accompanied by a fall of the levels of ATP and Pi to about one third to one fourth of their normal values (Mäenpää *et al.*, 1968). The depletion of ATP is explained by its high rate of utilization as phosphoryl-donor in the ketohexokinase reaction, while the subsequent mitochondrial rephosphorylation of ADP utilizes Pi (Fig.1). It is assumed that, since ATP is an inhibitor of 5'-nucleotidase and Pi an inhibitor of AMP deaminase, the depletion of both metabolites results in a degradation of AMP and a decrease of the total amount of adenine nucleotides. Consequently, IMP (Woods *et al.*, 1970), its degradation product uric acid and, in the rat, allantoin, accumulate (Mäenpää *et al.*, 1968).

The mechanism whereby the concentration of fructose 1-phosphate increases in normal liver has not been clearly defined. This accumulation could be easily explained by a fast phosphorylation of fructose by ketohexokinase, followed by a much slower splitting of fructose 1-phosphate by aldolase. The kinetic parameters of both enzymes have, however, been found to be similar in rat liver (Woods *et al.*, 1970). These authors have therefore proposed the hypothesis that fructose 1-phosphate accumulates as a consequence of the inhibition of aldolase by the increased concentration of IMP. Since the accumulation of IMP is the consequence of the depletion of ATP, which is itself the consequence of the accumulation of fructose 1-phosphate, it is difficult to explain the initial accumulation of the fructose ester by this hypothesis. The enzyme triokinase might thus be the limiting step of the metabolisation of fructose and the accumulation of fructose 1-phosphate result from the equilibrium of the aldolase reaction. The value for the mass-action ratio, calculated from the concentrations of glyceraldehyde, dihydroxyaceton phosphate and fructose 1-phosphate found by Heinz and Junghänel (1969) after the injection of fructose, is close indeed to the Keq of 2.8×10^{-6} M determined *in vitro* by Lehninger *et al.* (1955).

Large amounts of fructose are clearly toxic to normal animals since they result in disturbances of the synthesis of proteins (Mäenpää *et al.*, 1968; Mäenpää, 1972) and ultrastructural changes in the liver (Goldblatt *et al.*, 1970; Phillips *et al.*, 1970; Yu *et al.*, 1974), in impaired tubular function in the kidney (Morris *et al.*, 1973). The effects of the administration of large doses of fructose have not been studied as extensively in normal man as in experimental animals. The available data, in-

Table I. **Influence of fructose, L-sorbose and tagatose on the concentrations of hepatic metabolites.**

TREATMENT		ATP (mM)	P_i (mM)	KETOSE (mM)	KETOSE-1-P (mM)
NONE		2.82	4.57	–	–
FRUCTOSE					
	15 min	0.95	4.05	11.15	8.67
	25 min	1.16	6.03	8.42	9.46
	45 min	1.40	7.73	1.74	2.22
L-SORBOSE					
	15 min	1.43	4.50	14.50	3.60
	25 min	1.86	6.44	10.70	0.90
	45 min	1.97	6.76	2.70	0.50
TAGATOSE					
	15 min	0.64	2.97	15.45	9.40
	25 min	0.71	7.32	15.00	9.30
	45 min	0.61	7.56	6.14	9.90

Control mice and animals that had received the respective ketoses at the dose of 5 mg/g i.v. at zero time were killed at various times afterwards. Values shown are means of 4 animals. (Hue and Stalmans, unpublished experiments, by courtesy.)

crease of the level of uric acid in the blood (Perheentupa and Raivio, 1967), depletion of adenine nucleotides in the liver (Bode *et al.*, 1971), indicate that the same metabolic changes occur and could provoke a similar toxicity, which has been stressed by several authors (Hers, 1970; Woods and Alberti, 1972). It is thus likely, as concluded by Perheentupa *et al.* (1972), that the toxicity of small amounts of fructose in patients with hereditary fructose intolerance is due to the marked exaggeration of events taking place in normal individuals.

Metabolic effects, similar to those obtained with fructose, have also been observed in experimental animals with other ketones that can be phosphorylated by ketohexokinase namely sorbose (Raivio *et al.*, 1969) and tagatose (Hue and Stalmans, unpublished experiments). Table I gives a comparison of their influence on the concentration of certain liver metabolites. It can be seen that sorbose 1-phosphate, probably because of a lower affinity of sorbose for ketohexokinase (Sanchez *et al.*, 1971), does not accumulate to the same extent as the two other ketose phosphates. This is the reason why, in patients with hereditary fructose intolerance, sorbose does not provoke hypoglycemia (Wolf *et al.*, 1959). On the other hand, the effects of tagatose last much longer than those of fructose. This can be attributed to a slower metabolisation of tagatose 1-phosphate.

The administration of large doses of fructose or tagatose to normal animals constitutes therefore a very useful experimental model for the study of the pathogenesis of herediatry fructose intolerance, although it does not reproduce the characteristic hypoglycemia.

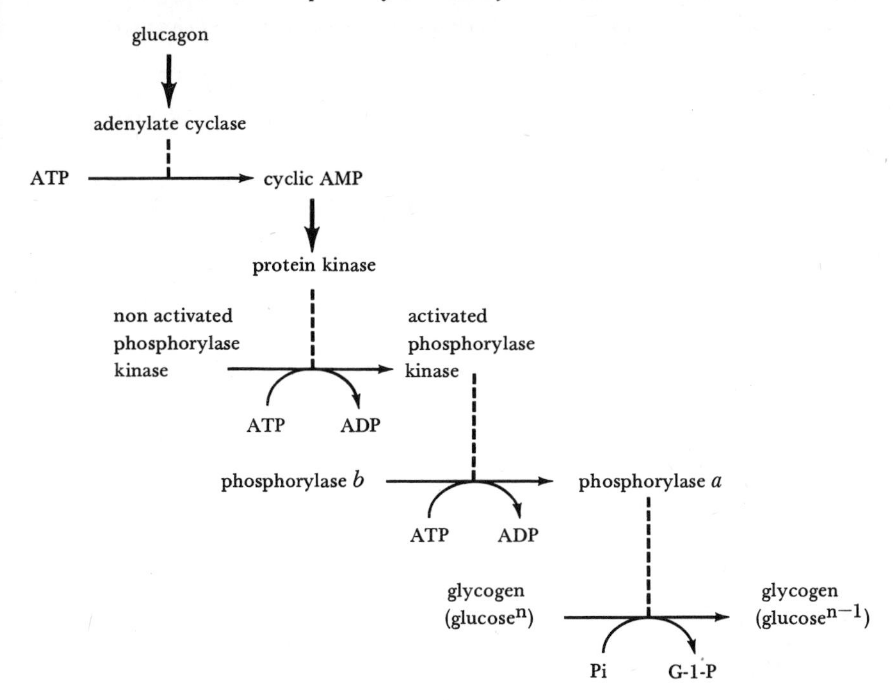

Fig. 2 The mechanism of the glycogenolytic action of glucagon. Explanations are given in the text.

5. THE MECHANISM OF THE GLYCOGENOLYTIC ACTION OF GLUCAGON

The mechanism whereby glucagon brings about its glycogenolytic effect has been worked out in great detail in recent years by the investigations of the groups of E.W. Sutherland and E.G. Krebs. The hormonal effect involves the following steps, which are described in Fig.2. (1) Glucagon stimulates liver adenylate cyclase causing a large increase in the concentration of cyclic AMP in the liver (for a review see Robison *et al.*, 1971); this increase in intrahepatic cyclic AMP is reflected in the plasma and the urine (Broadus *et al.*, 1970). (2) Cyclic AMP induces the conversion of phosphorylase from its inactive form *b* into the active form *a* through the successive stimulation of protein kinase and activation of phosphorylase kinase; this sequence has been established in muscle (for a review, see Krebs, 1972) and presumably also operates in liver. The action of both kinases is antagonized by that of two specific phosphatases (not shown in Fig.2) that inactivate active phosphorylase kinase and phosphorylase *a* respectively. (3) Phosphorylase *a* reacts with glycogen and Pi, producing glucose 1-phosphate, allowing the limiting step in the degradation of glycogen to glucose to proceed (Sutherland and Cori, 1951).

The inhibition of the glycogenolytic mechanism of glucagon by fructose in patients with hereditary fructose intolerance could thus be due to (1) decreased formation of cyclic AMP; (2) absence of conversion of the enzyme from its

Table II. Effect of a fructose load on the stimulation of the urinary excretion of cyclic AMP by glucagon in human subjects.

		Cyclic AMP excreted			
		Without fructose		After fructose	
Subject	Age	Before glucagon	After glucagon	Before glucagon	After glucagon
Controls					
VP	3 weeks	—	—	14.7	68.6
VDH	8 weeks	8.7	91.0	11.3	143.1
TA	18 months	11.8	69.3	16.6	81.6
VL	13 years	4.1	26.3	4.6	33.9
Hereditary fructose intolerance					
PN	7 weeks	16.3	41.5	16.8	16.0
TB	4 months	8.1	236.9	12.6	24.7
LJ	5 months	10.2	80.1	12.2	14.4
PT	3 years	6.3	69.5	7.8	20.5

Fructose, when given, was administered intravenously at the dose of 260 mg/kg 20 min before the injection of glucagon. Glucagon was injected intramuscularly at the dose of 0.1 mg/kg. A first urine sample was taken before the injection of glucagon, the bladder was emptied, and the second urine sample was collected during the 30-60 min that followed the injection of glucagon. Results are expressed as nmol of cyclic AMP/mg of creatinine. After Van den Berghe *et al.* (1973) by permission of the *Biochemical Journal.*

physiologically inactive *b* form to its active *a* form; (3) deficient interaction of phosphorylase *a* with its substrates glycogen and Pi. The effect of fructose on the different steps of the glycogenolytic mechanism has been evaluated by clinical investigations, enzyme studies *in vitro* and animal experiments *in vivo* that will be described in the following section.

6. THE EFFECT OF THE ADMINISTRATION OF FRUCTOSE ON THE GLYCOGENOLYTIC MECHANISM OF GLUCAGON

A. *The Formation of Cyclic AMP*

An indirect investigation of step (1) of the glycogenolytic mechanism of glucagon could be performed in patients with hereditary fructose intolerance by the study of the urinary excretion of cyclic AMP. Table II shows the concentration of cyclic AMP in the urine of control children and of children with hereditary fructose intolerance before and after the injection of glucagon. In both groups of subjects, the basal excretion of the cyclic nucleotide was similar. After the administration of glucagon the excretion of cyclic AMP was as a mean increased 10-fold. In control children, this effect of glucagon was not modified by the previous administration of a small dose of fructose (250 mg/kg) whereas in the children with hereditary fructose intolerance it was greatly diminished although not always completely suppressed by that treatment. The report by Broadus *et al.* (1970) that an infusion of glucagon in a hepatectomized dog produced no increase of the concentration of cyclic AMP in plasma or urine, strongly suggests that the diminished

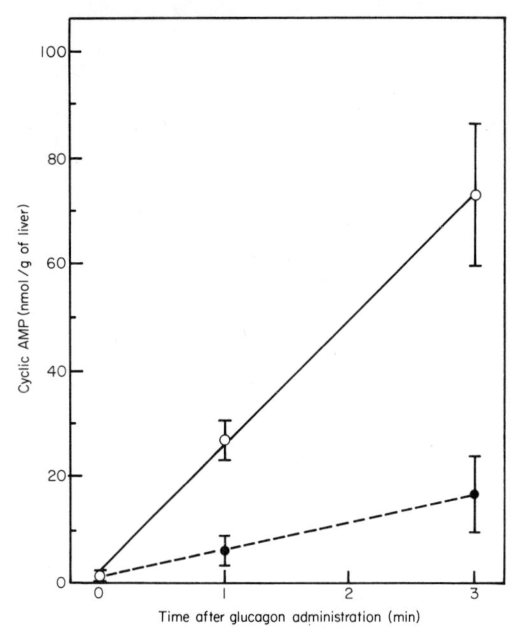

Fig.3 Effect of a fructose load on the action of glucagon on the level of cyclic AMP in the liver of mice. Control mice (○) and animals that had received a fructose load (5 mg/g i.p.) 20 min before (●) were injected with glucagon (0.1 μg/g i.v.) and killed at various times afterwards. There were 4-5 animals per experimental point. Vertical bars represent ± S.E.M.

urinary excretion observed reflects a decreased production of cyclic AMP in the liver.

This conclusion was confirmed by the animal experiment described in Fig.3. At 20 min after the administration of a large dose of fructose to mice, the concentration of cyclic AMP was the same as in the liver of control animals (about 0.6 μM). Within 3 min after the injection of glucagon, the concentration of cyclic AMP increased over 100-fold in the liver of control mice, whereas in the liver of mice treated with fructose the increase was only about 30-fold. Confirming the initial observation of Mäenpää *et al.* (1968), an important fall in the concentration of ATP in the liver was observed, from 2.43 ± 0.07 μmol/g in the control animals to 1.26 ± 0.08 μmol/g in the mice treated with fructose.

To study the mechanism of this decreased capacity to form cyclic AMP, the effect of other ATP-depleting agents on the ability of the liver to raise the concentration of the cyclic nucleotide on stimulation by glucagon was investigated. As shown by other authors, the administration of glycerol (Burch *et al.*, 1970) or of ethionine (Shull, 1962) can also lower the concentration of ATP in the liver. The mechanism of the action of these ATP-depleting agents is different since ethionine traps the adenosyl residue (Farber *et al.*, 1964) whereas glycerol like fructose binds to phosphate. In the experiment described in Fig.4, glycerol, ethionine, fructose or a combination of ethionine and fructose were administered to mice to obtain a wide range of ATP concentrations in the liver. The animals were killed 3 min after the administration of glucagon and the concentrations of ATP

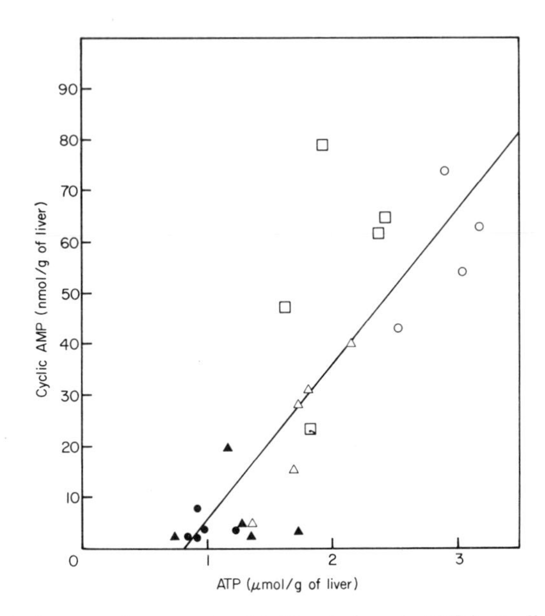

Fig. 4 Correlation between the contents of ATP and of cyclic AMP in the liver of mice 3 min after the administration of glucagon. ATP and cyclic AMP were measured in control mice (○) and in mice treated with glycerol (5 mg/g i.p. 20 min before) (□), ethionine (1 mg/g i.p. 4 h before) (△), fructose (5 mg/g i.p. 20 min before) (●) or ethionine and fructose (▲). Regression coefficient = 0.838 ($P < 0.001$). From Van den Berghe *et al.* (1973) by permission of the *Biochemical Journal.*

and cyclic AMP in the liver were determined. A linear correlation could be drawn between the concentration of ATP and of cyclic AMP, a half-maximal effect of glucagon being obtained at a concentration *in vivo* of 1.8 mM-ATP. This value is in close agreement with the Km of about 1.5 mM-ATP reported for the glucagon-stimulated adenylate cyclase present in plasma membranes of rat liver (Pohl *et al.*, 1971). It is therefore reasonable to assume that, both in human subjects and in mice, the lower responsiveness of adenylate cyclase to glucagon after a load of fructose is secondary to a decrease of the intrahepatic concentration of ATP.

B. *The Activation of Liver Phosphorylase*

In human subjects, a direct study of step (2) of the glycogenolytic mechanism could not be performed, but a harmless, overall investigation of steps (2) and (3) was realized by following the effect of the administration of cyclic AMP on blood glucose. The dibutyryl derivative was chosen as it has been reported to be better tolerated in man (Levine, 1970) and generally more potent in intact cells and organisms (Robison *et al.*, 1971).

Figure 5 shows that when the nucleotide was injected to a patient with hereditary fructose intolerance under control conditions, it produced a hyperglycemia comparable with that observed after the administration of glucagon. However, when dibutyryl cyclic AMP was given during the hypoglycemic phase of a fructose

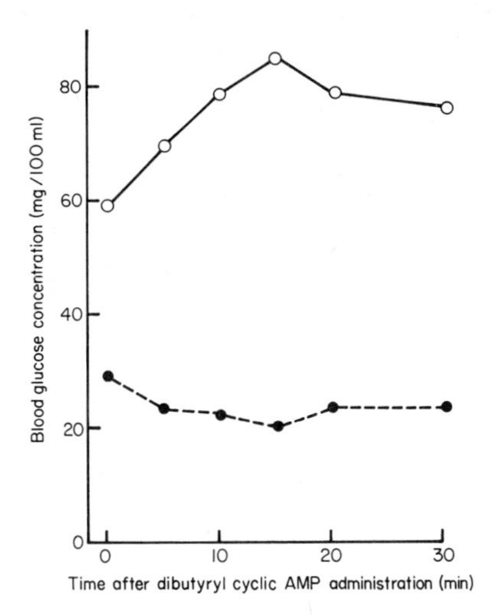

Fig.5 Blood glucose concentrations after administration of dibutyryl cyclic AMP to a patient with hereditary fructose intolerance. Dibutyryl cyclic AMP was injected at the dose of 2 mg/kg i.v. without prior administration of fructose (○) and after a fructose load (250 mg/kg i.v.), 30 min before zero time (●). From Van den Berghe *et al.* (1973) by permission of the *Biochemical Journal.*

tolerance test, no increase in blood glucose was recorded. In control children, the infusion of the same dose of fructose induced a slight hyperglycemia and the subsequent injection of dibutyryl cyclic AMP caused a further increase in blood glucose. These studies show that the fructose-induced hypoglycemia is characterized not only by a lower capacity to form cyclic AMP, but also by an insensitivity to the cyclic nucleotide.

A direct investigation of the effect of the administration of fructose on the conversion of phosphorylase from its inactive *b* form to its active *a* form was performed in animals. In the experiment shown in Fig.6, mice were treated with fructose and ethionine to obtain a marked decrease of the concentration of ATP and the increase in cyclic AMP observed after the administration of glucagon was consequently several times lower than in the experiment shown in Fig.3. An unexpected and hitherto unexplained finding was the pronounced decrease of the activity of phosphorylase *a* in the mice that had received fructose and ethionine without glucagon. Within the first minutes after the injection of glucagon, however, phosphorylase *a* reached the same activity in both groups. This experiment demonstrates that although the administration of fructose provokes an important decrease in the capacity of the liver to form cyclic AMP, the concentration of the cyclic nucleotide obtained after the injection of a pharmacological amount of glucagon is still large enough to cause the activation of phosphorylase. Indeed, the amount of cyclic AMP required to stimulate protein kinase is very small (Krebs, 1972). The same conclusion presumably applied to patients with hereditary fructose intolerance, since in some of them, the injection of glucagon during a fructose tolerance test still caused a slight increase in the urinary excretion of the cyclic nucleotide. It can thus be concluded that the administration of fructose blocks

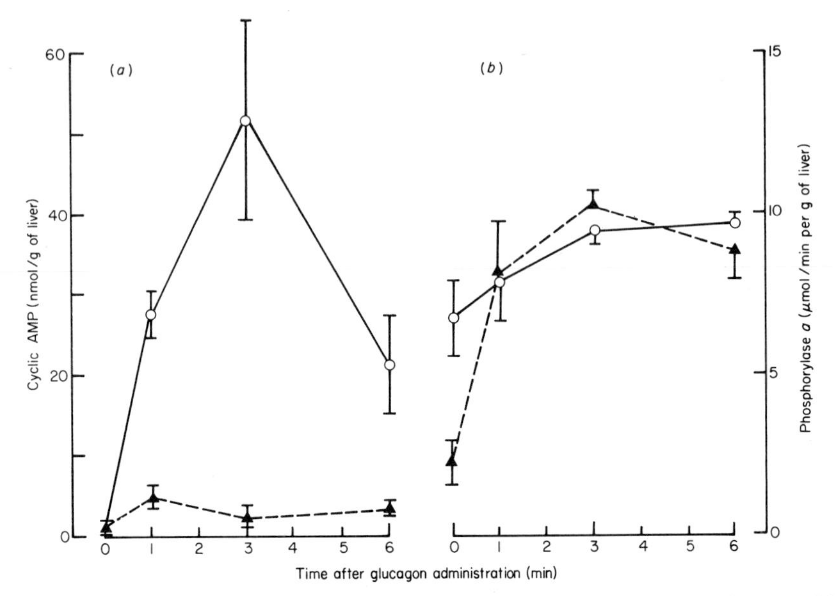

Fig.6 Action of glucagon on mice treated with fructose and ethionine. Phosphorylase *a* activity (*b*) and cyclic AMP (*a*) were measured in control mice (○) and in animals treated with ethionine and fructose as described in Fig.4 (▲). There were five animals per experimental point. Vertical bars represent ± S.E.M. From Van den Berghe *et al.* (1973) by permission of the *Biochemical Journal.*

step (3) of the glycogenolytic mechanism: the interaction of phosphorylase *a* with its substrates glycogen and Pi.

C. *The Activity of Phosphorylase a*

An inhibition of liver phosphorylase *a* by fructose 1-phosphate *in vitro* has been reported (Maddaiah and Madsen, 1966) and several authors have attributed the absence of the glycogenolytic effect of glucagon during fructose induced hypoglycemia to this phenomenon (Cotte *et al.*, 1968; Kaufmann and Froesch, 1973). It has also been hypothesized that the decreased concentration of Pi, observed after the administration of fructose, may explain the absence of phosphorolytic breakdown of glycogen after the injection of glucagon. The response to the hormone could, however, not be restored by the administration of phosphate salts (Desbuquois *et al.*, 1969), although the amount of phosphate used may not have been sufficient. In rats, a significant attenuation of the fructose-induced tubular dysfunction by the simultaneous administration of sodium phosphate has been reported (Morris *et al.*, 1973).

A quantitative evaluation of the decrease of the activity of phosphorylase *a* due to the accumulation of the inhibitor fructose 1-phosphate and the decrease of the substrate Pi is difficult to establish. As said before, very little data are available concerning the levels of these metabolites reached in the liver of patients with hereditary fructose intolerance after the ingestion of fructose. Table III shows the residual activity of liver phosphorylase *a* that can be expected after the administration of a large load of fructose, sorbose or tagatose in experimental ani-

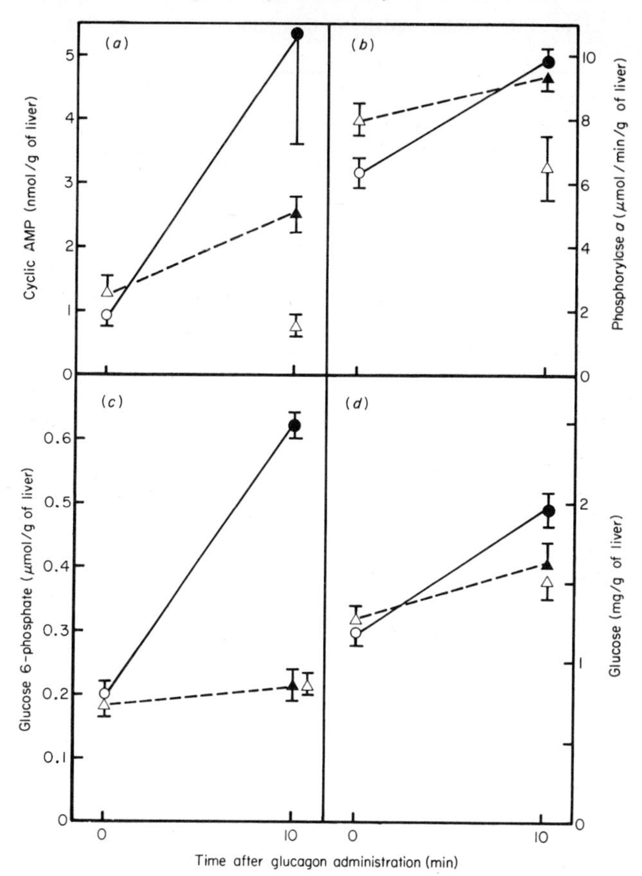

Fig. 7 Action of glucagon on rats treated with tagatose. Cyclic AMP (*a*), phosphorylase *a* activity (*b*), glucose 6-phosphate (*c*) and glucose (*d*) were measured in control rats (○,●) and in rats treated with tagatose (2 mg/g i.p. 10 min before zero time) (△,▲) that had not received glucagon (○,△) or were given the hormone (0.1 μg/g S.C.) at 0 time (●,▲). There were 5 animals per experimental point. Vertical bars represent ± S.E.M. (Van den Berghe *et al.*, 1974).

mals. These values have been calculated from kinetic data obtained with purified liver phosphorylase *a*, taking into account the accumulation of ketose 1-phosphate and the depletion of Pi observed with the respective ketoses. In so far as these data can be extrapolated to human liver, it is difficult to conclude that a decrease of the activity of phosphorylase *a* to about 30% of its initial value, can explain the complete unresponsiveness of the patients to glucagon.

The effect of the administration of a load of tagatose on the glycogenolytic response to glucagon was therefore studied *in vivo* in experimental animals (Fig. 7). Tagatose was chosen since its effects last longer (Table I). Tagatose, when given alone, provokes a small increase of the hepatic concentration of cyclic AMP, the activity of phosphorylase *a* and the glucose level. The glycogenolytic effect of glucagon was evaluated by its ability to increase the hepatic concentrations of glu-

Table III. Calculation of the residual activity of liver phosphorylase *a* after the administration of various ketoses.

KETOSE	[I] KETOSE-1-P (mM)	Ki KETOSE-1-P (mM)	RESIDUAL ACTIVITY (%)
Fructose	10	3.1	37
L-Sorbose	4	2.4	45
Tagatose	10	3.5	35

The residual activity of phosphorylase *a* is expressed as the % of the activity observed in the presence of a physiological concentration of Pi (4mM) and in the absence of inhibitors. The activities in the presence of the concentrations of the inhibitors shown have been calculated taking into account the lowest value of Pi observed at very short times after the injection of the ketoses (1.5 mM). The Km for Pi (1 mM) and the Ki values have been determined with purified liver phosphorylase *a* (from data reported by Van den Berghe *et al.*, 1973, and unpublished experiments by L. Hue).

cose 6-phosphate and glucose, the former index being more sensitive due to the smaller size of the glucose 6-phosphate pool. It can be seen that similarly to fructose, the administration of tagatose decreases the capacity of the liver to produce cyclic AMP on stimulation by glucagon but does not affect the activation of phosphorylase. The effect of the hormone on the concentrations of glucose 6-phosphate and glucose is, however, completely abolished by treatment of the animals with tagatose. This experiment conclusively demonstrates a complete block of the glycogenolytic response to glucagon in the conditions prevailing in the liver of experimental animals after the administration of tagatose. The discrepancy observed in comparison with the partial decrease of the activity of phosphorylase *a* shown in Table III may be due to several factors. Part of the total concentration of Pi, utilized in the calculations, may be bound and consequently not available for phosphorylase *a*. The activity of phosphorylase *a* might be overestimated. It is well known to students of liver phosphorylase that, for unexplained reasons, the production of glucose by the perfused liver is always much lower than expected from the measurement of the activity of the enzyme and that even the unstimulated liver contains an appreciable amount of phosphorylase *a* (Weintraub *et al.*, 1969). These findings emphasize the need to complement calculations based in part on *in vitro* data with *in vivo* experiments.

ACKNOWLEDGEMENTS

The author wishes to thank Professor H.G. Hers for reviewing the manuscript and continuous guidance. The personal investigations described in this paper were supported by the "Fonds de la Recherche Scientifique Médicale" and by the U.S. Public Health Service (grant AM 9235).

REFERENCES

Adelman, R.C., Ballard, F.J. and Weinhouse, A. (1967). *J. biol. Chem.* **242**, 3360-3365.

Baerlocher, K., Gitzelmann, R., Nüssli, R. and Dumermuth, G. (1971). *Helv. Paediat. Acta* **26**, 489-506.

Baker, L. and Winegrad, A.I. (1970). *Lancet* **ii**, 13-16.

Bode, C., Schumacher, H., Goebell, H., Zelder, O. and Pelzel, H. (1971). *Horm. Metab. Res.* 3, 289-290.

Bollman, J.L. and Mann, F.C. (1931). *Am. J. Physiol.* **96**, 683-695.

Broadus, A.E., Kaminsky, N.I., Northcutt, R.C., Hardman, J.G., Sutherland, E.W. and Liddle, G.W. (1970). *J. clin. Invest.* **49**, 2237-2245.

Bublitz, C. and Kennedy, E.P. (1954). *J. biol. Chem.* **211**, 951-961.

Burch, H.B., Max, P., Chyu, K. and Lowry, O.H. (1969). *Biochem. biophys. Res. Commun.* **34**, 619-626.

Burch, H.B., Lowry, O.H., Meinhardt, L., Max, P. and Chyu, K. (1970). *J. biol. Chem.* **245**, 2092-2102.

Chambers, R.A. and Pratt, R.T.C. (1956). *Lancet* **ii**, 340.

Cori, G.T., Ochoa, S., Slein, M.W. and Cori, C.F. (1951). *Biochim. biophys. Acta* **7**, 304-317.

Cornblath, M., Rosenthal, I.M., Reisner, S.H., Wybregt, S.H. and Crane, R.K. (1963). *N. Eng. J. Med.* **269**, 1271-1278.

Cotte, J., Mathieu, M., Nivelon, J.L., Bethenod, M., Kissin, C. and Collombel, C. (1968). *Clin. Chim. Acta* **19**, 215-226.

Desbuquois, B., Lardinois, R., Gentil, C. and Odievre, M. (1969). *Arch. Franç. Péd.* **26**, 21-35.

Dubois, R., Loeb, H., Ooms, H.A., Gillet, P., Bartman, J. and Champenois, A. (1961). *Helv. Paediat. Acta* **16**, 90-96.

Farber, E., Shull, K.H., Villa-Trevino, S., Lombardi, B. and Thomas, M. (1964). *Nature* **203**, 34-40.

Froesch, E.R. (1972). *In:* "The Metabolic Basis of Inherited Disease" (J.B. Stanbury, J.B. Wyngaarden and D.S. Frederickson, eds) 3rd edn., pp.131-148. McGraw-Hill, New York.

Froesch, E.R. and Ginsberg, J.L. (1962). *J. biol. Chem.* **237**, 3317-3324.

Froesch, E.R., Prader, A., Labhart, A., Stuber, H.W. and Wolf, H.P. (1957). *Schweiz. Med. Wochenschr.* **87**, 1168-1171.

Froesch, E.R., Prader, A., Wolf, H.P. and Labhart, A. (1959). *Helv. Paediat. Acta* **14**, 99-112.

Froesch, E.R., Wolf, H.P., Baitsch, H., Prader, A. and Labhart, A. (1963). *Am. J. Med.* **34**, 151-167.

Gentil, C., Colin, J., Valette, A.M., Alagille, D. and Lelong, M. (1964). *Rev. Franç. Études Clin. Biol.* **9**, 596-607.

Gitzelmann, R., Steinmann, B., Bally, C. and Lebherz, H.G. (1974). *Biochem. biophys. Res. Commun.* **59**, 1270-1277.

Goldblatt, P.J., Witschi, H., Friedman, M.A., Sullivan, R.J. and Shull, K.H. (1970). *Lab. Invest.* **23**, 378-385.

Günther, M.A., Sillero, A. and Sols, A. (1967). *Enzymol. Biol. Clin.* **8**, 341-352.

Heinz, F. and Junghänel, J. (1969). *Hoppe Seyler's Z. Physiol. Chem.* **350**, 859-866.

Heinz, F., Lamprecht, W. and Kirsch, J. (1968). *J. clin. Invest.* **47**, 1826-1832.

Hers, H.G. (1952a). *Biochim. biophys. Acta* **8**, 416-423.

Hers, H.G. (1952b). *Biochim. biophys. Acta* **8**, 424-430.

Hers, H.G. (1960). *Biochim. biophys. Acta* **37**, 120-126.

Hers, H.G. (1962). *In:* "Methods in Enzymbology" (S.P. Colowick and N.O. Kaplan, eds) *V*, pp.362-364. Academic Press, New York and London.

Hers, H.G. (1970). *Nature* **227**, 421.

Hers, H.G. and Kusaka, T. (1953). *Biochim. biophys. Acta* **11**, 427-437.

Hers, H.G. and Joassin, G. (1961). *Enzymol. Biol..Clin.* **1**, 4-14.

Holzer, H. and Holldorf, A. (1957). *Biochem. Z.* **329**, 283-291.

Hue, L. (1974). *In:* "Sugars in Nutrition" (Sipple, ed) *(in press)*. Academic Press, New York and London.

Hue, L. and Hers, H.G. (1972). *Eur. J. Biochem.* **29**, 268-275.

Ichihara, A. and Greenberg, D.M. (1957). *J. biol. Chem.* **225**, 949-958.

Kaufmann, U. and Froesch, E.R. (1973). *Eur. J. clin. Invest.* **3**, 407-413.

Kranhold, J.F., Loh, D. and Morris, R.C. Jr. (1969). *Science* **165**, 402-403.

Krebs, E.G. (1972). *Curr. Top. Cell. Regul.* **5**, 99-133.

Lehninger, A.L., Sicé, J. and Jensen, E.V. (1955). *Biochim. biophys. Acta* **17**, 285-287.

Leuthardt, F. and Testa, E. (1951). *Helv. Chim. Acta* **34**, 931-938.

Leuthardt, F., Testa, E. and Wolf, H.P. (1953). *Helv. Chim. Acta* **36**, 227-251.

Levine, R.A. (1970). *Clin. Pharmacol. Therap.* **11**, 238-243.

Levine, R. and Huddlestun, B. (1947). *Fedn Proc.* **6**, 151.

Maddaiah, V.T. and Madsen, N.B. (1966). *J. biol. Chem.* **241**, 3873-3881.
Mäenpää, P.H., Raivio, K.O. and Kekomäki, M.P. (1968). *Science* **161**, 1253-1254.
Mäenpää, P.H. (1972). *Acta Med. Scand. Suppl.* **542**, 115-118.
Mendeloff, A.I. and Weichselbaum, T.E. (1953). *Metabolism* **2**, 450-458.
Morris, R.C.Jr., McSherry, E. and Sebastian, A. (1973). *J. clin. Invest.* **52**, 57a-58a.
Nissel, J. and Linden, L. (1968). *Scand. J. Gastroent.* **3**, 80-82.
Nordmann, Y., Schapira, F. and Dreyfus, J.C. (1968). *Biochem. biophys. Res. Commun.* **31**, 884-889.
Pagliara, A.S., Karl, I.E., Keating, J.P., Brown, B.I. and Kipnis, D.M. (1972). *J. clin. Invest.* **51**, 2215-2123.
Perheentupa, J. and Raivio, K. (1967). **Lancet ii**, 528-531.
Perheentupa, J., Pitkänen, E., Nikkilä, E.A., Somersalo, O. and Hakosalo, J. (1962). *Ann. Paediat. Fenn.* **8**, 221-235.
Perheentupa, J., Raivio, K.O. and Nikkilä, E.A. (1972). *Acta Med. Scand. Suppl.* **542**, 65-75.
Phillips, M.J., Hetenyi, G.Jr. and Adachi, F. (1970). *Lab. Invest.* **22**, 370-379.
Pitkänen, E. and Perheentupa, J. (1962). *Ann. Paediat. Fenn.* **8**, 236-244.
Pohl, S.L., Birnbaumer, L. and Rodbell, M. (1971). *J. biol. Chem.* **246**, 1849-1856.
Raivio, K.O., Kekomäki, M.P. and Mäenpää, P.H. (1969). *Biochem. Pharmacol.* **18**, 2615-2624.
Reinecke, R.M. (1944). *Am. J. Physiol.* **141**, 669-676.
Robison, G.A., Butcher, R.W. and Sutherland, E.W. (1971). "Cyclic AMP", pp.234-240. Academic Press, New York and London.
Sanchez, J.J., Gonzalez, N.S. and Pontis, H.G. (1971). *Biochim. biophys. Acta* **227**, 67-78.
Saudubray, J.M., Dreyfus, J.C., Cepanec, C., Leloch, H., Trung, P.H. and Mozziconacci, P. (1973). *Arch. Franç. Péd.* **30**, 609-632.
Shull, K.H. (1962). *J. biol. Chem.* **237**, PC1734-PC1735.
Sidbury, J.B. (1959). *Helv. Paediat. Acta* **14**, 317.
Sillero, M.A.G., Sillero, A. and Sols, A. (1969). *Eur. J. Biochem.* **10**, 345-350.
Sols, A. and Crane, R.K. (1954). *J. biol. Chem.* **210**, 581-595.
Sutherland, E.W. and Cori, C.F. (1951). *J. biol. Chem.* **188**, 531-543.
Van den Berghe, G., Hue, L. and Hers, H.G. (1973). *Biochem. J.* **134**, 637-645.
Van den Berghe, G., Hue, L. and Hers, H.G. (1974). *Abstracts Annual Meeting of the European Society for Paediatric Research, Lausanne, 9-12 July*, p.55.
Vinuela, E., Salas, M. and Sols, A. (1963). *J. biol. Chem.* **238**, PC1175-PC1177.
Weichselbaum, T.E., Margraf, H.W. and Elman, R. (1953). *Metabolism* **2**, 434-449.
Weintraub, B., Sarcione, E.J. and Sokal, J.E. (1969). *Amer. J. Physiol.* **216**, 521-526.
Wick, A.N., Sherill, J.W. and Drury, D.R. (1963). *Diabetes* **2**, 465-468.
Wolf, H.P. and Leuthardt, F. (1953). *Helv. Chim. Acta* **36**, 1463-1467.
Wolf, H., Zschocke, D., Wedemeyer, F.W. and Hübner, W. (1959). *Klin. Wochenschr.* **37**, 693-696.
Woods, H.F. and Alberti, K.G.M.M. (1972). *Lancet* **ii**, 1354-1357.
Woods, H.F., Eggleston, L.V. and Krebs, H.A. (1970). *Biochem. J.* **119**, 501-510.
Yu, D.T., Burch, H.B. and Phillips, M.J. (1974). *Lab. Invest.* **30**, 85-92.
Zalitis, J. and Oliver, I.T. (1967). *Biochem. J.* **102**, 753-759.

DISCUSSION

Koster: It is nice to hear that you came to the same conclusion as we did on leucocyte phosphorylase. We used leucocyte phosphorylase for these experiments because we know that the deficiency of the liver phosphorylase is reflected in the leucocytes. You mentioned that the activity of aldolase for fructose-1,6-diphosphate is lowered in the patients with fructose intolerance. Did you measure the Km value for FDP?

Van den Berghe: No.

Koster: We have done it with three patients and found that the Km for fructose-1,6-diphosphate is increased about 10 times as compared to controls. That may be the reason why the activity for FDP-aldolase is diminished.

Hatzfeld: We found that in the normal human liver the Km for the most specific substrate, fructose-1-phosphate is about 4 mM, and in the patients with fructose intolerance a value between 15 and 30 mM was found.

Blass: Sidbury (*In:* "Molecular Genetics and Human Disease", L.E. Gardner, ed. C.C. Thomas, Springfield, Ill., p.61, 1960) reported that galactose-1-phosphate inhibits phosphoglucomutase by a mechanism which involves stripping the enzyme and the glucose-1,6-diphosphate cofactor of its phosphorous. Does fructose-1-phosphate have any effect on phosphoglucomutase?

Van den Berghe: Sidbury has indeed suggested that fructose-1-phosphate inhibits phosphoglucomutase on the basis of *in vitro* studies. This hypothesis has, however, been ruled out by Cornblath *et al.* (*New Eng. J. Med.* 269, 1271, 1963) who found that galactose could provoke a significant rise of blood glucose during a fructose-induced hypoglycemia.

Greengard: You mentioned that the residual phosphorylase activity was too high (about 50%) for you to consider the inhibition as being important. By the same token, are you not concerned whether the very small increase in phosphorylase activity in normal mice with glucagon or c-AMP actually accounts for the glycogenolysis?

Van den Berghe: The effect of glucagon on liver phosphorylase *a* which I have shown in normal mice is, indeed, small. One should take into account, however, the fact that this experiment was performed on unanesthetized animals; this results in an artefactual increase of the basal level of the enzyme, which is attributed to the stress of killing, resulting in a local release of catecholamines. It is possible to lower this basal level of phosphorylase *a* by anesthetizing the animals (Hornbrock and Brody, *Biochem. Pharmacol.* 12, 1407, 1963) and to demonstrate that way a more marked effect of glucagon. It should be mentioned, too, that both the basal and the stimulated activities of phosphorylase *a*, especially when measured in the synthetic direction as was done in this experiment, are much higher than the rates of glycogenolysis. This has been shown by Weintraub *et al.* (*Amer. J. Physiol.* 216, 521, 1969) in the perfused liver. The discrepancy is attributed to activation of phosphorylase during the homogenisation procedure. It is thus not possible to compare quantitatively the activities obtained *in vitro* in liver homogenates with rates of glycogenolysis *in vivo*.

Tager: How do you prevent any effect of fructose-1-phosphate on the phosphorylase *a* measurements in the experimental animals?

Van den Berghe: By a 33-fold dilution of the homogenate in the phosphorylase *a* assay.

Hommes: It has been demonstrated by Maddiah and Madsen (*J. biol. Chem.* 241, 3873, 1966) that hexosephosphates are inhibitory to phosphorylase *a*; but at least for glucose-6-phosphate it has been demonstrated that this inhibition does not occur in the presence of AMP. As you have such a decrease in the ATP level, you might as well expect an increase in the AMP level. The question is whether such an effect has also been demonstrated, or not, with respect to fructose-1-phosphate.

Van den Berghe: Mäenpää and co-workers (*Science* 161, 1253, 1968) have indeed shown an approximately 2-fold increase of the level of AMP after the administration of a load of fructose, but this lasts only a couple of minutes. The inhibitory effect of fructose-1-phosphate on liver phosphorylase *a* was described in the presence of 1 mM AMP by Maddiah and Madsen, but we have found a similar degree of inhibition in the absence of AMP.

Hatzfeld: One comment, which might have a therapeutic consequence. We have recently found that the Km can be normalized by β-mercapto ethanol treatment which lowers the Km for fructose-1-phosphate from 13 to 3.

INBORN ERRORS AND THE BRAIN VARIABILITY OF THE PATHOGENETIC PROCESS

L. Crome and J. Stern

Queen Mary's Hospital for Children, Carshalton, Surrey, England

Much progress has been made in the understanding of the biochemical basis of inborn errors of metabolism; in fact, as we have heard at this meeting, the metabolic pathways have in many cases been worked out in great detail. Much less well understood is the way in which biochemical aberrations lead to permanent mental handicap. In these circumstances it is sometimes opportune to return to the natural history of these disorders. I propose to review briefly 20 patients with four biochemically well-understood conditions in the hope that the data presented will help to provide a link between biochemical mechanism and neuropsychiatric manifestations.

MATERNAL PHENYLKETONURIA: MENTAL RETARDATION AND INTRAUTERINE GROWTH RETARDATION IN EIGHT SIBLINGS

The harmful effects on the foetus of untreated maternal phenylketonuria are well documented (Fisch and Anderson, 1971). In addition to microcephaly, mental retardation and intrauterine growth retardation a number of congenital abnormalities are frequently present (Table I). The view has been put forward that the risk to the (heterozygous) foetus increases dramatically if the maternal blood phenylalanine level exceeds about 1 mM. No fewer than 58 out of 59 non-pheneylketonuric offspring whose mothers had levels above 1 mM were mentally retarded, while 15 out of 16 offspring of mothers with blood phenylalanine levels below 1 mM were said to be normal (Hsia, 1970).

We have recently seen a phenylketonuric woman who has given birth to eight mentally retarded children seven of whom survive. The family was discovered in the course of an investigation of the genetic prognosis in mental handicap (Angeli and Kirman, 1971). The propositus, Gerald (Fig.1), had been seen over 10 years previously and had died soon afterwards in another hospital aged 5 years 4 months. He had microcephaly and Fallot's tetralogy and suffered from infantile spasms. The mother was described by a psychiatrist as 'obviously of normal intelligence but an inadequate mother and unreliable witness'. She was both nonco-operative and elusive but when eventually traced she was found to have given birth to seven

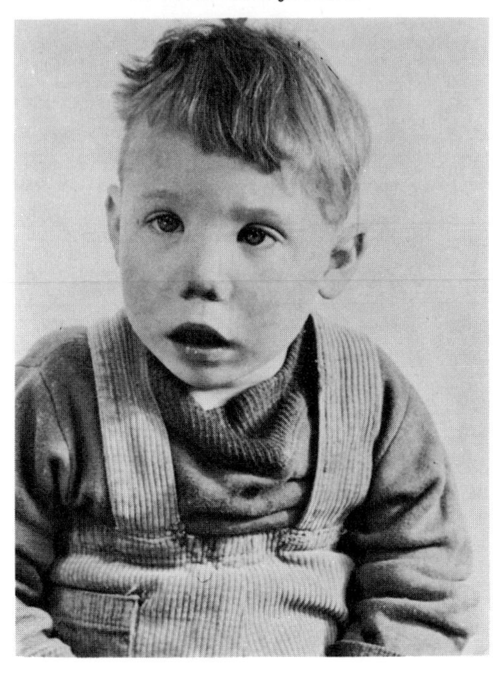

Fig. 1 Microcephalic and growth retarded son of a phenylketonuric mother at age 4 years 6 months.

Table I. Abnormalities observed in maternal phenylketonuria (Fisch and Anderson, 1971).

Mental retardation
Microcephaly
Growth retardation
Cerebral palsy
Congenital heart disease
Agenesis of the spleen
Extra lobe of lung
Oesophageal atresia
Eye defects including lipodermoids of cornea
Dislocation of the hip
Club foot
Anomalies of first rib
Spinal fusion
Missing phalanx of finger

children, all except the first fathered by her husband. Soon afterwards she became pregnant by a third man, divorced her husband and went to live in another part of the country. The eighth child is also retarded. A summary of the clinical findings is given in Table II; full details are being reported elsewhere (Angeli *et al.*, 1974). The reason for the long delay in diagnosis was that every one of the children

Table II. **Clinical findings in eight microcephalic and growth retarded children of a phenylketonuric mother.**

		Mother's age	Characteristic facies	Abnormalities of fingers, toes or ears	Fallot's tetralogy	I.Q.[a]	EEG (age in brackets)
1.	Teresa	24	–	+	–	57	Normal record. Good alpha (16 years).
–	Miscarriage	25					
2.	Gerald	26	+	+	+	< 45	Bilateral spike and wave discharges (4 years).
3.	Michael	28	–	+	–	36	Low voltage record. Very little alpha (12 years).
4.	Caroline	30	+	+	–	37	Mild excess of intermediate slow wave activity. Very little alpha (10 years).
5.	David	32	+	+	+	40	Normal record. Good alpha (8 years).
6.	Janet	34	+	+	–	36	Mild excess of intermediate slow wave activity (5 years).
7.	Colin	37	+	+	–	25[b]	Diffuse excess of intermediate slow wave activity (3½ years).
8.	Helen	42	+	–	–	45[b]	–

[a] intelligence quotient, Stanford-Binet, 1960 revision (form L-M)
[b] developmental quotient, Griffith scale

was born in a different hospital, that the mother failed to attend antenatal or welfare clinics, and refused almost without exception to cooperate in investigations. She also failed to persevere with dietary treatment offered her during her eighth pregnancy. Her blood phenylalanine was 1.09 mM (18 mg/100 ml) at age 40 and 0.85 mM (14 mg/100 ml) early in her eighth pregnancy.

The brain of the child who died had been preserved in formalin and over 10 years later a pathological examination was carried out by Dr. Crane. The fixed brain weighed 760 g (normal for age 1240 g), the cerebellum 130 g. The reduction in bulk was symmetrical in grey and white matter; the pattern of the gyri was normal. Histologically the brain showed no myelin loss and no abnormalities of myelin. Apart from micrencephaly there were no long-standing changes ascertainable by classical neuropathological methods.

A neurochemical examination of the brain by Professor Davison revealed a normal phospholipid pattern in both white matter and cortex apart from losses due to fixation. The sphingolipid pattern was also normal and there was no excess of cholesterol esters. Professor Davison concluded that there was no evidence of demyelination or dysmyelination. These findings in maternal phenylketonuria stand in contrast to those in classical phenylketonuria where microcephaly is usually accompanied by deficiency of myelin formation, demyelination and gliosis. A fuller account of this case will be published elsewhere (Angeli *et al.*, 1975).

There is good evidence that in phenylketonuria a number of metabolic processes may be adversely affected (Table III), a good example of such a process has been discussed at this meeting. However, the fact that mechanisms exist which are potentially dangerous or even fatal to the brain does not mean that these mechanisms are actually responsible for the brain damage. The same mech-

Table III. Some metabolic processes which may be adversely affected in phenylketonuria.

Serotonin synthesis
Catecholamine synthesis
GABA synthesis
Acetyl-CoA synthesis
Glycolysis
Gluconeogenesis
Amino acid transporr
Myelin protein synthesis
Brain polyribosome disaggregation
Microneurone formation
Dendritic arborisation

Table IV. Factors which may enhance the vulnerability of the nervous system in inborn errors of metabolism.

Malnutrition or inappropriate diet
Anoxia, hypoglycaemia, acidosis, brain oedema
Infection
Drugs
Suboptimal functioning of alternative pathways
Failure of protective transport mechanism

anism or toxic agent need not necessarily operate in every patient, or even in any one patient at all stages of the development of the brain. The variability of manifestations found in so many inborn errors suggests that we may look at many of these diseases as entailing enhanced vulnerability of the brain which may be further increased by a number of extraneous factors (Table IV).

In a complementary approach to the problem of pathogenesis Dobbing (1974) has emphasized the vulnerability of the brain at critical periods of rapid growth and differentiation. According to this view it is the severity and duration of growth restricting factors and the developmental stage of the brain which largely determine ultimate deficits rather than the precise nature of these factors. The offspring of a phenylketonuric mother is exposed to high phenylalanine levels during organogenesis in the early stages of pregnancy, during the period of rapid neuronal multiplication between ten and eighteen weeks of gestation and during the early, prenatal part (perhaps one sixth to one seventh) of the main brain growth spurt when glial proliferation, myelination, synaptogenesis and dendritic arborisation take place. The untreated phenylketonuric offspring of a heterozygous mother is exposed to the harmful effects of high phenylalanine levels after birth during most of the main brain growth spurt. Before birth the phenylketonuric infant is largely protected by the mother's ability to control the blood phenylalanine level, although the long-held view that he is normal at birth is being challenged (Saugstad, 1973).

It is then hardly surprising that myelination which in man is largely postnatal is less severely affected in maternal phenylketonuria than in untreated classical phenylketonuria. The effect of the maternal phenylalanine on the brain of the offspring was variable as shown by a fairly wide range of ability and considerable variability in the E.E.G. (Table II). The latter usually reflects the presence of epilepsy rather than the mental retardation (Harris, 1972), and, as might be expected, with the exception of Gerald's record, the observed abnormalities were only mild and two of the children actually had normal records. Although all the children probably shared a not dissimilar environment only two had cardiac malformation. While no malformations were found in Gerald's, the only brain so far examined morphologically, neural malformations might well occur in other cases.

The findings in this family suggest that phenylketonuric women who want to have children should ideally start dietary treatment before they become pregnant, and take a diet no less stringently controlled than that given to phenylketonuric infants. It should not be assumed that a blood phenylalanine level just below 1 mM is 'safe' during pregnancy.

Screening of all pregnant women for phenylketornuria has been advocated. However the number of undiagnosed adult phenylketonurics of near normal intelligence in the community is very small. It will further diminish as the proportion of the population screened in infancy increases. Furthermore some of these patients are psychiatrically disturbed and may fail to attend antenatal clinics or refuse to cooperate in treatment. The prospects of bearing a normal infant are probably more favourable for women who have been diagnosed by mass screening of infants and who after dietary treatment are mostly intellectually normal. It must be stressed that very few cases of successful dietary treatment of pregnant phenylketonuric women are on record (Farquhar, 1974).

HISTIDINAEMIA: DELAYED MANIFESTATIONS OF BRAIN LESIONS?

A lively discussion is still in progress about whether histidanaemia (deficiency of L-histidine ammonia lyase) is harmful or an essentially benign condition. Popkin *et al.* (1974) concluded from a review of published data that the risk of mental retardation was at least 40 per cent and recommended treatment of affected infants with a diet low in histidine. Neville *et al.* (1972) and Clayton (1974) on the other hand do not favour treatment of asymptomatic infants in whom the disorder has been diagnosed in the course of mass screening. These authors have observed normal development in 19 such infants whom they have followed for varying periods extending in some cases to over three years.

We recently came face to face with this problem. Urine from a healthy five-week old infant was submitted for screening when the child was being placed for adoption. Further tests point very strongly to the diagnosis of histidinaemia. It is at present extremely difficult in such cases to give parents a confident prognosis or to offer non-controversial advice on whether or not to treat the infant. Our experience of three other cases is perhaps of relevance in trying to assess the way in which pathogenic mechanisms operate in this disorder.

The first case is a little girl now eight years old whom we have followed for over six years. She has a W.I.S.C. I.Q. of 118 and is free from neurological signs. Her E.E.G. (Dr. Ruth Harris) showed occasional bursts of theta activity on a normal background especially during overbreathing. The record was judged somewhat abnormal but did not show definite epiletogenic activity. So far her development has been entirely normal.

The second and third cases are brother and sister. The brother was diagnosed

Table V. Differences between histidinaemia and phenylketonuria.

Higher renal clearance of histidine
Lower blood histidine level
Lower histidine content of diet
Slower uptake of histidine by brain
Less direct link to neurotransmitters
Benefits of dietary treatment not proven in histidinaemia

in a boarding school for epileptic children. He has a history of epilepsy starting at 9 months after diphtheria immunisation. His I.Q. is 80, but he has orientational and coordination difficulties, and an immature personality. At school he has done less well than expected from his test results. His E.E.G. (Dr. Ruth Harris) showed considerable generalised abnormality with an excess of slow activity and multiple discharges. Several potentially epileptogenic areas were present. The sister has a full scale Wechsler I.Q. of 100. After an apparently normal childhood she developed grand mal epilepsy at age 15 years. She has also serious personality problems, has twice attempted suicide, and has had several admissions to hospital for depressive illnesses. Her E.E.G. record was abnormal with a potentially epileptogenic area over the right temporal lobe.

We should like to stress two points which may be of more general significance. The first relates to the delayed manifestation of neurological defects. This is well appreciated in disorders such as Huntington's chorea and the late onset storage diseases but not so common in the extracerebral recessive disorders. It is indeed a general phenomenon that a lesion sustained at a given stage of development of the nervous system may have repercussions at almost any time during the subsequent development of the brain. Professor Gitzelmann has referred to this problem in galactosaemia, and acute late onset bilirubin encephalopathy in the Crigler-Najjar syndrome (Blaschke *et al.*, 1974) might in part be caused or exacerbated by pre-existing occult brain damage.

A clear implication of these findings is that follow-up of both treated and untreated histidinaemias should be continued at least into adolescence. Care will also be necessary in interpreting behavioural changes when treatment is discontinued. Apparent deterioration or failure to progress at school may be due to the relaxation of diet but can also be the late manifestation of an insult to the brain sustained much earlier or even before birth.

The second point which is worth noting is the wide spectrum of psychiatric symptoms found in both treated and untreated patients with inborn errors such as phenylketonuria, homocystinuria, and histidinaemia. Sir Martin Roth (1969) has pointed out that organic and functional psychiatric disorders occur together with a frequency well beyond chance expectation. A clear cut association has been demonstrated between cerebral disease and psychiatric illness in children, childhood autism in particular.

Do our data provide any evidence in favour of dietary treatment of histidinaemia? Arguments in favour of treatment are essentially based on the efficacy of the phenylalanine low diet in phenylketonuria and the similarities between histidinaemia and phenylketonuria. However, the two disorders also differ in important respects (Table V). In particular, mental handicap if present at all is of a dif-

Table VI. Findings in four cases of non-ketotic hyperglycinaemia.

	Age	Perinatal history	Main clinical signs	EEG	IQ	Course
Case 1	21	No problems	No fits, backward, no neurological signs	Minimally abnormal	55	Stable
Case 2	17	No problems	Fits, no neurological signs, 'irresponsible'	Abnormal	50	Stable
Case 3	15	No problems	Hypotonia, cerebral atrophy	Moderately abnormal	SSN[a]	'Improving'
Case 4	Neonate	Severely affected	Hypotonia, apnoea death at 72 hours	Grossly abnormal	—	Rapidly fatal

Blood glycine: 800-900 μmol/l in Cases 1, 2 and 3; 1350 μmol/l in Case 4. [a]SSN, severely mentally retarded.

ferent order of magnitude from that in untreated phenylketonuria, and the same is true of maternal histidinaemia (Neville *et al.*, 1971; Lyon *et al.*, 1974). Brain histidine levels are increased in rats fed excess histidine (Tyfield and Holton, 1974) and in a strain of histidinaemic mice (Bulfield and Kacser, 1974), but this does not appear to give rise to overt behavioural abnormalities. Female homozygous mutants do, however, produce offspring with vestibular damage and this suggests, tentatively, that treatment of histidinaemic women during pregnancy might be more rewarding than that of histidinaemic infants. The vulnerability of the nervous system is probably enhanced in histidinaemia. This in itself is not sufficient reason for instituting dietary treatment.

Semisynthetic diets give rise to a number of nutritional and social problems in infants under treatment and their families. The onus of providing evidence of the benefits of a histidine low diet must rest with those advocating its use. At present reliable evidence on when and how to treat histidinaemia is lacking. One way of testing the *potential* usefulness of the histidine low diet would be to feed it to pregnant histidinaemic mice to see if the vestibular damage in the offspring can be prevented.

DISORDERS OF NON-ESSENTIAL AMINO ACIDS : FAILURE OF PROTECTIVE TRANSPORT MECHANISMS?

In our hospital we have seen four cases of non-ketotic hyperglycinaemia (Table VI). One of these, case 4, followed a course very similar to that described, for example by Baumgartner *et al.* (1969), Bachmann *et al.* (1971), Koepp *et al.* (1973) and Trijbels *et al.* (1974) characterized by extreme hypotonia, hiccoughs, and respiratory but not metabolic acidosis. No excess of organic acids was detected in the urine (Dr. Gompertz). The EEG (Dr. Harris) showed intermittent bursts of large amplitude slow wave activity of variable duration interspersed with periods of low voltage activity.

The child died at 72 hours. Apart from subarachnoid and ventricular haemorrhages there were no noteworthy macroscopical abnormalities. Histologically it was found that many nerve cells had undergone severe degenerative changes, and in the brainstem nuclei had broken into numerous fragments - karyorhexis (Fig. 2). The brain also showed oedema and subpial glial proliferation. Many astrocytes

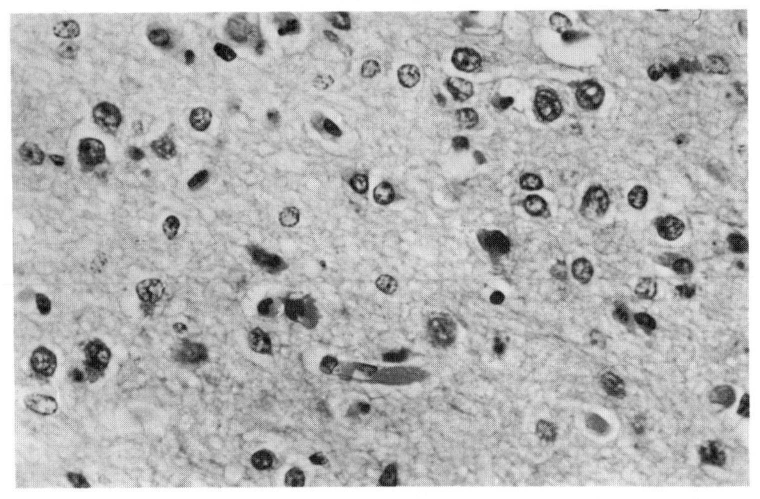

Fig. 2 Karyorhexis of nerve cells in the pontine tegmentum in a case of non-ketotic hyperglycinaemia. (Haematoxylin and eosin × 400.)

had degenerated into so-called Alzheimer 2 cells. The white matter showed spongiform degeneration. None of these changes are specific but are also found in the brains of infants dying from disorders unrelated to aminoacidurias.

The course of the disorder was completely different in cases 1, 2 and 3, three sisters who had the same defect in the glycine cleavage system demonstrated by Professor Nyhan who measured *in vivo* the formation of labelled CO_2 and serine from glycine labelled in the 1 and 2 positions respectively. In none of the sisters was there serious perinatal problems. Case 3, the most retarded of the sisters, is hypotonic and an air encephalogram at age two years was suggestive of cerebral atrophy. The second sister has epilepsy but no neurological signs. The eldest sister is least affected. A feature common to all three girls is the lack of deterioration. Their intelligence quotients have been stable, the eldest girl learnt to read at 15 years, the youngest and most severely affected, learnt to walk at 8 years, and has continued to acquire skills in our school for severely retarded children. Genetic heterogeneity might be invoked to explain, at least in part, the striking difference in involvement between the three sisters on the one hand and cases with severe neonatal course, but not the quite substantial variability within the sibship.

Citrullinaemia, deficiency of the enzyme argininosuccinate synthetase (L-citrulline: L-aspartate ligase, E.C. 6.3.4.5) is a very rare disorder of the Krebs urea cycle. Wick *et al.* (1973) reviewed eight cases including three of their own and concluded that the disorder could occur in three forms

(a) rapidly fatal
(b) intermediate with mental retardation and periodic exacerbations
(c) begign.

The severity of the disorder was closely related to the degree of ammonia intoxication, lethal levels being observed in the first group even in the absence of protein ingestion. In the second group ammonia tended to accumulate after protein ingestion or during catabolic states while in the single patient with a benign course blood ammonia did not reach toxic levels even after a protein load.

Table VII. Biochemical findings in eight citrullinaemic patients.

	Plasma citrulline mmol/l	Urine citrulline mmol/mmol creatinine	Course
Van Der Zee (1971)	4.26 – 4.57	21.2	Severe
Scott Emuakpor (1972)	1.10 – 1.47	10 – 15 (estimated)	Moderate
Wick (1973) Case 1	3.13		Severe
Wick (1973) Case 2	2.11	1.28	Severe
Wick (1973) Case 3	0.11 – 0.32	0.08 – 1.13	Benign
Roerdink (1973)	2.64	20 – 30 (estimated)	Severe
Danks (1974)	1.25 – 3.89	–	Severe
Case 1	–	11.3	Severe
Case 2	2.4	25.0	Benign

However, Roerdink *et al.* (1973) have described a case which did not fit into this pattern. This infant had the disease in a rapidly fatal form with high blood citrulline levels but no evidence of ammonia intoxication. Argininosuccinate synthetase was absent from liver but present in normal amounts in brain.

Very recently we have seen a case which was clinically similar to that of Roerdink *et al.* (1973) and cases 1 and 2 of Wick (1973). The first symptoms on the sixth day of life were irregular breathing and grunting followed by loss of consciousness, spasticity and opisthotonus which later changed to hypotonia, increasing numbers of apnoeic attacks and death on the eighth day. Autopsy failed to reveal an obvious cause of death. A sample of urine collected shortly before death was found to contain large amounts of citrulline. The parents have another child, a boy aged six years who is well and attending a school for normal children. He too has citrullinaemia. Biochemical findings in eight patients with the disease are shown in Table VII. The toxic effects in this disorder of high blood ammonia levels are well established. It now appears that high blood citrulline levels may be toxic by themselves in the absence of hyperammonaemia, but may also be compatible with normal or near normal development.

How are we to explain these devastating effects of what in hyperglycinaemia are comparatively modest increase in blood glycine levels, or of citrulline when in at least one case argininosuccinic acid synthetase was normal in brain. Oldendorf (1971) amongst others has demonstrated that blood brain barrier permeability is low to nonessential amino acids, particularly to those with roles as neurotransmitters. This suggests that failure of a protective transport mechanism might play a part in pathogenesis. Olney (1974) has studied the neurotoxic effect of glutamic acid when given to immature animals. To explain the toxicity of an amino acid normally present in high concentration in the brain this author suggests that it is only extracellular rather than intracellular glutamic acid which is implicated, fairly small changes in extracellular concentration leading first to neuronal excitation and subsequently to neuronal necrosis.

Elevated levels of glycine may well exert their toxic effects via a similar mechanism. Glycine is a putative transmitter in the cord and perhaps also supraspinally. Interference with transmission in the cord correlates well with the extreme muscular hypotonia and some clinical signs suggestive of Werdnig-Hoffmann's disease which are found in the severe form of non-ketotic hyperglycinaemia. On this view

Table VIII. Neuropathological findings in inborn errors.

Absence of congenital malformations
Usually structural changes
Usually lack of specificity
Common abnormal pathways
Often delayed clinical manifestation
Combined effects of
(a) Long term metabolic imbalance
(b) Acute exacerbations (e.g. fits)
(c) Terminal events

one would expect the toxic action of glycine to be similar in primary deficiency of the glycine cleavage system and in deficiency secondary to propionic acidaemia or methylmalonic acidaemia (Tada *et al.*, 1974) although in ketotic hyperglycinaemia the harmful effects of propionic or methylmalonic acid would be superimposed on those of glycine. Patients with non-ketotic hyperglycinaemia who survive early infancy will suffer from the effects of early insults to the brain but to the extent that glycine can subsequently be kept out of the nervous system should be protected from further deterioration. This appears to have happened in cases 1, 2 and 3 of Table VI.

SOME GENERAL COMMENTS

It is clear from this brief account that our knowledge about how pathogenetic mechanisms operate *in vivo* is still scanty and uncertain. In assessing pathogenetic mechanisms regard has to be paid to the developmental stage of the brain, bearing in mind regional differences in the rate of maturation and the stage of evolution of metabolic pathways. The concentration of potentially toxic metabolites in blood or even cerebrospinal fluid may be a poor guide to its topological distribution or compartmentation in the brain.

Important lessons may be learnt from the examination of the brains of affected patients (Table VIII). Inborn errors which give rise to mental and neurological abnormalities are nearly always associated with structural changes in the nervous system. The brain does not remain structurally normal if its metabolic activity is seriously impaired. In general lesions are not confined to any circumscribed part of the brain nor, on the other hand, are they evenly distributed, some areas and formations being more conspicuously involved, and others apparently spared. It must be stressed that most of the changes found in inborn errors of metabolism are non-specific; this applies, for example, to the hyperglycinaemias, maple syrup urine disease and disorders of the Krebs urea cycle. The repertory of structural change in the nervous system is limited in relation to the great variety of environmental and endogenous harmful factors. Diverse metabolic causes may give rise to common abnormal pathological pathways leading to similar structural changes. Ultimately, these structural changes reflect the underlying metabolic disease, the effects of acute exacerbations, such as fits, apnoeic episodes, and thromboses, and the terminal illness.

This work has been generously supported by the South West Metropolitan Regional Hospital Board.

REFERENCES

Angeli, E., Crome, L., Kirman, B.H. and Stern, J. (1975). *(In preparation)*

Angeli, E., Denman, A.R., Harris, R.F., Kirman, B.H. and Stern, J. (1974). *Develop. Med. Child Neurol. (in press).*

Angeli, E. and Kirman, B.H. (1971). *In:* "Proceedings of the Second Congress of the International Association for the Scientific Study of Mental Deficiency" (D.A.A. Primrose, ed) pp.692-697. Polish Medical Publishers, Warsaw.

Bachmann, C., Mihatsch, M.J., Baumgartner, R.E., Brechbühler, T., Bühler, U.K., Olafsson, A., Ohnacker, H. and Wick, H. (1971). *Helv. paediat. Acta* **26**, 228-243.

Baumgartner, R., Ando, T. and Nyhan, W.L. (1969). *J. Pediat.* **75**, 1022-1030.

Blaschke, T.F., Berk, P.D., Scharschmidt, B.F., Guyther, J.R., Vergalla, J.M. and Waggoner, J.G. (1974). *Pediat. Res.* **8**, 573-590.

Bulfield, G. and Kacser, H. (1974). *Arch. Dis. Childh.* **49**, 545-552.

Clayton, B.E. (1974). *In:* "Symposium of the Royal College of Pathologists" *(in press).*

Danks, D.M., Tippett, P. and Zentner, G. (1974). *Arch. Dis. Childh.* **49**, 579-581.

Dobbing, J. (1974). *In:* "Scientific Foundations of Paediatrics" (J.A. Davis and J. Dobbing, eds) pp.565-577. Heinemann Medical Books, London.

Farquhar, J.W. (1974). *Arch. Dis. Childh.* **49**, 205-208.

Fisch, O. and Anderson, A. (1971). *In:* "Phenylketonuria and some other Inborn Errors of Amino Acid Metabolism" (H. Bickel, F.P. Hudson and L.I. Woolf, eds) pp.73-80. Thieme, Stuttgart.

Harris, R.F. (1972). *In:* "The Brain in Unclassified Mental Retardation" (J.B. Cavanagh, ed) pp.225-240. Churchill Livingstone, Edinburgh and London.

Hsia, D. Y-Y. (1970). *Progr. med. Genet.* **7**, 29-68.

Koepp, P., de Groot, C.J., Grüttner, R. and Rybak, C. (1973). *Helv. paediat. Acta* **28**, 459-465.

Lyon, I.C.T., Gardner, R.J.M. and Veale, A.M.O. (1974). *Arch. Dis. Childh.* **49**, 581-583.

Neville, B.G.R., Bentovim, A., Clayton, B.E. and Shepherd, J. (1972). *Arch. Dis. Childh.* **47**, 190-200.

Neville, B.G.R., Harris, R.F., Stern, D.J. and Stern, J. (1971). *Arch. Dis. Childh.* **46**, 119-121.

Oldendorf, W.H. (1971). *Amer. J. Physiol.* **221**, 1629-1639.

Olney, J.W. (1974). *In:* "Drugs and the Developing Brain" (A. Vernadakis and N. Weiner, eds) pp.489-501. Plenum Press, New York.

Popkin, J.S., Clow, C.L., Scriver, C.R. and Grove, J. (1974). *Lancet* **1**, 721-722.

Roerdink, F.H., Gouw, W.L.M., Okken, A., van der Blij, J.F., Luit-de Haan, G., Hommes, F.A. and Huisjes, H.J. (1973). *Pediat. Res.* **7**, 863-869.

Roth, M. (1969). *Proc. roy. Soc. Med.* **82**, 765-772.

Saugstad, L.F. (1973). *Clin. Genet.* **4**, 105-124.

Scott-Emuakpor, A., Higgins, J.V. and Kohrmann, A.F. (1972). *Pediat. Res.* **6**, 626-633.

Tada, K., Corbeel, L.M., Eeckels, R. and Eggermont, E. (1974). *Pediat. Res.* **8**, 721-723.

Trijbels, J.M.F., Monnens, L.A.H., van der Zee, S.P.M., Vrenken, J.A.Th., Sengers, R.C.A. and Schretlen, E.D.A.M. (1974). *Pediat. Res.* **8**, 598-605.

Tyfield, L.A. and Holton, J.B. (1975). *In:* "Twelfth Symposium of the Society for the Study of Inborn Errors of Metabolism" *(in press).*

Van der Zee, S.P.M., Trijbels, J.F.M., Monnens, L.A.H., Hommes, F.A. and Schretlen, E.D.A. M. (1971). *Arch. Dis. Childh.* **46**, 847-851.

Wick, H., Bachmann, C., Baumgartner, R., Brechbühler, T., Colombo, J.P., Wiesmann, U., Mihatsch, M.J. and Ohnacker, H. (1973). *Arch. Dis. Childh.* **48**, 636-641.

DISCUSSION

Gaull: I would like to raise the paradox that by treating phenylketonuria, and in the case that we are unable to do anything about maternal phenylketonuria, we are going to end up with more diseased than before, because there will be more women treated and they will then produce progency who are almost foredoomed to a mental deficiency whether or not the father is a carrier.

Stern: Simple calculations can be done based on the assumption that the phenylketonuric mother will undergo adequate treatment during pregnancy (which in two or three cases has resulted in a normal child) and that the number of her children is close to the national average. These calculations show that the increase in the genetic burden of the community will be negligible over a large number of generations. (cf. World Health Organisation, 1972, Genetic Disorders: Prevention, Treatment and Rehabilitation. WHO Technical Report Series No. 497, Geneva, p.31).

Gaull: Provided that the experience on three cases is substantiated in the future!

Stern: Yes.

Hommes: I was particularly interested in your citrullinemia patient and his brother who had a high blood citrulline level. Our original suggestion that citrulline itself may be toxic to the brain, because of an inhibition of brain pyruvate kinase (K_i:3.2 mM) has been criticized by the New Zealand group (*Arch. Dis. Childh.* 49, 579 (1974)). A clue to this dispute could be whether or not high blood citrulline levels are indeed associated with high cerebrospinal fluid citrulline levels. Is the cerebrospinal fluid level of citrulline known in that patient?

Stern: No, this has not been done as yet. Oldendorf (*Amer. J. Physiol.* 221, 1629-1639 (1971)) measured the uptake by the brain of amino acids *in vivo* and found that nonessential amino acids, particularly those associated with neurotransmission are kept out very effectively, whereas the essential amino acids are allowed in. It may be that substances like citrulline and argininosuccinate which are usually intracellular are toxic when extracellular, particularly in the perinatal period when the brain may not be able to keep them out. An analogy is provided by animal experiments with glutamic acid, which at certain stages of development of the nervous system can be highly toxic and produce cellular changes as reported, for example, by Lemkey-Johnston and Reynolds (*J. Neuropath. exp. Neurol.* 33, 74 (1974)). The same may be true of glycine. The child who died in our hospital with hyperglycinaemia was first thought to be a severe case of Werdnig-Hoffman disease with complete paralysis of the respiratory muscles. One wonders whether in these cases glycine gets to places in the brain where it should not be and where it is toxic.

Sternowsky: During our recent survey of the literature on maternal phenylketonuria, (for which we have to thank Dr. Held of our department) it was found that it is not entirely clear that the institution of a low phenylalanine diet in the last trimester of gestation will prevent brain damage. Either essential amino acids in the L-form or protein hydrolysates are administered, but this administration of non-natural protein might do damage to the fetus. It is questionable whether one should do screening for maternal phenylketonuria in a maternal care unit because we cannot offer a real treatment.

Stern: I essentially agree with you; treatment in the third trimester may well be too late. The safety and effectiveness of the treatment with a phenylalanine low diet of phenylketonuric women during pregnancy cannot be regarded as proven. I know of only a handful of cases with a successful outcome. On the other hand, a few phenylketonuric mothers have produced normal children without treatment.

Gaull: I can only agree with what Dr. Sternowsky said about screening for these mothers when we are not quite sure what we have to offer therapeutically. With regard to the point raised about amino acid toxicity for the fetus, Olny *et al.*

(*Exp. Brain Res.* 14, 61 (1971)) have shown in mice that massive doses of cysteine will give a lesion which very much resembles the lesion induced by glutamate. It is of interest in this regard that the human fetus prior to birth is protected against large amounts of cysteine by two mechanisms of which we know. One is the absence of cystathionase and the inability to convert methionine sulfur to cysteine and the other is related to the transfer of cysteine across the human placenta. It turns out that all the amino acids that are normally present in plasma are higher in the human fetal plasma than in the maternal plasma, with the exception of cysteine. Although there is a transport system for cysteine across the placenta, it is apparently a facilitated transfer and not an active transfer against the gradient.

Wadman: Do you think it to be necessary to treat by dietary restriction pregnant histidinaemic women?

Stern: This is a very difficult question to answer. I know of four offspring of histidinaemic mothers. Our case (Neville *et al., Arch. Dis. Childh.* 46, 119 (1971)) was normal in every way when seen by neurologist, psychologist and paediatrician at the age of 4½ years. Another family has recently been described (Lyon *et al., Arch. Dis. Childh.* 49, 581 (1974)) in which three children of a histidinaemic mother had a mean I.Q. of about 90 compared to a midparent value of about 110. However, the intelligence of all the children was within the normal range. Again one has to balance the risks of the dietary treatment against the risk to the fetus of the uncorrected maternal disorder. I do not think that we shall have the data to answer this question until more cases are discovered.

Wadman: But we come in a somewhat difficult position. On the one hand we can diagnose histidinaemia easily and on the other hand we have the means for dietary treatment. Can we take the risk of giving no treatment?

Stern: I agree this is a serious problem to which, I fear, there is at present no answer.

Sternowsky: In the first paper of this session, it was inferred that the pathogenesis of brain damage in phenylketonuria was not due to phenylalanine but rather to phenylpyruvate. The level of phenylalanine in the ammionsac is high, or as high as in the blood, but phenylpyruvate has not been found, so you may not even be sure what the damaging factor in the fetus is, if he is not by himself a phenylketonuric child.

SUBJECT INDEX

G

H

K

L

M

T

U

THE LIBRARY